CLIMATE
PROCESS
&
CHANGE

To Dianne, Mark and Kate

Climate
Process
&
Change

EDWARD BRYANT

School of Geosciences
University of Wollongong

 CAMBRIDGE
UNIVERSITY PRESS

PUBLISHED BY THE PRESS SYNDICATE OF THE UNIVERSITY OF CAMBRIDGE
The Pitt Building, Trumpington Street, Cambridge CB2 1RP, United Kingdom

CAMBRIDGE UNIVERSITY PRESS
The Edinburgh Building, Cambridge CB2 2RU, United Kingdom
40 West 20th Street, New York, NY 10011–4211, USA
10 Stamford Road, Oakleigh, Melbourne 3166, Australia

First published 1997

Printed in Hong Kong by Colorcraft

Typeset in Sabon 11/12 pt

National Library of Australia Cataloguing in Publication data
Bryant, Edward.
Climate process & change.
Bibliography.
Includes index.
ISBN 0 521 48189 9.
ISBN 0 521 48440 5 (pbk.).
1. Meteorology. 2. Climatic changes. I. Title.
551.5

A catalogue record for this book is available from the British Library

ISBN 0 521 48189 9 hardback
ISBN 0 521 48440 5 paperback

Contents

Illustrations

Note: Unless otherwise stated, a McBryde-Thomas flat-polar quartic projection is used for all world maps presented in this text.

Tables

Preface

I have been sitting at my home computer for the last few months absorbed in finishing the drafts of this text. The writing has not been onerous. Outside through the open doors into my garden, I have been witnessing one of the best autumns I have experienced: clear blue skies, a light breeze and temperatures up to 23–25°C each day for weeks on end. Unfortunately, I have not been able to venture far to enjoy the weather more. The weather is overdue because southeastern Australia underwent the coldest summer in thirty-three years. It was certainly the coldest summer I have experienced. While many of the vagaries in temperature were due to the fluctuations in the state of tropical ocean temperatures in the Pacific Ocean, between the El Niño–Southern Oscillation (ENSO) and La Niña, this was not the complete picture. Neither were the abnormal temperature changes due to enhanced 'greenhouse' warming, as trumpeted by the media when it was too warm. Nor did they represent the demise of the Earth's present, warm interglacial climate.

In the clear autumn sky, every time a northeast sea breeze came up, a white haze wafted across my view of the Illawarra Escarpment in the distance. When I have managed to walk along my local beach, I can trace the white haze back to an orangish smog above Sydney, 40 kilometres to the north. The haze does not officially exist. At least publicly the state Environmental Protection Agency does not seem overly concerned about it in Wollongong. The haze has existed in some form or other for the twenty-five years that I have lived in the region. Unlike most North American cities, Sydney has a limited air pollution monitoring network, and weak laws restricting industrial or vehicular emissions. The skies over Sydney can be as dirty as any seen over American cities, including Los Angeles. Between the cold summer, in a supposedly 'greenhouse'-warming world, and the excursions of urban air pollution during a bonzer autumn, I can't help but think that climate change is not simplistic. This text is partially my attempt to understand the climate change that the Earth has undergone over the latter half of the twentieth century, to understand the unbearably cold winters I experienced as a teenager living in Canada, the smogs that I saw in an Arctic summer, the amazing rains and floods witnessed along the Sydney–Wollongong coast during the 1980s, and the everlasting drought that has set in over eastern Australia in the 1990s. Further, this text is an attempt to combine into one source book all the far-ranging descriptions of climate process and change that students so diligently try to seek out when writing essays, yet fail to discover, either because of their acquired bias from previous instruction, or the inadequacies of a modern, fund-strapped university library.

Unless otherwise stated, a McBryde-Thomas, flat-polar, quartic projection has been used for all world maps presented in the text. This is an equal-area, pseudocylindrical projection with pole-lines one-third as long as the equator. The central meridian lies at 0°, and degrees of longitude and latitude are plotted at 15° intervals. This map reproduces continents with a recognisable shape, and it is similar to projections such as the Mollweide, Hammer, Robinson and Hoelzel commonly used in atlases. The projection appears in many geography texts. Much of the mapped climatic data from satellite monitoring or computer simulation models, especially on the topic of anthropogenically enhanced 'greenhouse' warming, are biased, not because of any deficiency in the science, but because they are not plotted as equal-area projections.

Units of measurement used throughout the text follow the International System of Units. All abbreviations and conversions are based upon Rocke (1984). In order to convey viewpoints and arguments, unobstructed by copious referencing, strict adherence to formal, academic referencing has been relaxed. Usually, each section begins by listing the relevant journal articles or books that have either influenced my thinking, or are central to the topic. I apologise to anyone who feels that their crucial work has been ignored, but the breadth of coverage precluded a complete review of the literature on many topics. The full reference to publications can be found at the end of each chapter. Academic accreditation has been maintained only where published material is specifically referred to in a diagram or table. In addition some articles and data were acquired from the Internet. The Internet address in these cases is also referenced. Such material may not be readily sourced, due to changes in Internet addresses, or the lack of an archival tradition for this new resource medium. Where material is not available in the literature, or through these forums, it has been acknowledged at the beginning of the text.

The reporting of geological time can be confusing because of the terminology, and the fact that radiocarbon ages are younger than calendar ones by 22–27%. Unless otherwise stated all ages are reported in calendar years. Radiocarbon years are only used where dates older than 18,000 years BP have been quoted in the literature, and no correction exists to convert them to calendar years. All non-historical ages before 0 AD are referred to as years before the present (BP). The only exceptions are historical events, such as wars and epidemics, where the term BC is used. Ages after 0 AD are simply quoted as the year, without appending 'AD.'

I have communicated over the years with most of the people at the extremes of polarised climate views. I am struck by a quote by H. Tazieff, a world renown volcanologist:

> I had the surprise – oh, how pleasant – of receiving the approbation of numerous scientists, especially specialists in these matters. On the other hand, I have attracted innumerable enmities, some naive and some from certain people of doubtful honesty. But a small number of friends of quality is worth more than a bunch of fans or a bunch of foes.

TED BRYANT

REFERENCES

Rocke, F.A. 1984. *Handbook of Units and Quantities*. Australian Atomic Energy Commission, Lucas Heights, 270p.

Snyder, J.P. 1993. *Flattening the Earth*. University of Chicago Press, Chicago, 365p.

Acknowledgments

Many of the data sources and diagrams are derived from United States government publications which are not copyrighted. This book could never have been written without that policy. Individual contributors for that data are acknowledged separately in the text. Specifically, I would like to thank Dr Tom Boden of the Carbon Dioxide Information Analysis Center for making available data collated by that centre and figures appearing in Trends'93; and Dr Tom Karl of the National Climatic Data Center for Figure 1.4.

Some data, reported in Chapter 7 in the discussion of cloud, sulphates and satellite temperature measurements, are taken from anonymous articles in the *World Climate Review*, a publication of the Department of Environmental Sciences, University of Virginia, Charlottesville, USA, Dr P.J. Michaels, Chief Editor.

Other material comes from the Internet, including World Wide Web sites. The monthly sunspot numbers between 1749 and 1995 plotted in Figure 2.3 are part of a United States National Geophysical Data Center file regularly updated by the Data Support Section, Scientific Computing Division, National Center for Atmospheric Research (NCAR) at **ftp://ncardata.ucar.edu/datasets/ds834.0/monthly_data**. The Southern Oscillation index plotted in Figure 4.9 is partially derived from data compiled by the Joint Institute for the Study of Atmosphere and Ocean, University of Washington in Seattle at **http://tao.atmos.washington.edu/pacs/additional_analyses/soi.html**. The description of precipitation changes over historical timescales, specifically dealing with flooding, presented in Chapter 5, includes facts summarised on the Greenpeace web site at **http://www.greenpeace.org/climate/flood_report/index.html**. Milankovitch insolation values plotted in Figure 6.1 were downloaded from the NOAA web site at **ftp://ngdc1.ngdc.noaa.gov/paleo/insolation/**. The original source of data is cited in the figure caption. The Halley Bay ozone values for 1994 presented in Figure 7.16 were taken from the British Antarctic Survey ozone information web site at **http://**www.acd.ucar.edu/gpdf/ozone/science/bas.html**. Values are used with the permission of the British Antarctic Survey. Updated October ozone values for Halley Bay plotted in Figure 7.17 were compiled by Dr J.D. Shanklin, British Antarctic Survey and downloaded from Gregory P. Dubois-Felsmann's Internet web site at **http://www.acd.ucar.edu/gpdf/ozone/science/bas.html**. An inventory of historical values was taken from FAQ on ozone assembled by Dr Robert Parson, Department of Chemistry and Biochemistry, University of Colorado at **rparson@rintintin.colorado.edu**. The latter were checked for accuracy against graphs published in original sources.

Much of the material on climate impacts on health was obtained through collaborative work with Professor Christine Ewan, University of Wollongong between 1990 and 1993. Background material was compiled by John Marthick, Department of Geography, University of Wollongong, and Deanne Condon-Paoloni. John Marthick also kindly permitted Figure 8.2 to be used. Dr Paul Fraser, CSIRO Division of Atmospheric Research, Melbourne, provided valuable information on the health effects of enhanced ultraviolet radiation. Estimates of the medical cost of enhanced ultraviolet radiation in Australia presented in Table 8.3 were calculated by Professor Don Lewis, Department of Economics, University of Wollongong. Dr Keith Bentley, Environmental Health Unit, Australian Department of Community Services and Health, initiated much of the research, and facilitated funding through the National Health and Medical Research Council. Where this work is accessible to the general public, it has been referenced in the text. However the section on the health effects of ozone depletion comes from a limited-circulation report (Bryant, E., Lewis, D., Calvert, D., Ewan, C. and Fraser, P. 1992. *Estimation of the health costs, to 2030 AD, of enhanced ultraviolet radiation due to climatic change.* Final Report to the Department of Health Housing and Community Services and the Department of Arts, Sport, The Environment and

Territories, 519p.). Some background material on the biological consequences of enhanced 'greenhouse' warming was provided by Martin Gregory, a research assistant working for the University of Wollongong Environmental Research Institute on a project sponsored by the Department of the Environment, Sport and Territories.

This text is very much the effort of people associated with Cambridge University Press, in particular: Dr Robin Derricourt, who promoted the concept for this text; Jane Farago, Phillipa McGuinness and Catherine Flack, who professionally accelerated the pace of publication; and Kathleen Gray, who meticulously improved the manuscript. This book also benefited from the skill of David Martin, who drew many of the maps, thus relieving me of a difficult task. However, the responsibility for all other material, and any errors in the text, rests solely with me.

Finally, I would like to thank many colleagues, who through difficult years, sustained my enjoyment for academia and my enthusiasm for scholarly research.

1

Climate History of the Earth & Background Concepts

RATIONALE

A cursory examination of historical climate records, media archives or the recent climate literature will indicate that the Earth's climate, as characterised by temperature, precipitation and extreme hazard events, is undergoing change. Unfortunately, many recent scientific and popularised reports imply that this change is dramatic and related to global warming caused by the enhanced 'greenhouse' effect. This is not necessarily correct, and today in the late 1990s enhanced 'greenhouse' warming has not been conclusively substantiated. The debate is not necessarily over the question of 'greenhouse' warming, but rather over the question of whether or not the signatures of that warming have risen irrefutably beyond the background level of noise inherent in the Earth's climate. There is very little material in the current literature that the inquisitive student can find presenting alternative reasons for the observed climate change of the past century. Certainly, little appears in recent published literature about the uncertainties of enhanced 'greenhouse' warming. This is unusual given the fact that, as recently as twenty years ago, a similarly emotional debate was raging over the question of global cooling due to increased atmospheric dustiness. It is odd that climatic evidence over such a short space of time should support these two contrasting hypotheses.

In teaching climate change, I am continually amazed by the naivety of many people's arguments concerning basic climatic processes, the nature of past climate change, and society's response to that change. Many students do not know that the most important 'greenhouse' gas in the atmosphere is water vapour, or that a single solar flare event on the sun can destroy, in a few days, more ozone in the stratosphere than has disappeared in the past thirty years because of ozone depletion. Popularised reports concerning enhanced 'greenhouse' warming also conjure up apocalyptic disaster scenarios in a significant number of cases. If society is unwilling to accept the fact that it has to adapt to climate change, then it is also unwilling to accept the fact that the inherent noise within the Earth's climate system continually generates hazardous extremes. Many of the changes forecast for enhanced 'greenhouse' warming pale into insignificance when compared to the extreme climatic events that occurred in Europe during the initial warming of the Middle Ages, and the subsequent cooling leading to the Little Ice Age. The Great All Saints Day storm, on 1–6 November 1570 in northern Europe, killed an estimated 400,000 people. Not even the most extreme 'greenhouse' scenario would contemplate such a death toll in Europe within the next fifty years.

This book presents, in simple terms, the processes that drive the Earth's present climate system which are necessary for an understanding of the nature of climate change. The record of that change is presented so that the magnitude and impact of any future shifts can be kept in perspective. It will be shown that temperature changes on the order of 5–10°C have been a natural component of global climate over the past two million years, dominated by the Ice Ages. The reasons for

1

glaciations will be clarified with the caveat that our present science cannot give an unequivocal reason for those extreme temperature fluctuations. Against this background, the nature of twentieth century climate change will be discussed. The existing shift in climate is being forced by enhanced 'greenhouse' warming, ozone depletion, urban growth, increased dust and increased atmospheric pollution. While each of these causes is related to human activity, there are natural components forcing climate change such as random fluctuations, volcanic activity and critical ocean–atmosphere interactions. Finally, future climate change will be discussed for two major implications, on human health and ecosystems. The rest of this chapter describes the evolution of the Earth's climate and defines the basic terminology used in describing the temporal and spatial nature of climatic data.

EVOLUTION OF THE EARTH'S CLIMATE
(Kasting et al., 1988; Kasting, 1993)

The evolution of the Earth's climate is related very much to the history of the development of the sun and the degassing of the Earth. The sun 4.6 billion years ago was 25–30% dimmer than it is now. Since then, it has progressively increased its radiation output. If the Earth's atmosphere had had the same composition then as now, this lower solar radiation output would have allowed the Earth to be covered in ice for the first two billion years of its existence. But the geological record indicates that the Earth had liquid oceans as early as 3.8 billion years ago, and no glaciation until 2.7 billion years ago. This dichotomy is known as the 'faint sun' paradox. At present the Earth's surface receives only half of the solar radiation impinging upon its disk as it orbits the sun. Reflectance and atmospheric interference account for the rest. If the Earth was not ice-locked in its early history, then its surface would have needed to receive most of the solar radiation emanating from a fainter sun. This could have been possible only if there were unrealistically fewer clouds and less moisture in the atmosphere than at present. So the only viable mechanism to negate the consequences of a 'faint sun' had to be a global warming caused by an enhanced 'greenhouse' effect that was dependent upon large amounts of carbon dioxide in the Earth's early atmosphere.

Geological evidence supports this enhanced carbon dioxide-induced 'greenhouse' effect. The amount of carbon dioxide (CO_2) present in carbonate rocks in the Earth's crust is equivalent to sixty times (60 bars) present atmospheric pressure. At present the partial pressure of carbon dioxide in the atmosphere is only 0.0003 bars. Only a few tenths of a bar would have been needed to compensate for the 'faint sun' paradox. However the carbon dioxide in the Earth's atmosphere must have decreased at just the right rate over time to compensate for the steadily increasing radiation output of the sun. Such a delicate balance was not easily maintained for long periods. A mechanism was needed to modulate periods of excessive carbon dioxide in the atmosphere and variability in solar activity.

The modulator can be found in the geochemical and tectonic nature of the Earth. The carbonate-silicate geochemical cycle can turn over 80% of the Earth's carbon dioxide in 500,000 years or more. Too much carbon dioxide causes acid rain that dissolves calcium through the weathering of igneous bedrock. Calcium-rich water flows into the oceans where it is used by organisms to form calcium carbonate ($CaCO_3$) in skeletal material. When these organisms die, skeletal material settles and accumulates on the ocean floor. The theory of plate tectonics hypothesises that the ocean floor moves outwards from the oceanic ridges to be consumed by the Earth's mantle at subduction zones. Calcium carbonate moves on this oceanic conveyor belt to the edges of continental plates, where it is subducted into the mantle and subjected to extreme pressures and heat. In this process, calcium carbonate releases carbon dioxide into magma. The magma returns to the Earth's surface via volcanoes where the carbon dioxide is released back into the atmosphere. The warmer the Earth's atmosphere, the faster the Earth's hydrological cycle operates, and the more it rains. Increased rain dissolves more carbon dioxide and removes this gas faster from the atmosphere. Lesser amounts of carbon dioxide in the atmosphere lead to cooler surface temperatures because of a reduction in the carbon dioxide-induced 'greenhouse' effect. Cooler temperatures produce less acid rain and less carbon dioxide is removed from the atmosphere, permitting this gas eventually to accumulate to levels that restore a warm regime. Even if the oceans froze over, cutting off the carbon dioxide cycle, carbon dioxide degassing from the Earth via volcanoes would eventually raise the partial pressure of carbon dioxide in the atmosphere to the point where atmospheric temperatures would melt this ice cover. Degassing of carbon dioxide from volcanoes can increase atmospheric pressure at the rate of one bar per twenty million years. An increase in carbon dioxide of one bar would raise surface temperatures of the Earth by 50°C, melting any ice, and restoring the Earth's benign temperatures.

Biological activity aids this process, but is not crucial to it. The Earth could still have had habitable temperatures without any biota; however, plant life accelerates the removal of carbon dioxide from the atmosphere. Without biological activity, the carbon dioxide content of the atmosphere would be several times greater than it is at present. Water vapour played no major role in this buffering. While water vapour may have reinforced warming or cooling, the carbon dioxide balance could have been driven solely by chemical weathering and subduction of calcium carbonate-rich sediments beneath the Earth's crust. As the sun warmed over the eons, carbon dioxide levels in the atmosphere progressively decreased to preserve the surface temperature of the Earth close to its present levels. Carbon dioxide levels today are only 280 parts per million, in the preindustrial present era, and have dropped as low as 150 parts per million during recent glacials. This latter figure is close to the lower limit needed to support plant life as we know it. In the future, as the sun continues to warm, carbon dioxide will cease to be a significant constituent of the Earth's atmosphere, and all plant life will reach a crisis point for its existence.

CLIMATE DIFFERENCES AMONGST EARTH, MARS AND VENUS
(Kasting et al., 1988; Ramanathan et al., 1989; Haberle, 1990; Kasting, 1993)

The Earth is delicately positioned within the solar system in terms of a climate required to sustain life. If the Earth were 5% closer to the sun, its ocean would boil off to form a dense atmosphere similar to that of Venus. If the Earth were 1% further from the sun, its oceans would permanently freeze, forming an Antarctic-like planet such as Mars. And yet both Venus and Mars have, or have had, atmospheres which for part of their geological history were similar to Earth's. All three planets formed by accretion of solid material from condensation within the solar nebula. Any primary atmospheres that might have existed were lost, to be replaced by secondary atmospheres generated by volatile compounds making up their interiors,

and released to their atmospheres through volcanic degassing. Mars appears to have had running water until 3.8 billion years ago, as evidenced by fluvial mega-channels, buried permafrost and ocean strandlines. However, Mars required 150–800 times more carbon dioxide than it presently has in its atmosphere, to raise temperatures above freezing. This carbon dioxide must have been replenished in some manner, otherwise these high levels could not have been maintained in the presence of weathering for longer than ten million years. However, Mars never developed a replenishment mechanism. There was no continental drift to recycle carbon dioxide in a manner similar to that evident throughout Earth's history. Because Mars is a smaller planet, its interior cooled down earlier, slowing down the volcanic activity needed to add carbon dioxide to its atmosphere. What carbon dioxide was present disappeared through weathering, while Mars' water became buried beneath the surface under dust stirred up by subsequent meteorite impacts. Mars is an arid planet, thinly veiled in carbon dioxide and plagued by dust storms generated by strong winds of 20–360 kilometres per hour.

Venus evolved differently. The top of its atmosphere presently receives almost twice as much solar radiation as the Earth's (2,620 watts per square metre ($W\ m^{-2}$) for Venus versus 1,368 $W\ m^{-2}$ for the Earth). However, Venus is completely covered in clouds of sulphuric acid that reflect back to space 80% of incoming solar radiation, compared to the Earth which reflects back to space 31%. As a result, Venus absorbs only 130 $W\ m^{-2}$ of radiation, or 54% of that absorbed by the Earth. Venus has a much hotter surface temperature than the Earth because it has an exaggerated 'greenhouse' effect due to carbon dioxide. With a mean surface temperature of 462°C, Venus emits 17,900 $W\ m^{-2}$ of longwave energy, of which less than 1% escapes to space. Venus thus recycles two orders of magnitude more longwave radiation within its atmosphere than the Earth either receives from the sun, or recycles in its atmosphere. Whereas Mars has lots of water frozen under its surface, Venus is waterless. It is doubtful whether Venus was always dry, because it evolved geologically in a similar fashion to the Earth. On Venus, the solar flux of radiation was so great that the surface water evaporated and was transferred to the upper atmosphere, where it was disassociated into its ionic components by high energy particles. The lighter hydrogen ions were then slowly torn from the gravitational field of the planet by the solar wind. Sulphur gases reacted with any remaining moisture to form sulphuric acid clouds. Without water, it was impossible for carbon dioxide to dissolve and form acid rain. Thus carbon dioxide accumulated in the atmosphere over time through planetary degassing. This process still continues today. Without the 'greenhouse' effect of carbon dioxide, Venus would be much colder than the Earth and only slightly warmer than Mars.

PLANETARY 'GREENHOUSE'
(Idso, 1989)

The 'greenhouse' warming effect on Venus is approximately 500°C, while that on Mars is 5–6°C. Yet, both planets are similar to each other in that their atmospheres contain 96–100% carbon dioxide. Venus has a runaway 'greenhouse' effect, while Mars does not have enough. This dichotomy appears at first sight odd, if it assumed that any carbon dioxide-induced 'greenhouse' effect depends upon the proportion of carbon dioxide making up a planet's atmosphere. But it does not. A carbon dioxide-stimulated 'greenhouse' effect depends only upon the amount, or partial pressure, of carbon dioxide in a planet's atmosphere. Venus has an atmospheric pressure of 93 bars, while Mars has one of only 0.007–0.010 bars. As a result, the planetary 'greenhouse' effect on Venus raises temperature one hundred times more than it does on Mars. This effect is shown poignantly in Figure 1.1, which plots the linear relationship between partial pressure of carbon dioxide and

temperature. For the Earth, the present partial pressure of carbon dioxide (0.0003 bars) results in only a 1°C 'greenhouse' warming effect. Under a 'faint sun', the partial pressure required to maintain the temperature of the Earth around present day temperatures needed to be 0.08 bars. Note that according to Figure 1.1, the projected doubling in carbon dioxide by anthropogenic enhancement over the next forty years should result in only a 0.4°C increase in present air temperatures.

THE CLIMATE RECORD OF THE EARTH
(Frakes, 1979)

Despite the 'faint sun' paradox, the Earth's climate has varied dramatically throughout geological time, departing by +5°C to -10°C from present values. This record is presented schematically in Figure 1.2. The exact nature of climate before 1,000 million years is indeterminable. Overall, it was wetter and warmer than at present. The evolution of life had a dramatic effect upon the composition of the Earth's atmosphere. Early plant photosynthesisers consumed carbon dioxide but locked the poisonous oxygen waste into iron compounds, producing extensive banded iron formations between 3,000 and 1,500 million years ago. A second wave of photosynthesisers after this period released their oxygen waste directly into the atmosphere, leading to intense oxidisation across the surface of the Earth. Reactions between oxygen and iron formed extensive red beds. The first indication of glaciation occurred around 2,300 million years ago, and represents a period when the amount of carbon dioxide in the atmosphere was lowered quickly because of the proliferation of blue-green algae forming stromatolites. Other periods of extensive glaciation occurred during the late Precambrian between 950 and 615 million years ago, and the Permian around 250 million years ago. The Earth underwent a general cooling following the Cretaceous warm peak when temperatures were 10°C above present values. Permanent icesheets formed over the Antarctic around 25–30 million years ago; but cyclic glaciation developed globally only within the last two million years. Currently, the Earth's average surface air temperature occupies the upper 10% of values recorded over this geological time. Our present climate is still controlled by the mechanisms that can generate glaciation.

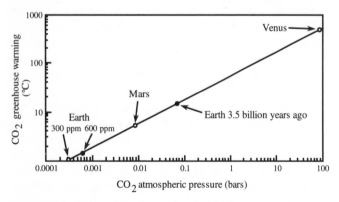

Figure 1.1 The relationship between the partial pressure of carbon dioxide (CO_2) and surface temperature for the inner planets (Idso, 1989).

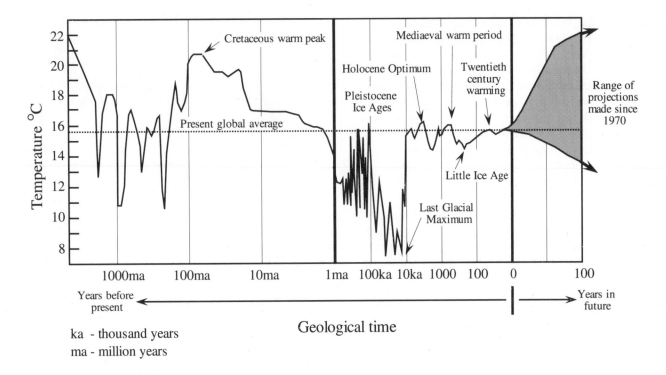

Figure 1.2 The temperature record of the Earth over geological time. A logarithmic scale is used for time, but there are breaks in scale at one million years and the present.

THE NATURE OF PRESENT RECORDS OF CLIMATE CHANGE

Trends and variability
(Bryant, 1991)

Geophysical variables such as temperature or rainfall, when plotted as time series or maps, evidence a number of fundamental characteristics. Figure 1.3 presents the annual temperature record for the northern hemisphere since 1854. Temperatures are plotted as deviations from the 1951–1979 average. This record consists of a trend or a signal over time which can be determined statistically. In Figure 1.3, the signal is one of warming at the rate of 0.40°C per century. This accounts for 51% of the variability in the data and is significant at the 0.05 level of significance. Simulation studies have shown that it is extremely unlikely to get warming trends of this magnitude for intervals longer than seventy years. This rise has been attributed to enhanced 'greenhouse' warming, but could be ascribed just as easily to climate recovery from the Little Ice Age which peaked in the seventeenth century.

The trend has not been consistent over time. There were periods, for example 1960–1978, when warming was reversed and the data did not fit the trend line. In addition, this rate of change is not consistent globally. Temperatures in the southern hemisphere warmed at a different rate than did those in the northern hemisphere. These spatial differences are suggestive of another controlling variable upon global warming, or of an additional mechanism in the Earth–atmosphere system that modulates solar radiation. Thus a trend or signal can be obscured by noise, variability or scatter in the data. This may be due to measurement error, additional factors or randomness. For temperature, some of these additional factors include volcanic eruptions and solar variability. In Figure 1.3, 49% of the variability in the data is still unaccounted for after the removal of the trend (detrending). Often the amplitude of this variability has been greater than the trend, especially over timescales of one or two centuries. For instance, in Figure 1.3, the amplitude of the variability after detrending is as high as 0.4–0.6°C, or about twice the magnitude of the trend. Any interpretation that temperatures have warmed in recent years must be qualified, because unless the signal of this warming exceeds the amplitude of noise inherent in the temperature record (the signal-to-noise ratio), then the change could be due simply to chance.

Variability may also fluctuate over time. For

instance, in the late 1800s, the amplitude of fluctuations in temperature amounted to 0.6°C; but during the 1940s, it was as low as 0.3°C. If the trend of a time series is determined using statistical techniques such as linear regression analysis, then a change in variability weakens, or even invalidates, the analysis. Statistically, this change in variability is said to violate the assumption of homoscedascity or equal variance over time.

Coherence

In spatial terms, time series may show strong similarities in the timing of major fluctuations. This is termed coherence. At timescales of 3 to 5 years, temperature changes in the northern and southern hemispheres tend to parallel each other. Usually, there is a reason for this globally induced change, such as a large volcanic eruption. Coherence is quite common and exists in a wide range of climatically controlled variables. For instance, the climatic control on sea-levels can be illustrated by the fact that about 50% of the change in sea-level from year to year, at such diverse locations as Sydney, Venice, London and New York, is coherent. This coherence has an amplitude of 10 to 15 centimetres and is caused by changes in the global hydrological cycle, induced by short-term climate change produced by trade wind fluctuations in the central Pacific Ocean.

Extremes

When the variation about some trend line is examined year by year, there will always be a few individual years where temperatures are exceedingly large or small. Confidence limits can be fitted to a trend line to reveal values that do not follow the signal. For the 140 years of data shown in Figure 1.3, about 7 to 10 values should lie outside the 95% confidence limits to the trend line. In actual fact, only five values (1862, 1878, 1917, 1976, 1990) lie outside the 95% confidence limits for the trend. These values are labelled extremes and, because their occurrence is difficult to explain, they pose a problem in climatic modelling, description and interpretation. Extreme values may exist for a plethora of reasons, but often they appear to be just random. One of the best climatic examples of extreme values occurred in the Burgundy region of France in the 1420s, one of the periods of the Middle Ages when temperatures were cooling rapidly in northern Europe. This single decade witnessed six of the coldest years in one thousand

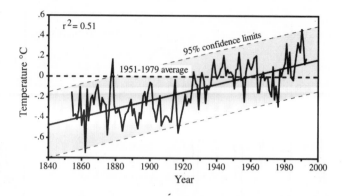

Figure 1.3 Northern hemisphere annual temperatures 1854–1993 referenced to the mean for the period 1951–1979. The amount of variability explained by the straight line is given by r^2, the regression coefficient squared. If all the data fit the straight line (no variability), this value would equal 1.0. Data from Jones et al. (1994).

years, in this region. However, embedded within this cooling was the warmest summer experienced for a millennium. This summer was not just slightly above normal, but exceptionally so. The grape harvest in this year occurred at the beginning of August instead of during the second or third week of September.

Extremes may not always be random events. Figure 1.4 plots the occurrence of extreme climatic events in the coterminous United States, between 1909 and 1993, using the summation of five different climatic parameters: maximum and minimum temperatures, moisture deficit–surplus, extreme rainfall above 50 millimetres in one day,

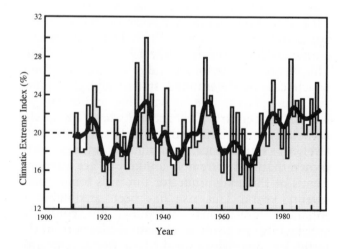

Figure 1.4 Times series of an aggregated index of climatic extremes in temperature and precipitation for the United States, 1909–1993 (from Karl et al., 1995). The solid line shows the smoothed or filtered record.

and the number of days with precipitation. For each parameter, the percent area of the country registering in the upper or lower 10% of values is used. In Figure 1.4, a value of 0% for the Climate Extreme Index means that no portion of the United States is subject to temperature or precipitation extremes. This situation never occurs. At any one point in time, about 15% or more of the United States is subject to some sort of climatic extreme. Similarly, it is very unusual for more than 25% of the country to be subject to climatic extremes. When smoothed, the time series shows clusterings, five years in length, when climate is more extreme than normal. While no reason can be given as to cause of these events, it can be said that once extreme events begin to occur, they tend to recur.

Probability of exceedence
(Bryant, 1991; Smith, 1992)

For very long time series, it is possible to construct distributions to visualise the number or frequency of observations over the range of the data. The shape or *skewing* of these distributions can reveal the presence of climate change. For instance, if temperatures have been sampled randomly from the optimum range of temperatures available, if there is no long term warming or cooling trend, and if the variability in the record remains constant (statistically known as stationarity), the resulting frequency distribution should be bell-shaped. This particular distribution type is relatively common and is known as a normal distribution. Such a distribution can be generated randomly. Figure 1.5A simulates the frequency distribution of 1,500 years of temperature data assuming present day averages and variability in annual global temperature. This distribution has been labelled 'Time 1'. If the globe warms by 0.5°C, then the simplest manifestation of this change will be a shift in the mean of the distribution. The resulting distribution has been labelled 'Time 2'. However, some researchers believe that such a shift towards a warmer climate will also lead to more variability. For instance, if the standard deviation in temperature rose from 0.25 to 0.5°C, then there would be a greater spread of temperatures about the new mean. This is shown in Figure 1.5B. Note that a simple shift in the mean of a distribution shifts all frequencies about that mean towards higher values. However, if the variability also increases, then it is quite possible to have temperatures just as cold, after warming, as before. (Compare the number of times

that temperatures are below 14.5°C at Time 1 and Time 2 in each diagram.)

It is of interest to risk assessors to know the likelihood or probability of occurrence of these extremes. For example, global warming may make frosts rarer, but have disastrous consequences for ski resorts. At the other extreme, excessively high temperatures could lead to more heat stress, and be a health risk. The probability of rare events of high magnitude can be ascertained by constructing a probability of exceedence diagram, that depicts the rarity of an event in terms of the percentage of time that such an event will be exceeded. If the magnitude of an event can be measured at regular time intervals, the value at each interval can then be ranked in magnitude relative to each other. The probability of exceedence of each interval is determined using the following formula:

$$\text{Exceedence probability} = \frac{M}{N+1} \times 100\% \qquad \textbf{1.1}$$

where N = the number of ranks
M = the rank of the individual event (highest = 1)

The results are then plotted on normal probability paper. On this paper, normal distributions plot as a straight line. The percentage scales are designed such that the probability of occurrence of rare events is enhanced visually. The scales on each y-axis are also the inverse of each other. Thus it is possible to read off not only the probability of exceedence of any particular value, but also the percentage of time that observations will be less than this value. In addition, if the frequency distributions before and after climate has changed are plotted on the same graph, it is possible to determine the change in exceedence of rare values. Probability of exceedence diagrams are plotted in Figure 1.5 for each temperature distribution. In the case of simple warming (Figure 1.5A), the value at which annual temperatures will be exceeded 1% of the time occurs at 15.6°C. After an overall warming of 0.5°C, this same extreme value has the likelihood of being exceeded 44.1% of the time. Thus an annual temperature that recurs once in 100 years (1:100), will be exceeded eleven times in 25 years after warming. The degree of change can be illustrated further by calculating the probability of exceedence of the mean temperature (15.0°C) for 'Time 1', at 'Time 2'. After warming, this mean will be

A Warmer, same variability

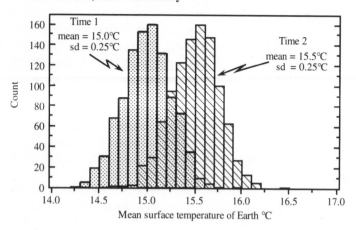

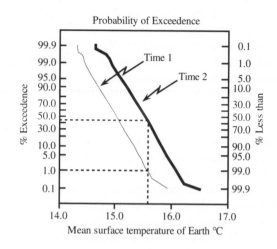

B Warmer, greater variability

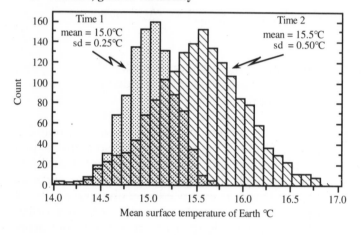

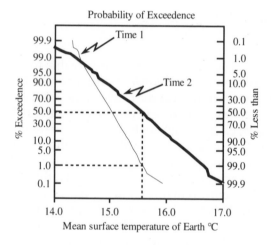

Figure 1.5 Simulated frequency distributions of global annual temperature. Reference is the present climate, 'Time 1', having a mean of 15.0°C and a standard deviation of 0.25°C. (A). Temperature increases overall by 0.5°C, but the variability remains the same. (B). Temperature increases overall by 0.5°C but the standard deviation increases to 0.5°C. The probability of exceedence curves for both sets of data are plotted to the right of each panel.

exceeded 96.7% of the time. In the second case, where variability also increases with warming, the mean temperature will be exceeded 84% of the time. Note, that this latter scenario also includes just as many, if not more, extreme cold events after warming, as before.

Persistence and feedback

In some time series, noise and extreme values can cluster systematically in time due to climatic persistence. For instance, up until 1993, the greatest flood ever recorded on the Mississippi River was one that occurred in 1973. As flood records had only been collected for two hundred years, this 1973 event was considered at the time to be the 1:200 year flood event. But the next 1:200 year flood event on this river occurred the very next year, in 1974! The reason for this is that the climatic conditions that produced the heavy winter snowfall and spring rains leading to flooding in 1973, persisted beyond the calendar year into 1974. Similar persistence is also present in the time series of climatic extremes in the United States presented in Figure 1.4. In terms of short-term climate change, persistence represents a temporary stability in the conditions controlling that change.

Persistence is usually controlled by feedback mechanisms. Take the simple example of heavy rainfall. Once a weather system bringing that rainfall has disappeared and clearer skies have returned, higher evaporation can be produced because of an abundant ground-moisture supply. This evaporated moisture can condense to form cloud, leading to atmospheric instability that brings more rain. If rain-bearing systems are repetitive, they need only top up ground moisture to maintain a season of wet conditions. Any process where a shift in climate is reinforced, is termed positive feedback. In some cases, this positive feedback may lock climate into a stable, but altered state. This often occurs in temperate latitudes of the northern hemisphere with the first snowfalls of winter. Snowfall exacerbates the reflectance of solar radiation from the ground surface, leading to cooler temperatures. The cooler ground temperatures then favour the formation of more snowfall when precipitation forms again. Negative feedback occurs if the interaction amongst variables cancels them out. For example, in spring in the northern hemisphere, increased warming melts snow, thus minimising the snow's ability to reflect incoming sunlight. The process does not enhance the warming effect of increased solar radiation, rather it weakens the cooling effect of snow.

Serial correlation, lags and response times

The temperatures plotted in Figure 1.3 are not distributed randomly about the trend line. In the 1940s temperatures were warmer than the trend for about a decade, while in the 1970s, temperatures were cooler for a similar length of time. This persistence or bias in a detrended time series is termed serial correlation. Implied is the notion that temperatures can be predicted as being warmer or colder than the trend simply by knowing where in time the first value in the sequence occurs. This trait weakens the statistical significance of a regression line fitted to any time series. The problem is not a major one with most climatic time series and can be corrected statistically. However, many reported analyses fail to acknowledge this limitation.

The ability to show the interrelationships or cross-correlation between two time series is a powerful technique in unravelling the causes of global climate change. There may not necessarily be a cause-and-effect relationship, but often a logical association can be made. For instance, near the ocean, rainfall is frequently correlated to sea surface temperature. Warmer ocean waters have greater evaporation rates, and the resulting condensation and precipitation over land should be greatest closest to this moisture source. However, the interaction may not be contemporaneous. It may take time for instability to develop in the atmosphere over warmer water, before increased rainfalls are measured over land. Where coastal rainfall and sea surface temperatures are cross-correlated, the strongest correlation is more than likely to show lagging of increases in rainfall behind increases in sea temperature. Lags between climatic variables are quite common, and often reflect inertia against change. For instance, air temperatures along a coast lag those further inland in spring, because bodies of water have a larger thermal capacity, and thus take longer to warm up. This thermal inertia effect can delay coastal warming by two or three months, and controls the response time of this climatic process to change.

Cycles
(Lamb, 1972; Dewey, 1987; Burroughs, 1992)

The noise in a time series can also vary systematically or cyclically. The timespan of a cycle may be constant or quasi-periodic. A cycle can be defined as any periodic phenomenon that repeats at a fixed interval. Figure 1.4 tends to show an 18 to 22-year periodicity in the occurrence of extreme climate in the United States, with greatest severity peaking in 1918, 1934, 1954 and 1976. The temperature record in Figure 1.3 is also quasi-periodic at a 40 to 50-year interval. A cycle can usually be characterised by a sine wave whose wavelength defines the periodicity. Alternatively, the cycle can be described by the number of repetitions per unit time. This is labelled frequency. The strength of a cycle is determined by the magnitude of the sine wave. While cycles may be apparent in a time series they may be so weak as to be irrelevant.

Many climatic variables display cyclicity over time (Table 1.1). The most common periodicities occur at 9.6, 11.2, 18.6 and 22 years, and reflect either features of sun processes or the lunar orbit. Whether or not such cycles are significant climatic determinants is presently subject to debate. Such cycles form intriguing and alluring alternatives for predicting future climate change. Generally, significant cycles are ones where a substantial amount of the noise remaining in a time series after detrending can be explained by some regular periodicity. Noise is categorised as 'white', spread equally across a range of frequencies; or 'red', restricted to lower frequencies or longer cycles. Significantly, the characteristics defining a cycle, or the importance of a cycle, cannot be determined accurately if the length of a time series is short relative to the periodicity of the cycle. This aspect is often overlooked in many studies purporting to explain climatic phenomenon as cyclic.

More than one cycle may also be embedded in a record. Terms such as filtering, smoothing, prewhitening, spectra analysis and harmonics are all part of the terminology of statistical techniques used in the the search for cycles. The techniques for defining cycles are complicated and involve complex computer algorithms. Cycles can be complicated by harmonics and interaction. Harmonics occur when a strong cycle displays a weaker cycle at one-half of its wavelength. For example, the 5.6 year cycle in Table 1.1 appears to be a simple harmonic of the 11.2 year sunspot cycle. The tendency for cycles to interact in an additive or subtractive manner is even more complicated. For two cycles, the resulting cycle length can be determined as follows:

additive: $\qquad x = mn/(n+m)$ **1.2**

subtractive: $\qquad x = mn/(n-m)$ **1.3**

where $\qquad x =$ the periodicity of the resulting cycle
m & $n =$ the periodicities of the forcing cycles

For example, there are three primary cycles controlling the timing of glacial Ice Ages: ~96,000 years (eccentricity), ~40–41,000 years (obliquity) and 19–23,000 years (precession). The details of these cycles can be found in Chapter 6. The additive interaction of the 96,000 and 41,000 year cycles can produce a cycle of 28,700 years, while the subtractive interaction of obliquity and precession can produce cycles of 35,000 and 54,000 years. This type of interaction is also discussed in Chapter 6. Finally, one of the benefits of a cycle is the fact that if its periodicity, magnitude and starting time are known, it is possible to predict the effect of the cycle at any point in the future. This aspect may be complicated by the tendency for some cycles to suddenly switch direction after

showing long-term regularity. This aspect will also be described in more detail later.

Chaos theory
(Gleick, 1988; Burroughs, 1992)

Many geophysical systems or time series involve stochastic or random components. Rainfall and temperature time series, for example, appear to behave as chaotic systems, and may be explained better using chaos theory. The use of the word chaos does not imply that the behaviour of these parameters is without order. Chaotic systems often have a high degree of order; however, it is impossible to predict the future trend or state of a chaotic system no matter how regular it might appear. Chaotic systems are the result of the interaction of two or more non-linear processes. For instance, the amount of rain that will fall during a thunderstorm is dependent upon the supply of moisture, the degree of uplift in the atmosphere, the rate of condensation and agglomeration of water particles, and the rate at which droplets fall. All of these processes change in a non-linear fashion and interact with each other. For this reason, it is virtually impossible to predict the amount of rain that will fall from a thunderstorm at any one location.

A linear system mathematically has the following form:

$$y = a + bx^n$$ **1.4a**

or

$$y = bx^n$$ **1.4b**

where $n = 1.0$

The first form of this equation is that normally used to define a straight line. The second form is a special case where the line intercepts the y-axis at the point where x equals 0. In linear regression analysis, equation 1.3a describes the relationship between a dependent variable y and an independent variable x. The trend line in Figure 1.3 is an example of this.

A non-linear system has the following form:

$$y = bx^n$$ **1.5**

where $n \neq 1.0$

Many climatic processes behave like this and have

Table 1.1 Reported cycles for climate-related phenomena.

Phenomenon	Periodicity (in years)	Phenomenon	Periodicity (in years)
Meridional winds in Europe	2.1–2.2	Baltic sea-ice, 1900–1950	11–14
Tropical stratospheric winds	2.2	Drought in the Canadian Prairies, 1583 ff.	18.6
Strength of northern hemisphere upper westerlies	2.2	Drought in the US Great Plains, 1805 ff.	18.6
El Niño events	2.2	Drought in northern China, 1582 ff.	18.6
Pressure fields in the north Atlantic, 1871–1974	2.2	Drought in the Patagonian Andes, 1606 ff.	18.6
Temperatures in Europe, 1760 ff.	2.2	Drought in the Nile valley, 622 ff.	18.6
Temperatures in eastern North America, 1900 ff.	2.2–2.5	Latitude world subtropical highs	19
Pleistocene ice varves	2–3	Rainfall in South Africa	20
Rainfall in eastern North America	2–3	Sunspots (Hale cycle)	22
Blocking in the northern hemisphere	2–3	Droughts in China, 1440 ff.	22
Rainfall in Europe, 1800 ff.	2–2.5	Floods in India	22
Baltic sea-ice, 1900–1950	3	El Niño events	22
Pressure fields in the north Atlantic, 1871–1974	3.4	Baltic sea-ice, 1900–1950	21–24
Pressure fields in the north Atlantic, 1871–1974	5	Seasonal pressure difference in the United Kingdom	18–23
Rainfall in the United Kingdom, 1896–1975	5	Air pressure in the northern hemisphere	18–24
Sunspot numbers	5.5	Drought in the Dnieper basin, 1650 BC ff.	20–25
El Niño events	5.5	Land temperatures in the northern hemisphere	40–50
Baltic sea-ice, 1900–1950	5-6	Greenland sea-ice	71–77
Rainfall in the United Kingdom, 1727–1927	9.5	Drought in the Nile valley, 622 ff.	77
Nile floods	9.5	Greenland ice accumulation $\partial^{18}O$	78
Latitude world subtropical highs	9.5	Latitude highs in Siberia	80–85
Atmospheric ozone	9.6	Sunspots (Gleissberg cycle)	80–90
Storm tracks in North America	9.6	Latitude highs in the north Atlantic	85–110
Barometric pressure in Paris	9.7	Severity of European winters, 1215–1905	90
Abundance of Canadian mammals	9–10	Sunspots	178
Abundance of North American birdlife	9–10	Greenland ice accumulation $\partial^{18}O$	181
Rainfall in South Africa	10	Rainfall in England	170–200
Pressure fields in the north Atlantic, 1871–1974	11	Radioactive carbon	200
Global thunderstorms	11	Southwesterly winds in England, 1340–1965	200
Droughts in China, 1440 ff.	11	Night cloudiness in China, 2300 BC ff.	400
Air pressure in the New Zealand region	11	Radioactive carbon	2300
Droughts in India	11	Holocene aridity in the tropics and subtropics	2300
Davis Strait pack ice	11	Greenland ice accumulation $\partial^{18}O$	2500
El Niño events	11	Ice Ages	~19–23,000
Sunspots	11.2	Ice Ages	~40–41,000
Arctic stratospheric ozone	11.2	Ice Ages	~96–100,000
Arctic stratospheric temperature	11.2		
Air pressure in the northern hemisphere	11–12		
Pleistocene ice varves	10–12		

ff. – following

a value of n between 1.0 and 2.0. Climatic phenomena are often the product of two or more, simple, interacting non-linear processes. As a result, chaotic processes are extremely sensitive to small disturbances. Small, almost undetectable, variations in the perturbations can result in very different outcomes. If this is the case, then it becomes impossible either to measure the system accurately or to predict its future state.

This aspect was first discovered by Ed Lorenz in the 1960s. He was the first to realise that climatic systems are non-linear and highly subject to the effects of small changes. The simulation of climatic processes in a computer model can lead to very different results between the real world and the mathematical model because of rounding errors in the iterative computer calculations. For this reason, it is impossible to predict climate more than a few days in advance using computers. Events like storms, that are highly changeable, are particularly difficult to simulate for this reason. Climatic systems behave as chaotic systems and hence always contain an unpredictable component.

Time series of chaotic systems display many of the characteristics described above. They can manifest a trend in which the signal-to-noise ratio is low, display sudden changes or steps, have variability that changes over time, incorporate extreme values, show cyclicity, and contain persistence. Logical explanations can be formulated to explain the behaviour of chaotic time series, with many explanations incorporating the concepts of positive and negative feedback. Fundamentally, time

series appear to oscillate between one or two defined end states which are the opposite of each other. For example, the time series in Figure 1.3 tends to gravitate towards temperatures that are either -0.4°C below the 1951–1979 average, or slightly above it. These particular states or stages in chaos theory are called 'strange attractors', and it is this attribute that makes chaotic systems unpredictable. The variability in a time series may simply represent the oscillation of the system between these two states. Many other climatic phenomena that will be discussed in this text display 'strange attractors' as well: cold glacials alternating with warm interglacials; widespread droughts associated with El Niño (described in Chapter 4) alternating with cool, wet La Niña years; and cold winters followed by summer heat waves. Chaotic systems also inherently contain elements of surprise. This aspect is most applicable to short-term time series of between 100 and 200 years. Chaos appears to be the dominant feature of climate and of climate change. Throughout this book it should always be remembered that the explanations being presented to describe the behaviour of climate, while valid and logical, may not be warranted.

Random or stochastic effects
(Moss and Wiesenfeld, 1995)

The time series in Figure 1.3 assumes that both the signal and noise are part of the same process or system, and that the noise is actually superimposed

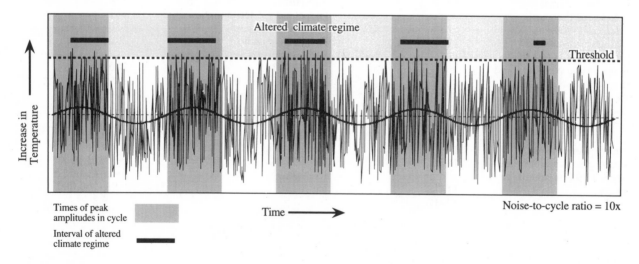

Figure 1.6 Simulated temperature time series illustrating the effect of stochastic resonance of a low amplitude cycle in the presence of high background noise. The amplitude and length of the cycle is 10% and 200 times, respectively, that of the noise.

upon the signal. It is, therefore, a relatively simple process to detect the signal and to define whether or not it is statistically significant. However, in some climatic systems this technique fails. For instance, the cycle of Ice Ages over the last 700,000 years appears to follow a 96,000 year cycle. There are features of the Earth's orbit that can explain this cycle, but the variation in solar energy received across the Earth's surface is very small, too small apparently to generate Ice Ages with intervening warm periods at this periodicity. Two additional factors are required. First, the amplitude of the 96,000 year cycle must be amplified, and second some threshold must be crossed which locks the climate system into an Ice Age through positive feedback.

These conditions can be produced by stochastic resonance. 'Stochastic' is a statistical term meaning randomness. A system that resonates is one in which some low frequency is slowly amplified until it dominates the system. Consider the time series in Figure 1.6 representing global temperature change over time. It has been generated using random numbers and is dominated by noise. The noise has been added to a low amplitude cycle having only one-tenth the amplitude and two hundred times the wavelength of the noise. The threshold drawn across the top of the diagram represents a global temperature at which positive feedback is triggered. This may be a temperature where enhanced 'greenhouse' warming begins to dominate global climate, or a warm interglacial sets in. The Earth's climate is most likely to reach this threshold at times when randomness is superimposed upon peaks in the underlying cycle, even when the amplitude of the cycle is barely noticeable. Clustering of these events above the threshold triggers positive feedback processes which can lock the system into a new state, such as glaciation (or deglaciation at troughs in the cycle). The crucial parameters in stochastic resonance are the level of the threshold, the nature of the positive feedback, and the periodicity of the cycle. The amplitude of the underlying cycle does not appear to be important. With stochastic resonance, long cycles, even of low amplitude, can emerge as a dominant signature in a resulting climatic time series. These concepts can be applied to many of the climate change phenomena described throughout this book.

REFERENCES AND FURTHER READING

Bryant, E.A. 1991. *Natural Hazards*. Cambridge University Press, Melbourne, 294p.

Burroughs, W.J. 1992. *Weather Cycles Real or Imaginary?* Cambridge University Press, Cambridge, 201p.

Dewey, E.R. 1987. 'Cycles and periodicities'. In Oliver, J.E. and Fairbridge, R.W. (eds) 1987. *Encyclopedia of Climatology*. Van Nostrand Reinhold, New York, pp. 371–375.

Frakes, L.A. 1979. *Climates throughout Geologic Time*. Elsevier, Amsterdam, 310p.

Gleick, J. 1988. *Chaos: Making a New Science*. Heinemann, London, 352p.

Haberle, R.M. 1990. 'The climate of Mars'. *Scientific American*, Special Issue v. 2 pp. 70–80.

Idso, S. 1989. *Carbon Dioxide and Global Change: Earth in Transition*. Institute for Biospheric Research, Tempe, 292p.

Jones, P.D., Wigley, T.M.L. and Briffa, K.R. 1994. 'Global and hemispheric temperature anomalies-land and marine instrumental records'. In Boden, T.A., Kaiser, D.P., Sepanski, R.J. and Stoss, F.W. (eds) *Trends'93: A Compendium of Data on Global Change*. ORNL/CDIAC-65. Carbon Dioxide Information Analysis Center, Oak Ridge National Laboratory, Oak Ridge, Tennessee, USA, pp. 603–608.

Karl, T.R., Knight, R.W., Easterling, D.R. and Quayle, R.G. 1995. 'Trends in U.S. climate during the twentieth century'. *Consequences* (http://www.gcrio.org/CONSEQUENCES/introCON.html) v. 1 no. 1.

Kasting, J.F. 1993. 'Earth's early atmosphere'. *Science* v. 259 pp. 920–925

Kasting, J.F., Toon, O.B. and Pollack, J.B. 1988. 'How climate evolved on the terrestrial planets'. *Scientific American* v. 258 no. 2 pp. 46–53.

Lamb, H.H. 1972. *Climate: Present, Past and Future*: v. 1&2. Methuen, London, 613p.

Moss, F. and Wiesenfeld, K. 1995. 'The benefits of background noise'. *Scientific American* v. 273 pp. 50–53.

Ramanathan, V., Barkstrom, B.R. and Harrison, E.F. 1989. 'Climate and the Earth's radiation budget'. *Physics Today*, May, pp. 22–32.

Smith, K. 1992. *Environmental Hazards: Assessing Risk and Reducing Disaster*. Routledge, London, 324p.

I

PROCESSES

$$2$$

Climatic Processes

ELEMENTS OF THE SUN

The solar spectrum
(Miller, 1971; Griffiths, 1976; Linacre and Hobbs, 1977; Willson, 1993)

The energy driving the Earth's atmospheric circulation and ocean currents is derived originally from the sun. The sun is a common form of gaseous star about 1.4×10^6 kilometres in diameter with a core temperature reaching $10\text{--}15 \times 10^6$ degrees centigrade. Energy is generated in the core by thermonuclear fusion, and radiates over a one- to two-million-year period towards the surface of the sun where convection and conduction become important. At its surface, the sun has a temperature of 5,770 degrees kelvin (°K) and emits 3.85×10^{26} joules of energy per second. The sun radiates its energy across the electromagnetic spectrum from longwave radio frequencies to shortwave gamma and x-rays. The electromagnetic spectrum classifies radiation according to wavelength (Figure 2.1). The longest wavelengths correspond to radio and television waves, with lengths between one and ten metres. Heat energy, the heat we can feel or sense, has a wavelength of about 10 micrometres (μm). Visible light has wavelengths between 0.4 and 0.8 μm, and dominates solar radiation output. For this reason, solar radiation is termed shortwave radiation, whereas heat radiating from the Earth is known as longwave radiation.

Figure 2.2 displays the amount of solar radiation

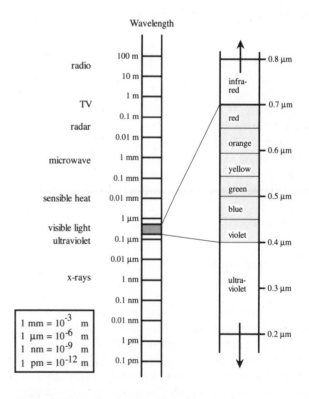

Figure 2.1 The electromagnetic spectrum.

emitted at various wavelengths. The sun behaves as a blackbody, which can be defined as any object which both absorbs and emits radiation with maximum efficiency. The theoretical distribution of energy from the sun as a blackbody is also shown on Figure 2.2. The slight discrepancies between the sun as a blackbody and as an imperfect radiator are due to absorption by gases in the atmosphere of the

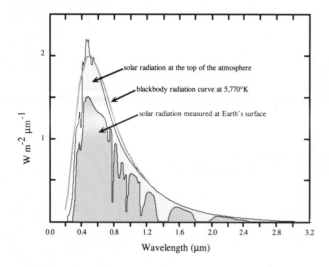

Figure 2.2 The spectrum of solar radiation (measured as Watts per square metre per micro metre) received at the top of the Earth's atmosphere and at the Earth's surface. The difference between the two curves is due to gases in the Earth's atmosphere.

sun, and by fluctuations in the output of the sun in the ultraviolet part of the spectrum. About 7% of solar radiation exists in the ultraviolet band at wavelengths below 0.4 μm, 46% within the visible range at wavelengths between 0.38-0.72 μm, and 47% in the infrared band at wavelengths above 0.72 μm. While human vision is restricted to the visible part of the spectrum, many insects can see ultraviolet light. Ultraviolet is a very important component of sunlight. In the amounts emitted by the sun, it is hazardous to all life.

Sunspots and solar flares
(Giovanelli, 1984; Shove, 1987; Willson, 1993)

Sunspots are regions on the sun's surface where magnetism increases 100–1,000 times above average. The intense magnetic field inhibits convection, and as a result 400 to 500-kilometre-deep pockmarks, 4,000–20,000 kilometres in diameter, form on the sun's surface. These depressions radiate energy less efficiently at a temperature of 4,200–4,500°K. Compared to the adjacent surface, a sunspot appears darker. Sunspots can occupy up to 1 or 2% of the sun's surface. Bright areas known as faculae surround sunspots, and actually increase overall solar irradiance when sunspots are numerous. Satellite measurements since 1978 indicate that average solar radiation can vary by as much as 0.4% on a daily basis, and about 0.1% over a sunspot cycle. This translates into a temperature fluctuation

of 0.15–0.30°C at the Earth's surface, a value of the same magnitude as the forecast temperature change per decade under anthropogenic 'greenhouse' warming. Changes in solar irradiance over the past hundred years may have ranged between 0.2–1.0%. The upper value is equivalent to 3.4 watts per square metre (W m^{-2}) at the top of the Earth's atmosphere, and can account for temperature changes measured over the last century.

Sunspots occur in groups with magnetic polarity alternating throughout a cycle lasting on average 11.2 years. Sunspots have reversed polarity between the sun's hemispheres, and also change polarity between alternate cycles. For example, if the sunspots developing in the northern hemisphere of the sun in the first eleven-year cycle show a north–south polarity, they will display a south–north polarity in the following cycle. Thus the sun has a fundamental twenty-two-year magnetic cycle, termed the Hale cycle, consisting of two 11.2 year sunspot cycles with reversed polarity. Figure 2.3 plots the number of sunspots recorded since 1500. These are termed Wolf or Zurich sunspot numbers after the first person and observatory systematically recording their occurrence. The 11.2 year cycle dominates, taking four years to rise to a maximum, and seven years to fall to a minimum. This periodicity is quasi-cyclic, and can be as short as six years and as long as seventeen years. Every second cycle has slightly more sunspots. This is the twenty-two year Hale cycle. The greatest number of sunspots in historical times was 254, occurring in October 1957. Also embedded in the record are cycles of 80–90 and 180 years, the latter correlating with the 179-year cyclic alignment of planets. While sunspot cycles appear to be a consistent feature over time, there were few sunspots and weak cycles between 1645 and 1715, during the Maunder Minimum, at the peak of the Little Ice Age. Additionally, the low numbers during the early part of the nineteenth century, termed the Dalton Minimum, may be an aberration. Reexamination of records indicates that the preceding fifty years may have had about 50% fewer sunspots than shown.

Solar flares, representing ejection of predominantly ionised hydrogen from the sun's atmosphere at speeds in excess of 1,500 kilometres per second, often develop in sunspot regions. Flares, consisting of electrically neutral, ionised hydrogen (protons and electrons), enhance the solar wind. Accompanying any solar flare is a pulse of electromagnetic radiation that takes eight minutes to reach the Earth. This radiation, in the form of soft x-rays (0.2–1.0 nanometre wavelengths), interacts with the Earth's

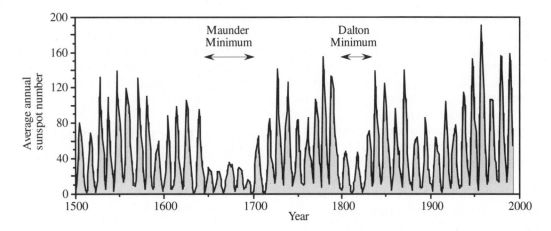

Figure 2.3 Zurich sunspot numbers between the years 1500 and 1993. Values between 1500 and 1749 are from Shove (1987); between 1749 and 1993, from the Data Support Section, Scientific Computing Division, National Center for Atmospheric Research (NCAR).

magnetic field, increasing ionisation in the lowest layer of the ionosphere at altitudes of 65 kilometres. The enhanced solar wind arrives one or two days after this magnetic pulse, and disturbs the Earth's geomagnetic field. Enhanced magnetic and ionic currents during these periods can heat and expand the upper atmosphere. Very large solar flares are called proton events. A proton event can destroy 20% of the Earth's ozone in a matter of a few days.

There are correlations between sunspot cycles and climatic effects. For instance, it is known that thunderstorm and lightning activity increase worldwide during peak periods of sunspot activity. Ultraviolet radiation also increases at times of more sunspots and during solar flares. Recent research has shown that ultraviolet radiation may have a disproportionate effect upon temperature in the upper atmosphere. Additionally, periods of low or non-existent sunspot activity over the past 5,000 years correlate with cold periods in historical records. The major periods with decreased sunspots were the Oort Minimum (1010–1050), Wolf Minimum (1280–1340), Spörer Minimum (1420–1530) and Maunder Minimum (1645–1715). The climatic implications of sunspots are discussed in more detail in Chapter 6.

SOLAR RADIATION INTERCEPTED BY THE EARTH
(Griffiths, 1976; Ramanathan et al., 1989; Houghton, 1994)

It is possible to calculate theoretically what the Earth's surface temperature should be because of solar heating. The energy output of any blackbody radiator can be calculated from its temperature using the Stefan–Boltzmann Law as follows:

$$T = (E/k)^{1/4} \qquad \textbf{2.1}$$

or

$$E = k\,T^4 \qquad \textbf{2.2}$$

where T = temperature in °K
E = energy
k = 5.67×10^{-8} W m^{-2} K^{-4}, the Stefan–Boltzmann constant

For a temperature of 5,770°K, the sun's energy output is 6.3×10^6 W m^{-2}. The average distance of the Earth from the sun is 149.6×10^6 kilometres. From simple geometry, the Earth should receive 0.46×10^{-7}% of the sun's output. Spread over the area of the Earth's disk facing the sun, the Earth should receive at the top of its atmosphere 1353 W m^{-2} of solar radiation. This is termed the solar constant. Satellites since 1978 have confirmed this value, measuring an input of 1,368 W m^{-2}. To average this value over the whole Earth, the amount of energy received by the Earth's disk (πr^2) needs to be divided by the surface area of the Earth's sphere ($4\pi r^2$). This results in an average value of 342 W m^{-2} of solar energy at the top of the Earth's atmosphere. Not all this energy is available to heat the atmosphere. The Earth and its atmosphere behave like a partial mirror, and reflect 31% of the incoming solar radiation back to space. This amount is termed the planetary albedo. Because of the planetary albedo, the Earth and atmosphere together receive 239 W m^{-2} of solar energy to drive all atmospheric and ocean circulation. This is still a large amount of energy. The amount of solar energy received by the Earth in one hour is equivalent to that generated by forty tropical cyclones, or in one second to that generated by 1,200 thunderstorms (Table 2.1).

If the Earth is assumed to be a blackbody, then its average surface temperature, using equation 2.1, should be 254.8°K. (Note that technically, with a planetary albedo of 31%, this value should actually lie some distance above the Earth's surface.) The measured surface temperature of the Earth is approximately 15°C, or about 33°C warmer than it should be. Two conclusions can be drawn from this result. First, the Earth does not

Table 2.1 Comparison between the energy generated by common phenomena and the amount of solar radiation received by the Earth.

Phenomenon	x Factor
In one day, solar input equals	
total world's coal reserves	0.4
total world's oil reserves	2.9
total world's gas reserves	3.5
annual melting of northern hemi-sphere snow	10
world's annual energy consumption in 1989	50
In one hour, solar input equals:	
tropical cyclone	40
large earthquake	60
major volcanic eruption	60
In one second, solar input equals:	
average thunderstorm	1,200
atomic bomb at Nagasaki in 1945	1,200
In one millisecond, solar input equals:	
average forest fire	12
average tornado	1,200
average lightning strike	120,000

Amount of solar energy received (in joules)
in one day = 1.5×10^{22}
in one hour = 6.3×10^{20}
in one second = 1.7×10^{17}
in one millisecond = 1.7×10^{14}

behave as a blackbody absorbing and emitting radiation efficiently. Second, the composition of the atmosphere must have important positive enhancement effects on the heating of the Earth's surface. This enhancement effect is known as the planetary 'greenhouse' effect.

ATTRIBUTES OF THE EARTH–ATMOSPHERE SYSTEM

Orbital effects
(Linacre and Hobbs, 1977)

The Earth follows an elliptical orbit about the sun with its closest approach of 146.9 × 10^6 kilometres occurring on 3 January, and its farthest approach of 152.1 × 10^6 kilometres occurring on 4 July (Figure 2.4). This inequality is such that the

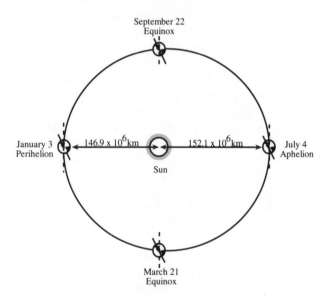

Figure 2.4 Orbital attributes of the Earth about the sun. Note that the diagram is not uniformly scaled.

southern hemisphere receives 6.7% more solar radiation than the northern hemisphere. The solar constant is only applicable at the top of the atmosphere where the rays of the sun impinge at right angles to the atmosphere. This location, termed the solar equator, moves seasonally because of the tilt of the Earth (Figure 2.5). Presently the angle of tilt of the Earth is 23.47° (23° 27'). At the solstices, either 22 December or 22 June, alternate poles are exposed to continual sunlight and darkness. The angle of the sun to the Earth's atmosphere and surface, as well as the path length of solar radiation through the atmosphere, increases towards

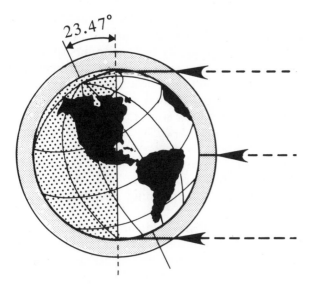

Southern Hemisphere solistice
December 22

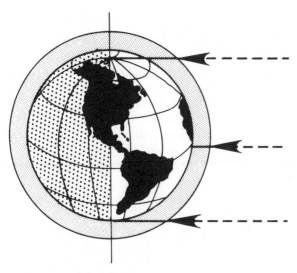

Equinox
March 21 or September 22

Figure 2.5 Effect of tilt upon input of solar radiation at the Earth's surface.

of years. Except for the latter, these orbital movements are unimportant in terms of short-term climate change at timescales of centuries, but as shown in Chapter 6, have a significant control on the timing of Ice Ages.

Composition of the atmosphere
(Linacre and Hobbs, 1977; Houghton, 1977; Plumb, 1989; Houghton et al. 1990; Stolarski et al., 1991)

Spatial variations in solar radiation between the equator and the poles drive atmospheric air circulation. However, not all of this radiation is available to heat the Earth's surface. As shown in Figure 2.2, a significant proportion of solar radiation is absorbed by molecules of gas, cloud and dust in the Earth's atmosphere. Table 2.2 summarises the present composition of the Earth's atmosphere. (A more detailed account of manufactured or anthropogenic gases appears in Chapter 7.) Of these gases, ozone, carbon dioxide, water vapour, methane, nitrogen oxide and chlorofluorocarbons (CFCs, which are chlorine-containing chemicals used as refrigerants, degreasers and foaming agents) are 'greenhouse' gases. The bulk of the atmosphere consists of the virtually inert gas nitrogen. Oxygen is a highly reactive gas, and its presence in the Earth's atmosphere is a clear sign of a life based upon photosynthesis. Carbon dioxide makes up only 0.036% of the Earth's atmosphere. And even though this is now increasing, biological activity has kept this concentration exceedingly low over geological time. The proportion of carbon dioxide has increased 29% since the Industrial revolution beginning in 1790. At present, the 'greenhouse' forcing effect of carbon dioxide on longwave emission from the Earth's surface amounts to 1.5 W m^{-2}, which is equivalent to 0.93% of the total amount of solar energy reaching the Earth's surface. This value is expected to increase to 4.0 W m^{-2} with a doubling of carbon dioxide above its preindustrial value, in about forty years' time. A complete discussion of the role of carbon dioxide and other anthropogenic 'greenhouse' gases in the actual warming of the Earth is found in Chapter 7.

Water vapour or H$_2$O is a clear gas with significant absorption properties. Its 'greenhouse' forcing effect in the atmosphere is greater than for any other gas. When it condenses into its liquid phase, it produces cloud, mist or fog which also have substantial effects on the radiation budget of the Earth's atmosphere. Water vapour is the most important 'greenhouse' gas in the atmosphere. Its

the poles. These orbital variations give rise to seasonal variations in solar radiation at any point on the surface of the Earth, and a slight discrepancy in the amount of radiation received at any one latitude between the two hemispheres. Finally, it should be pointed out that the tilt of the Earth, the degree of eccentricity of the Earth's orbit about the sun, and the time of the year that the Earth is closest to the sun, vary over periods of thousands

Table 2.2 Present gaseous composition of the Earth's atmosphere.

Constituent	Chemical formula	Concentration by volume
Nitrogen	N_2	78.08%
Oxygen	O_2	20.95%
Argon	Ar	0.93%
Water Vapour	H_2O	3.0% at the equator 0.2% near the poles
Carbon dioxide	CO_2	364.0 ppmv
Neon	Ne	18.2 ppmv
Helium	He	5.2 ppmv
Methane	CH_4	1.7 ppmv
Krypton	Kr	1.1 ppmv
Nitrous oxide	N_2O	310.0 ppbv
Xenon	Xe	100.0 ppbv
Hydrogen	H_2	50.0 ppbv
Ozone	O_3	>100.0 ppbv in stratosphere 10–100.0 ppbv in troposphere
Chlorofluorocarbons (CFCs)	Various	600–800.0 pptv

ppmv = parts per million by volume
ppbv = parts per billion by volume
pptv = parts per trillion by volume

concentration can vary between 0.2% and 2.5%, reaching 3% in extremely humid environments in the tropics. The amount of water vapour in the atmosphere is expressed as absolute humidity, and is temperature dependent. For instance, saturated air at 0°, 10° and 30°C can hold 4.0, 10.0 and 30.5 grams per cubic metre of water respectively. This attribute accounts for the wide variation in moisture content in the atmosphere between the poles and the equator.

Ozone is one of the most important gases in the atmosphere. Ozone is discussed in detail in Chapter 7, and only a brief overview is presented here. Ozone is measured by its total thickness in the atmosphere using Dobson units (1,000 Dobson units equal 1 centimetre thickness of pure ozone gas at normal temperature and pressure at sea-level). If all the ozone in the atmosphere was concentrated into a single layer, it would be only 1–5 millimetres thick. Ozone is created in the stratosphere by photolysis of molecular oxygen, and is

destroyed through reactions with atomic oxygen, usually in the presence of a catalyst. Destructive reactions are catalysed by trace amounts of hydrogen, nitrogen and halogen (chlorine and bromine) radicals. These reactions show large periodic and aperiodic variations related to the 11.2 sunspot cycle (1–2% variation), solar proton events (20%), volcanic eruptions (2–20%) and other climatic cycles (3–5%). It is now well established that anthropogenic activities have significantly increased chlorofluorocarbons and halons (fire-fighting chemicals containing bromine) in the stratosphere.

Ozone is both benevolent and noxious. It not only acts as a 'greenhouse' gas to outgoing long-wave radiation, but also intercepts and absorbs solar ultraviolet radiation. The biologically significant component of ultraviolet radiation is divided into three segments: UVA at 315–400 nm, UVB at 280–315 nm, and UVC at 100–280 nm. Ozone has a peak absorption at 250–255 nm; however, in conjunction with atmospheric particles and clouds, ozone reduces harmful UVB radiation to levels where life is possible on the Earth's surface. As an anthropogenic pollutant in the lower atmosphere, ozone in trace amounts can cause respiratory failure in humans. Yet, even here, ozone plays an ambivalent role. It reacts, in the presence of ultraviolet light, with water molecules to produce hydroxyl (OH−) radicals that cleanse air of pollutants and trace 'greenhouse' gases.

About 90% of ozone is located in the stratosphere at altitudes of 10 to 50 kilometres (Figure 2.6). The lower atmosphere or troposphere has a height of 10 kilometres at the equator, declining to 5 kilometres at the poles. Air temperature in the troposphere cools adiabatically with elevation. The lower boundary of the stratosphere starts where ozone and ultraviolet absorption becomes significant. This reversal in temperature change with elevation is quite abrupt, and forms an inversion with warm air overlying cooler air below in the troposphere. The location where the rate of change in temperature changes direction is known as the tropopause. The inversion makes the atmosphere above very stable in terms of vertical air movement. Thus, the tropopause constitutes a barrier to air circulation between the troposphere and the stratosphere.

Important processes and circulation occur in the stratosphere that affect short-term climate change in the lower atmosphere. For instance, air in the tropics at an elevation where pressure

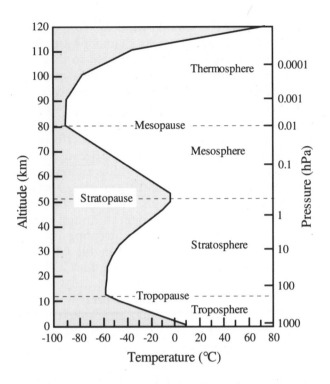

Figure 2.6 Temperature structure of the Earth's atmosphere.

is only 30 hectopascals (hPa), alternates between westerly or easterly flow about every 2.2 years. This phenomenon is termed the Quasi-Biennial Oscillation (QBO). The impact of the QBO on climate change is discussed in more detail in Chapter 6.

The final category of gases having significance in the atmosphere are chlorofluorocarbons or CFCs. CFCs do not occur naturally and are anthropogenic compounds that were virtually non-existent before 1950. The term halocarbons is now being used to include a wider range of industrial manufactured, carbon-based gases. Most of these gases are used as refrigerants, foam-blowing agents, solvents and fire retardants. Their inertness and non-conductivity make them extremely useful in the electronics industry. The most common CFCs are CCl_3F (CFC-11) and CCl_2F_2 (CFC-12), $C_2Cl_3F_3$ (CFC-113) and CCl_4, with present atmosphere concentrations of 280, 503, 82 and 132 pptv respectively. There are over thirty halocarbons, not all of which are noted as deleterious in the atmosphere. CFCs have been linked to destruction of ozone over both poles in spring, and over large industrial cities in the northern hemisphere. All CFCs and halocarbons are also significant 'greenhouse' gases with potential radiative forcing in the atmosphere up to eight thousand times greater

than carbon dioxide on a molecule-for-molecule basis. A more detailed discussion appears in Chapter 7.

EFFECT OF THE ATMOSPHERE AND EARTH'S SURFACE
(Miller, 1971; Griffiths, 1976; Linacre and Hobbs, 1977; Daly, 1989; Idso, 1989)

Absorption

There are three processes which affect radiation in the Earth's atmosphere: absorption, scattering and reflectance. Many of the so-called 'greenhouse' gases in the atmosphere are primarily absorbers of radiation across a wide bandwidth in the electromagnetic spectrum. This bandwidth includes shortwave solar radiation entering the Earth's atmosphere, as well as longwave radiation emitted from the Earth's surface. The effect of absorbers on solar radiation is shown broadly in Figure 2.2. An absorber is any gas molecule or particle which intercepts radiation at a particular wavelength, and then re-emits that energy at a longer wavelength. In the process of absorption, the gas or particle itself heats up. Many gases are selective absorbers at very narrow bandwidths. For example, ozone and oxygen absorb almost all solar radiation at wavelengths below 0.29 μm, but hardly any radiation between 0.3 to 9 μm. The degree of absorption for the main 'greenhouse' gases, together with the total absorptivity of the Earth's atmosphere, are plotted in Figure 2.7. In this diagram, if the atmosphere is completely open to radiation transmission, then the absorptivity index equals 0. If the atmosphere is completely closed, then the index equals 1. Note that this diagram plots absorptivity across both the shortwave spectrum between 0.1 and 2.5 μm and the longwave spectrum for the Earth between 4.0 and 40.0 μm.

Three features stand out. First, ozone (O_3) and oxygen (O_2) close the Earth's atmosphere to ultraviolet radiation transmission. Ozone is also a 'greenhouse' gas and closes the atmosphere to the transmission of longwave radiation between 9 and 9.7 μm. Because the bandwidth or window at which ozone absorbs radiation is totally closed, additions of ozone to the atmosphere have little further effect on its absorptivity in the atmosphere. Thus the forcing of ozone as a 'greenhouse' gas is relatively insensitive to change. Second, because of

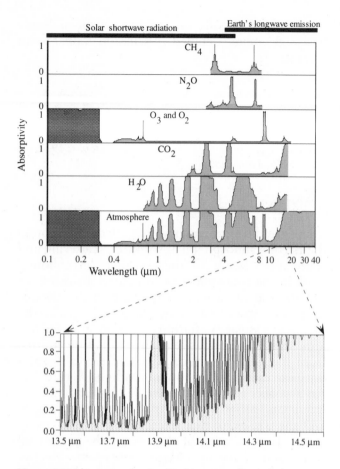

Figure 2.7 Absorption by the Earth's atmosphere of shortwave and longwave radiation across the electromagnetic spectrum from 0.1 to 40 μm. The inset diagram shows the detail of absorptivity of the Earth's atmosphere to longwave transmission between 13.5 and 14.7 μm (based upon Fleagle and Businger, 1963 and Houghton, 1977).

mainly water vapour (H₂O), the Earth's atmosphere is closed to longwave emission at wavelengths between 5.3 and 7.7 μm and above 15 μm. While ozone and oxygen block out 2% of incoming solar radiation, water vapour absorbs 15% of outgoing radiation and a significant amount of incoming solar radiation. Third, if water vapour increases because of the effects of increased anthropogenic 'greenhouse' gases, the effect will be partially counterbalanced by absorption of incoming solar radiation into the atmosphere.

The situation is slightly more complicated for the other 'greenhouse' gases. Nitrous oxide (N_2O) absorbs longwave radiation totally in two narrow windows, but only partially below 4 μm. A similar situation applies for methane (CH_4). Its dominant window around 7–8 μm is partially transparent, although it overlaps the absorption window for

nitrous oxide. The main window for carbon dioxide (CO_2) occurs between 13 and 17 μm. This window above 14.5 μm is also closed, so that any increase in carbon dioxide would have a minimal effect intercepting more outgoing longwave radiation. The windows at 2 and 4 μm for carbon dioxide are in the region where longwave emission from the Earth is small. The absorption effects for CFCs are not plotted in Figure 2.7. Because these gases occupy a virtually open window to longwave transmission between 8 and 13 μm, their forcing on climatic warming through absorption of longwave emission is at least one thousand times greater than for carbon dioxide. Specifically, absorption peaks occur at 11.9 μm for CFC-11, 8.6 and 10.9 μm for CFC-12, 8.9 μm for CFC-113 and 9.9 μm for methyl chloride.

The above is a broad, simplistic presentation of radiation absorptivity in the atmosphere. The bottom panel in Figure 2.7 presents the actual absorption spectrum for the atmosphere in the 13.5 to 14.7 μm range, where carbon dioxide and water vapour become very effective at absorbing outgoing longwave radiation. The low resolution graphs in the top panel smooth out reality. In actual fact, there are many substantial gaps in the absorption spectrum. Gas molecules collect radiation by their vibrational energy states, forming bands through rotational splitting. The absorption windows are formed by the averaging of a series of narrow bands. If one of these narrow bands is still partially open, then absorptivity will increase linearly as the concentration of the 'greenhouse' gas involved increases. If the band is saturated or closed, then an increase in the concentration of the gas will begin to increase absorption at the edge of the band (termed band wings). Absorption in the band wings or broadening can also be increased by resonance, temperature and pressure increases. Band wing broadening appears to be insensitive to present temperature increases. In order for absorption band broadening to occur to any degree, the partial pressure of 'greenhouse' gases in the atmosphere would have to increase by a factor of one thousand times.

Scattering

The second process affecting radiation transfer through the atmosphere is scattering. Scattering affects shortwave radiation, and involves the interference and deflection of radiation, travelling downwards through the atmosphere, by gas molecules and particulates such as dust and pollution. At present, the Earth's atmosphere scatters about 18%

of all incoming solar radiation. The effect redirects about 7% of radiation upwards and 11% of it downwards. The degree of scattering depends upon the size of the molecule and the wavelength of radiation. Longwave emission from the Earth's surface has wavelengths which are too large to be scattered in the atmosphere. Gas molecules scatter shorter wavelengths more effectively. Thus, when the sun is at a high angle in the sky, almost all blue light is scattered, making the sky appear blue. As the sun sets lower in the sky, radiation travels through a greater thickness of atmosphere, scattering longer wavelengths. Thus sunsets appear red and then become purple as the sun drops below the horizon. Dust particles, smoke and crystals of frozen water are bigger than air molecules, and tend to scatter all wavelengths of visible light. This causes the sky to become whitish, a sign that the atmosphere has become contaminated by dust from arid regions or by pollutants emanating from cities.

Reflectance (albedo)

The third process affecting radiation transference through the atmosphere is reflectance. Reflectance is the amount of shortwave radiation that is not absorbed or scattered in the Earth–atmosphere system, but returned directly to space unobstructed after hitting a molecule, particle or object. Reflectance can occur both within the atmosphere and at the Earth's surface. Table 2.3 lists some of the reflectance or albedo values for various surfaces. The total amount of short radiation reflected by the Earth-atmosphere system is termed the planetary albedo. As already mentioned, this value amounts to 31% of solar radiation.

Reflectance depends upon cloud thickness, type of vegetation, soil wetness, freshness of frozen water and sun angle. Clouds on average reflect 45% of the incoming solar radiation that they intercept. Low-level clouds associated with rainfall reflect between 25 and 84% of solar radiation; they are an important factor in explaining the decrease in solar radiation between the top of the atmosphere and the Earth's surface. The precise degree of reflectance by various types of clouds at different altitudes is not yet known, however such information is needed to explain fully the effect of enhanced 'greenhouse' warming. Reflectance from vegetated surfaces is very sensitive to vegetation type. The albedo of eucalypt forest in Australia is very low at 5%, whereas the figure for dry grassland is as high as 22%. This has obvious environmental implications regionally and globally. For instance, the conversion of forests

Table 2.3 Albedos of various surfaces.

Surface	Reflectance (%)
Clouds:	
average	45
cirrus	2–3
fair-weather cumulus	30
low-level:	
<150 metres thick	25–63
150–300 metres thick	45–75
300–450 metres thick	59–84
thick cumulonimbus	90
Ground surface:	
asphalt	5–10
concrete	17–27
sand	30–60
soil:	
dark	5–15
light and dry	25–30
moist	14–17
snow:	
fresh	75–90
old	45–70
Vegetation:	
forest:	
conifer	4–18
deciduous	8–23
tropical	4–18
grassland	5–25
green crop	5–25
tundra	15–35
Water surfaces:	
new ice	>90
water:	
sun angle 3°	60
sun angle 10°	38
sun angle 25°	9
sun angle 90°	4
Planets:	
Earth	31
Earth's moon	7
Mars	16
Venus	80

to grassland or cropland, as occurred in the nineteenth century in westernised countries, should have produced significant regional increases in albedo with a consequent decrease in local temperatures. Similarly, the large-scale clearing of rainforest for grazing land should produce a cooling effect in developing tropical countries. The exposure of grasslands to desertification or to droughts should also display increases in reflectivity. The simple wetting of large areas of land during prolonged rain-fall periods, and the drying out of these tracts during droughts, leads to changes in microclimate at the Earth's surface for similar reasons.

Both snow- and ice-covered landscapes have the highest albedos of any surface. If snowfall increases areally because of a colder than usual winter, especially in the northern hemisphere which contains the greater proportion of the Earth's landmasses, then regional air temperatures should decrease substantially. Reflectance from water surfaces is low when the sun lies directly overhead, but can increase to 38% when the sun angle falls to 10°, and to 60% when the sun angle is only 3° above the horizon. This difference should not be important globally, because the sun angle for any latitude varies consistently throughout the year. However, during Ice Ages, sea-levels were up to 130 metres lower than at present, and large expanses of the world's continental shelves would have been exposed to sunlight. The lifting of reflectances from an average of 8% to as high as 25 or 30%, as continental shelves turned from being water-covered to soil-covered, would have had a positive feedback on the lowering of global temperatures. The effects of both decreased areas of water in the world's oceans, and increased areas of snow and ice, had important implications for the onset and development of glaciations. These effects are discussed further in Chapter 6.

GLOBAL RADIATION BUDGET
(Ohring and Gruber, 1983; Ramanathan et al., 1989; Graedel and Crutzen, 1993; Harrison et al., 1993)

The balance between incoming solar radiation and outgoing longwave radiation defines the Earth's radiation budget. An idealised model of these radiation components, averaged annually for the whole Earth, is presented in Figure 2.8. Budget calculations have been performed for the Earth's surface, the atmosphere and at the top of the atmosphere. Note, that the amount of radiation added to the atmosphere or Earth's surface must be exactly balanced by emissions. In addition, the 100 units of shortwave energy originally entering the Earth–atmosphere system must be balanced by reflected, or upwardly scattered, shortwave radiation, plus upward longwave emissions.

Of the 342 W m^{-2} of solar radiation coming into the atmosphere, 74% is intercepted in some way by gases, particles or clouds in the atmosphere. Approximately 39% of incoming solar radiation is intercepted by cloud. Of this, 20% is reflected back to space, 5% is absorbed by cloud and 14% is diffused downwards to be absorbed by the Earth's surface. The 5% absorption by clouds (this value is subject to a high degree of uncertainty) is important, because it becomes a significant component of the transport required to remove excess heat out of equatorial regions. Absorption also allows a temperature inversion to develop at cloud level. This can form a pollution trap over cities or industrial areas, unless accompanied by wind or substantial convection. Gases such as ozone, water vapour and carbon dioxide, as well as any dust or acid particles, absorb 17% of all incoming radiation. Of this only 6.9 W m^{-2} is absorbed by ozone in the stratosphere. Gases and dust scatter 18% of incoming radiation, of which 11% is directed downwards, and 7% is reflected back to space as part of the planetary albedo. Only 4% of radiation reaches the Earth's surface directly and is reflected unimpeded back to space. This latter contribution to the planetary albedo is relatively small, amounting to 17.9 W m^{-2}. In terms of climate change, changes in land use on the Earth's surface play a secondary role to changes in the composition of the atmosphere. Only 22% of all solar radiation passes unobstructed through the atmosphere, and is absorbed directly at the Earth's surface. In total, the Earth's surface receives 47% (161 W m^{-2}) of solar radiation, while half as much (75.5 W m^{-2}) is absorbed within the atmosphere. Heating in the atmosphere provides a considerable amount of energy available for the generation of atmospheric circulation.

The longwave radiation budget is very different from the shortwave one. First, scattering and reflection processes do not affect radiation at long wavelengths. Second, the longwave radiation flux from the Earth's surface is very much affected by 'greenhouse' gases, of which water vapour and carbon dioxide are the major components. Clouds also play a major role in this 'greenhouse' process. Finally there are two important fluxes, beside the radiation flux, that transfer heat from the Earth's surface to the atmosphere. The lesser of these fluxes involves the physical transfer of heated air from the Earth's surface to the atmosphere. This transfer, termed sensible heat, is due to convection and turbulence. Convection is the heating of air at the Earth's surface with the subsequent movement

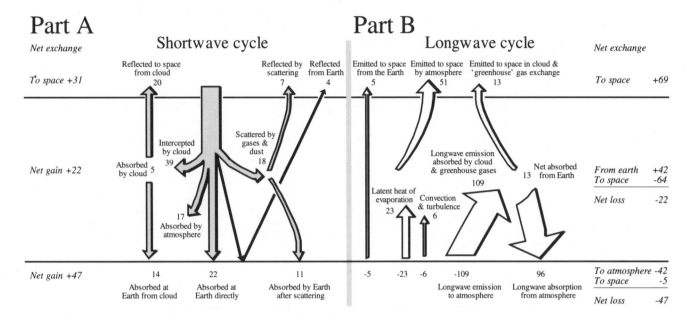

Figure 2.8 Schematic diagram of the radiation budget for (A) shortwave and (B) longwave radiation for the Earth–atmosphere system. Values are expressed as percentages where 100% equals 342 W m⁻², the average annual input of solar radiation at the top of the atmosphere. The values represent estimations substantiated by satellite information obtained through the Earth Radiation Budget Experiment. Individual figures vary by ±2%, and in the case of clouds by 10% overall.

of that air upwards into the atmosphere because of its lower relative density. Turbulence is the tendency for air to be turned over as it moves. While convection induces turbulence, air moving laterally across the Earth's surface can also become turbulent because of ground obstructions. The flux due to convection and turbulence amounts to 20.5 W m⁻², proportional to 6% of the initial amount of solar radiation impinging upon the Earth–atmosphere system. The greater of these fluxes is due to latent heat of evaporation at the surface, and subsequent condensation of water vapour as fog, mist and cloud in the atmosphere. A gram of water at 20°C requires 2.45 kilojoules of energy to evaporate. There is no sensible heat involved in this change of phase from liquid water to vapour, so the energy can be considered latent. The amount of heat used in evaporation of water from the Earth's surface becomes a very significant factor in the radiation budget. Because water vapour can be transported by winds, the latent heat process is a significant mechanism for transferring heat laterally across the Earth's surface. The process of evaporation and condensation amounts to 78.7 W m⁻², proportional to 23% of the initial amount of solar radiation impinging upon the Earth–atmosphere system.

Only 17.1 W m⁻² of the longwave radiation flux,

proportional to 5% of the total incoming solar radiation, is lost directly to space from the Earth's surface without any interference by the atmosphere. More important is longwave emission of radiation from the atmosphere itself. This flux is equivalent to 51% of the total solar radiation input. These fluxes are both dwarfed by the longwave radiation flux from the Earth's surface, absorption of that radiation by cloud and 'greenhouse' gases, and transmission of that radiation back to the Earth's surface. The Earth radiates to the atmosphere, and the atmosphere reradiates back, energy equivalent to 109% and 96% respectively of the total solar radiation input to the Earth–atmosphere. In this exchange, 44.5 W m⁻² is reradiated upwards to space. This process involves the recirculation of energy seven times between the Earth and the atmosphere before it is lost to space.

The effectiveness of this process in heating the Earth's surface can be illustrated by comparing the differences in surface temperature between a clear and cloudy night. The cloudy night will stay warmer for longer, because of the continual 'trapping' by clouds and recycling of longwave radiation emanating from the Earth's surface. Cloud is a crucial factor in the Earth's 'greenhouse' process. The debatable questions are whether an

Part A

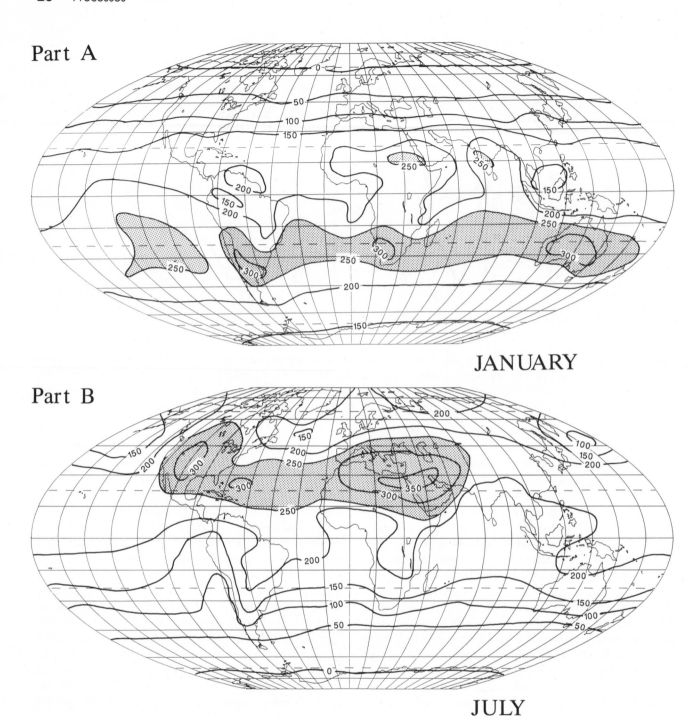

JANUARY

Part B

JULY

Figure 2.9 Solar radiation in watts per square metre reaching the Earth's surface for (A) January and (B) July (modified from Linacre, 1992). Areas with values greater than 250 W m⁻² are shaded.

increase in cloudiness cools the Earth's surface temperature, because the added cloud reflects more incoming solar radiation back to space, or whether it warms the Earth's surface, because it enhances the 'greenhouse' effect. Possible answers to these questions are found in Chapter 7.

TIME- AND SPACE-AVERAGED RADIATION BUDGET EFFECTS
(Graedel and Crutzen, 1993; Harrison et al., 1993)

Figure 2.8 presents schematically the average state of the Earth's radiation budget. Because of tilt,

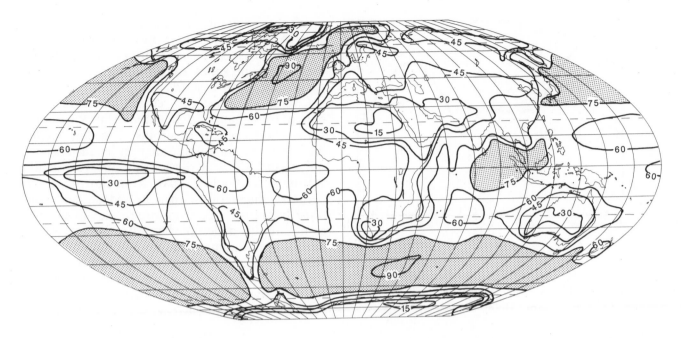

Figure 2.10 Global distribution of annual mean cloud amount expressed as percent cover (modified from Rossow, 1993). Areas with greater than 75% cloud are shaded.

landmasses and spatial variability in cloud, the pattern of net radiation varies seasonally across the Earth. Figure 2.9 plots the effect of these variations across the Earth's surface for the months of January and July. The effect of cloud on this pattern is shown in Figure 2.10 which plots the average annual amount of cloud measured over two years by satellites. Logically, equatorial regions should have more incoming solar radiation than the poles, because the sun is positioned overhead in the tropics more of the time. While there is a zonal increase towards the solar equator seasonally, there are many exceptions to this expected pattern. For instance, both the Canadian Arctic and the Antarctic, in summer, receive as much radiation as Malaysia situated on the equator. While part of this difference is due to the presence of 24 hours of sunlight near the poles in summer, some is due to the low amounts of solar radiation reaching Malaysia because of extensive cloud cover (Figure 2.10). Cloud covers Malaysia 75% of the time, but only 30% of the time does it cover the Antarctic. Elsewhere, the Southern Ocean around 50° S is cloud-covered for over 90% of the time, while large sections of the ocean south of Africa and east of Newfoundland in North America are cloud-covered for 90 to 100% of the year.

The highest inputs of solar radiation are received over the semi-arid belts of the northern and southern hemisphere around 20–30° north and south of the equator. The region from the Sahara desert in Africa through to the Arabian peninsula, receives an average of over 300 W m^{-2} of solar radiation per day in July. This is also a region which has less than 15% cloud cover throughout the year. Other regions of high solar input occur over the Kalahari desert in South Africa, the semi-arid area of eastern Australia, and the southwestern arid zone of the United States.

The broad latitudinal effect of cloud upon both long- and shortwave radiation is summarised in Figure 2.11. While this diagram plots cloud only for December, January and February (summer in the southern hemisphere), the zonal pattern for the whole year is well represented. Overall, cloud covers 59% of the globe. Generally, total cloud cover is higher over the oceans than over landmasses. Three peaks appear in the data. Two peaks occur over both land and ocean around 60° north and south of the equator. These are regions where intense storms form off the east coasts of Asia, North America and South America. These storms decay over landmasses downwind. The third peak appears over land in equatorial regions, predominantly in the southern hemisphere. This peak overlies most of the world's tropical rainforests, and the Indonesian–

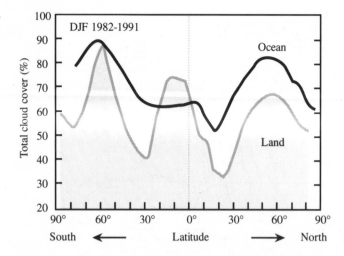

Figure 2.11 The total amount of cloud for the months of December, January and February by latitude over land and ocean (modified from Hahn et al., 1994).

Australian tropical landmass. In the latter case, land is surrounded by numerous shallow seas which are dominated by clusters of intense, complex, convective storms. The lack of cloud over the ocean in the tropics, reflects the presence of a large pool of cool water extending westwards, along the equator from the coast of South America. The reason for this pattern, and its importance in climate change, are described in greater detail in Chapter 4.

The annual latitudinal balance between shortwave radiation input and longwave emission is presented in Figure 2.12. This plot is based upon satellite measurements. The area under each curve represents the total amount of incoming shortwave and outgoing longwave radiation at the Earth's surface. The difference between the two curves represents surpluses or deficits in radiation across the globe. The sum of these values must equal zero in order for the Earth to maintain an average surface temperature of 15°C. Solar radiation reaching the Earth's surface varies more with latitude than does longwave emission. The difference indicates the dependence of shortwave radiation upon the sun's angle and the tilt of the Earth. Despite the fact that cloud dominates over equatorial landmasses, the tropics receive the most solar radiation, simply because of their greater exposure to sunlight at high sun angles throughout the year. The southern hemisphere receives slightly more solar radiation than the northern hemisphere, because of the fact that the Earth is closest to the sun during the southern

hemisphere summer (Figure 2.4). Longwave emission is greatest at 25° north and south of the equator. These areas are more cloud-free because they occur in zones of subsiding air associated with subtropical deserts.

Overall the poles are regions where there is a deficit in radiation, while the tropics occupy a zone where there is a surplus. The points where incoming and outgoing radiation balance occur at 35° N and 40° S. If the Earth's surface is neither warming nor cooling, apart from short-term temperature fluctuations, then heated air and water must be transported from the tropics to the poles to negate this deficit in the radiation budget. Similarly, the latitudinal imbalance in radiation can be rectified by transporting cold air and water from the poles to be warmed in the tropics. These transfer mechanisms drive the Earth's observed air circulation and ocean currents.

CONCLUDING COMMENTS

The facts and figures summarised in this chapter encompass a degree of uncertainty. For instance, Figures 2.8 and 2.9 should be treated with caution. The Earth has undergone substantial short-term climate change since 1975, so that even if the satellite measurements were accurate, there is a degree of variation or noise in the measurements of albedo, longwave emission and cloud that reduces the accuracy of the plotted values. In addition, satellite measurements are subject to error and to sampling deficiencies. Both of these problems increase towards the poles. At most, the radiation estimates are accurate to $\pm 2\%$. In addition, measurements of the average global longwave emission vary between 229 and 244 W m^{-2}. While this range approximates the value of 239 W m^{-2} required to balance incoming solar radiation absorbed by the Earth–atmosphere system, the 15 W m^{-2} uncertainty is substantial, being equivalent to the longwave energy transfer between the Earth's surface and the atmosphere due to either convection or turbulence. These errors must be considered, or at least acknowledged, when modelling radiation transfers in the Earth–atmosphere system.

Finally, this chapter has emphasised the important role that clouds play in intercepting both incoming and outgoing radiation. This process is called cloud forcing. Clouds globally force warming ('greenhouse' effect) and cooling (albedo) by amounts equivalent respectively to

13% and 20% of the total incoming solar radiation flux. On average, clouds have a net cooling effect on the Earth throughout the year of 17.3 W m⁻². Computer simulations of the Earth's radiation budget, involving cloud, often do not replicate these measured values. In some cases, the error is as much as 100%. Recently, the United States space agency, NASA, measured radiation transfers simultaneously at altitudes of 8 and 19 kilometres, and consistently found that the intervening clouds were absorbing substantially more solar radiation than the 5% theoretical estimate plotted in Figure 2.8. While the effects of cloud upon radiation would appear to be an easy aspect to measure, the processes are more complex than presently conceived.

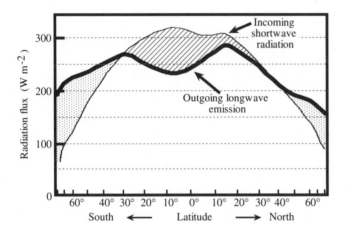

Figure 2.12 Net incoming solar radiation versus outgoing longwave emission by latitude (based on Graedel and Crutzen, 1993 and Harrison et al., 1993). The scale for latitude is proportional to area. Deficits are shaded with stippling; surpluses with slanted lines.

REFERENCES AND FURTHER READING

Daly, J.L. 1989. *The Greenhouse Trap*. Bantam, Sydney, 192p.

Fleagle, R.G. and Businger, J. 1963. *An Introduction to Atmospheric Physics*. Academic Press, New York, 346p.

Giovanelli, R.G. 1984. *Secrets of the Sun*. Cambridge University Press, Cambridge, 116p.

Graedel, T.E. and Crutzen, P.J. 1993. *Atmospheric Change: An Earth System Perspective*. Freeman, New York, 446p.

Griffiths, J.F. 1976. *Climate and the Environment:* *The Atmospheric Impact on Man*. Paul Elek, London, 148p.

Hahn, C.J., Warren, S.G. and London, J. 1994. *Climatological data for clouds over the globe from surface observations, the total cloud edition*. Carbon Dioxide Information Analysis Center, Oak Ridge, CDIAC Dataset NDP026A.

Harrison, E.F., Minnis, P., Barkstrom, B.R. and Gibson, G.G. 1993. 'Radiation budget at the top of the atmosphere'. In Gurney, R.J., Foster, J.L. and Parkinson, C.L. (eds) *Atlas of Satellite Observations related to Global Change*. Cambridge University Press, Cambridge, pp. 19–38.

Houghton, J. 1977. *The Physics of the Atmospheres*. Cambridge University Press, Cambridge, 203p.

Houghton, J. 1994. *Global Warming: The Complete Briefing*. Lion Publishing, Oxford, 192p.

Houghton, J.T., Jenkins, G.J. and Ephraums, J.J. (eds) 1990. *Climate Change: The IPPC Scientific Assessment*. Cambridge University Press, Cambridge, 365p.

Idso, S.B. 1989. *Carbon Dioxide and Global Change: Earth in Transition*. Institute for Biospheric Research Press, Tempe, 292p.

Linacre, E. 1992. *Climate Data and Resources: A Reference and Guide*. Routledge, London, 366p.

Linacre, E. and Hobbs, J. 1977. *The Australian Climatic Environment*. Wiley, Sydney, 354p.

Miller, A. 1971. *Meteorology*. 2nd ed. Merrill, Columbus, 154p.

Ohring, G. and Gruber, A. 1983. 'Satellite radiation observations and climate theory'. *Advances in Geophysics*, v. 2 pp. 237–304.

Plumb, R.A. 1989. 'Atmospheric ozone-Physics and Chemistry'. *Transactions of the Menzies Foundation* v. 15 pp. 3–13.

Ramanathan, V., Barkstrom, B.R. and Harrison, E.F. 1989. 'Climate and the Earth's radiation budget'. *Physics Today*, May, pp. 22–32.

Rossow, W.B. 1993. 'Solar irradiance'. In Gurney, R.J., Foster, J.L. and Parkinson, C.L. (eds) *Atlas of Satellite Observations related to Global Change*. Cambridge University Press, Cambridge, pp. 141–163.

Shove, D.J. 1987. 'Sunspot cycles'. In Oliver, J.E. and Fairbridge, R.W. (eds) *Encyclopedia of Climatology*. Van Nostrand Reinhold, New York, pp. 807–815.

Stolarski, R., Bloomfield, P., McPeters, R. and Herman, J. 1991. 'Total ozone trends deduced from Nimbus 7 TOMS data'. *Geophysical Research Letters* v. 18 pp. 1015–1018.

Willson, R.C. 1993. 'Solar irradiance'. In Gurney, R.J., Foster, J.L. and Parkinson, C.L. (eds) *Atlas of Satellite Observations related to Global Change*. Cambridge University Press, Cambridge, pp. 5–18.

3

Scales of Heat & Mass Transfers in the Atmosphere

BASIC TRANSFERS
(Griffiths, 1976; Harrison et al., 1993)

The transfer of energy to maintain thermal equilibrium over the globe is carried out by four different types of fluxes in the oceans and the atmosphere: sensible heat in air, latent heat, ocean currents and ice. In the atmosphere, energy can be transported by the movement of sensible heat contained in air masses or by latent heat that is locked up in evaporated moisture. In the oceans, the thermal capacity of water is much higher than in the atmosphere, and heat can be transferred by ocean currents. In addition, ice drifting from the poles uses up latent heat of fusion in the process of melting. One gram of ice requires 334.72 joules of heat at 0°C to melt. The magnitude of the first three fluxes are presented in Figure 3.1. The flux for ice melting is uncertain but it is small compared to the other three parameters and can be ignored. The total energy transport zonally across the globe is based upon measurement techniques. Variability in this latter value is due to year-to-year fluctuations between 1974 and 1986, and to different satellites. Energy transport is maximum at 35° N and 40° S, the same points in Figure 2.12 where incoming solar radiation balanced outgoing longwave radiation. While the net energy transfer in each hemisphere must be towards the poles, individual components can vary. Both the sensible heat flux in the atmosphere and that involved in ocean currents follow the net transfer curve. There are slight variations between hemispheres due to the latitudinal variations in the amount of land.

Sensible heat transfer reaches a peak between 50 and 60° N because the bulk of Earth's landmasses lie there. In the southern hemisphere, the sensible heat flux is greatest between 20° and 30° S, again over landmasses. Ocean currents transfer more heat poleward in the southern hemisphere because this hemisphere contains more ocean. The latent heat flux follows the average pattern of surface air movement, with energy being transferred towards the equator between 20° N and S, and towards the poles poleward of these latitudes. The latent heat flux decreases abruptly within 20° of the poles because air holds little moisture at the cold temperatures found at these latitudes.

CORIOLIS FORCE, VORTICITY AND ROSSBY WAVES
(Miller, 1971; Linacre and Hobbs, 1977; Bryant, 1991)

The energy transfers shown in Figure 3.1 occur on a rotating sphere and are thus subject to deflection because of Coriolis force. The degree of deflection is accounted for by the following equation:

Coriolis force = 2 ω sin ø V   **3.1**

where ω = rate of spin of the earth
    sin ø = latitude
    V = wind speed

The rate of spin of the Earth is a fixed quantity except for very small scale fluctuations on the order of milliseconds, induced by wind variations

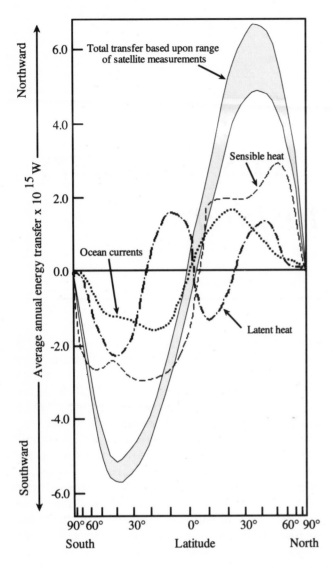

Figure 3.1 Annual distribution of components of latitudinal energy flux. The scale for latitude is proportional to area (based on Griffiths, 1976 and Harrison et al., 1993).

across the globe. Coriolis force varies across the surface of the globe because of latitude, and wind or current velocity. Coriolis force is zero at the equator ($\sin 0° = 0.0$), and maximum at the pole ($\sin 90° = 1.0$). The force exists simply because of the rotation of the Earth and is illusionary. For example, a person standing perfectly still at the pole would appear to an observer, viewing the Earth from the moon, to turn around in a complete circle every 24 hours. However, the same person standing perfectly still at the equator would always be facing the same direction, and not rotating about himself. As a result of Coriolis force, air and water movement at the equator is dominated by zonal flow (flow parallel to degrees

of latitude). The equator is also a barrier to inter-hemispheric movement because vortices rotate in opposite directions in each hemisphere. For example, tropical cyclones cannot cross the equator because they rotate clockwise in the southern hemisphere and anti-clockwise in the northern hemisphere. Away from the equator, any movement, involving air or water, whether on the surface of the Earth, at depth in the ocean, or at some height in the atmosphere, is affected by Coriolis force. In the northern hemisphere deflection is always to the right, while in the southern hemisphere deflection is to left.

The atmosphere has weight which can fluctuate at the surface of the Earth because of a change in elevation, or because air may be converging towards or diverging away from a point. The weight of air above a location on the surface of the Earth is termed barometric pressure. Spatial variations in pressure lead to alternating patterns of high and low pressure. Points of equal pressure across the Earth's surface are joined together on a weather map by isobars usually spaced at intervals of 4 hectopascals (hPa) or millibars (mb). Mean pressure for the Earth is 1,013.6 hPa. Wind is generated by air moving from high to low pressure, simply because a pressure gradient exists due to the difference in air density. The pressure gradient lies perpendicular to isobars (Figure 3.2). The stronger the pressure gradient, the closer the spacing between isobars and the stronger the wind. In reality, wind does not flow down pressure gradients but parallel to isobars because of Coriolis force (Figure 3.2). The resulting wind due to a combination of Coriolis force and the pressure gradient is termed the geostrophic (turning with the earth) wind. Air spirals outwards from high pressure anti-clockwise in the southern hemisphere and clockwise in the northern hemisphere. As a corollary, air spirals into low pressure clockwise in the southern hemisphere and counterclockwise in the northern hemisphere.

Deflection of airflow (or water) tends to form a vortex. The stronger the wind, the smaller and more intense the resulting vortex. Very strong vortices of air form tropical cyclones and tornadoes. Vortices cannot develop at the equator. The rate at which a cell of air rotates is termed vorticity. If a vortex forms at a particular latitude, it obtains a degree of vorticity conditioned by Coriolis force at that latitude. The vorticity of an air mass tends to be conserved, an attribute that is not important unless a vortex changes latitude. This can occur because of deflection by topography, such as

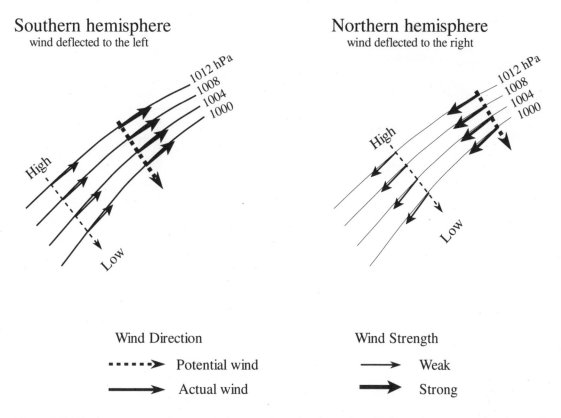

Southern hemisphere
wind deflected to the left

Northern hemisphere
wind deflected to the right

Wind Direction		Wind Strength	
┅┅┅▸	Potential wind	⟶	Weak
⟶	Actual wind	⟹	Strong

Figure 3.2 Wind movement relative to isobars, with and without Coriolis force, (i.e. potential and actual).

mountains, or displacement by moving masses of different air. Conservation of vorticity dictates that the path of a vortex moving poleward will curve (clockwise in the northern hemisphere and anticlockwise in the southern hemisphere) deflecting the air back towards the latitude where it acquired its vorticity (Figure 3.3), for the northern hemisphere. However, moving air also has momentum, and rather than stop at its original latitude, it will tend to overshoot this location and proceed towards the equator. The need to maintain vorticity, this time with the vortex in a more equatorial position, tends to deflect its path in an anticlockwise direction (northern hemisphere) or a clockwise direction (southern hemisphere) back towards the pole. Hence, any rotating air mass that is shifted from its original latitude tends to trace out a path of alternating cyclonic and anticyclonic loops, or waves, downwind.

In the upper atmosphere, these waves, termed Rossby waves, become quite prominent and are associated with the jet stream at the top of the troposphere (Figure 3.3). The looping in the upper atmosphere also generates pressure cells at the surface. Air moving as a stream towards the equator tends to undergo convergence which piles up air. Convergence of upper air leads to increased pressure and subsidence in the atmosphere, with the attendant formation of stable high pressure cells at the Earth's surface. Air moving poleward in the upper atmosphere undergoes divergence. Air is drawn in from below, leading to atmospheric instability and the formation of low pressure at the surface. This latter process is termed cyclogenesis. Sometimes the looping is so severe that winds in the jet stream simply take the path of least resistance, and proceed zonally (parallel to latitude), cutting off the loop. In this case, high and low pressure cells can be left stranded in location for days or weeks, unable to be shifted easterly by the prevailing westerly airflow. High pressure cells are particularly vulnerable to this process and form blocking highs. These can deflect frontal lows to higher or lower latitudes than normally expected, producing extremes in weather. Blocking is more characteristic of winter than summer, especially in the northern hemisphere. It also tends to occur over abnormally warm sea surface temperatures, on the western sides of oceans.

The jet stream consists of a zone of strong winds, no more than one kilometre deep and 100 kilometres wide, flowing downwind over a distance of

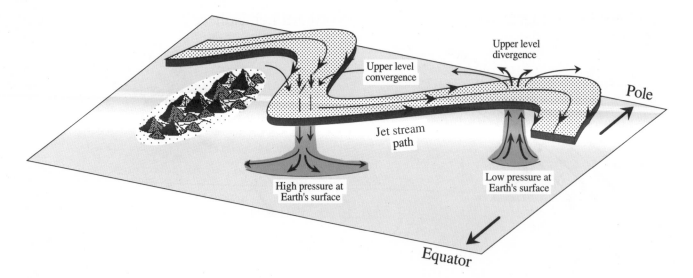

Figure 3.3 Formation of Rossby waves in the jet stream of the northern hemisphere and hypothesised generation of high and low pressure at the Earth's surface.

1,000 kilometres or more. Wind speeds can reach in excess of 250 kilometres per hour. The polar jet stream is most prominent and continuous in the northern hemisphere, where it tends to form Rossby waves encircling the globe. This jet is topographically controlled by the Tibetan Plateau and the Rocky Mountains. The polar jet tends to form in the region of the Tibetan Plateau, where its position is linked with the seasonal onset and demise of the Indian monsoon. The jet then swings northward over Japan and crosses the west coast of North America, normally north of the Rocky Mountains. It then swings southerly across the Great Plains of the United States, northeastward parallel to the Appalachian Mountains, and exits North America off Newfoundland, breaking up over Iceland.

Both the Tibetan Plateau and the Rocky Mountains produce a resonance effect in wave patterns in the northern hemisphere, locking high pressure cells over Siberia and North America, and a low pressure cell over the north Pacific. Barometric pressure, measured along the 60° N parallel of latitude through these cells, reveals a quasi-stable Rossby or planetary wave. The jet stream loops to the equatorial side of the highs and to the poleward side of the low. This looping is most enhanced in the northern hemisphere (boreal) winter, where deviations from this pathway contribute significantly to short-term climate change. As the looping strengthens, so does the pressure differential across this planetary wave. Strengthening high pressure in the lee of the Rocky Mountains, coupled with more intense looping in the jet

stream, favours drought on the United States Great Plains. There is evidence that both the intensification of the looping, and the amplitude of the planetary wave, are affected by either the 18.6-year lunar tide (the M_N component) or the 22-year Hale sunspot cycle. These aspects are examined in more detail in Chapter 6.

THE PALMÉN–NEWTON MODEL OF GENERAL AIR CIRCULATION
(Miller, 1971; Linacre and Hobbs, 1977; Bryant, 1991)

Rising air motion is a prerequisite for vortex development, and about 10% of all air movement takes place vertically. As air rises, it spirals because of Coriolis force, and draws in adjacent air at the surface to replace it. The speed at which air rises is a function of its changing temperature, relative to the environmental lapse rate existing in the atmosphere at the time. (The concept of lapse rates is not covered here; readers interested in more detail should refer to one of the reference texts.) Rising air is unstable if it is warmer than adjacent air, and instability is favoured if latent heat can be released through condensation. The faster air rises, the greater the velocity of surface winds spiralling into the centre of the vortex. Of course, in the opposite manner, descending air spirals outwards at the Earth's surface. Subsiding air also warms adiabatically and has the potential for evaporating moisture.

These concepts can be combined to account for

air movement in the troposphere. The Palmén–Newton general circulation model is one of the more thorough models in this regard (Figure 3.4). Intense heating by the sun, at the equator, causes air to rise and spread out polewards in the upper troposphere. As this air moves toward the poles, it cools through longwave emission, begins to descend back to the Earth's surface around 20–30° north and south of the solar equator, forming high pressure at the ground. Upon reaching the Earth's surface, this air either moves polewards or returns to the equator to form a closed circulation cell, termed the Hadley cell. Because of Coriolis force, equatorial moving air forms two belts of easterly trade winds astride the equator. Air tends to converge towards the solar equator, and the uplift zone here is termed the intertropical convergence. Over the western sides of oceans, the tropical easterlies pile up warm water, resulting in convection of air that forms low pressure and intense instability. In these regions, intense vortices known as tropical cyclones preferentially develop. At the pole, air cools and spreads towards the equator along the Earth's surface. Where cold polar air meets relatively warmer subtropical air at mid-latitudes, a cold polar front develops with strong uplift and instability. Tornadoes and strong westerlies can be generated near the polar front over land, while intense extratropical storms develop near the polar front especially over water bodies. A belt of strong wind and storms, dominated by westerlies poleward of the polar front, forms around 40° latitude in each hemisphere.

Other strong winds exist in the upper atmosphere adjacent to the tropopause boundary. The more significant of these is the polar jet stream on the equatorial side of the polar front. In the northern hemisphere, looping in this jet occurs, with the formation of Rossby waves (Figure 3.3). In the southern hemisphere, the polar jet stream tends to track around 45–50° S, with strengthening over the main oceans. Another major jet stream forms above subsiding air on the poleward side of the Hadley cell. In the northern hemisphere, this subtropical jet tends to form a continuous stream around 30° N, with maximum winds located at 70° W, 40° E and 150° E. Other jet streams exist regionally above the intertropical convergence during the Indian monsoon, and at the boundary with very cold air lying over the Antarctic.

The Palmén–Newton model accounts for most of the observed zones of pressure, wind and instability around the globe. There are two significant areas of the world where rising air motion predominates, leading to the formation of vortices and low pressure cells: at the intertropical convergence, and along the leading edge of the polar front. High pressure forms at the poles and beneath two bands of subsiding air on each side of the solar equator. Because of Coriolis force, winds moving towards the equator blow from the east, forming the easterly trades. The zone of calm air at the intertropical convergence, where air motion is mainly vertical, is termed the doldrums. Air moving away from each pole forms the polar easterlies. Air moving towards the poles blows from the west. Thus air moving poleward from the Hadley cell and meeting the polar front forms the roaring forties or westerlies. These are especially prominent in the southern hemisphere where winds blow unobstructed by land or, more importantly, by significant mountain ranges. Instability and rainfall is greatest above the intertropical convergence and along the polar front. Clear skies and stable air exist under the descending arms of the Hadley cell and beneath polar highs. In these regions, evaporation is enhanced by adiabatic warming of descending air.

Surface air pressure patterns partially support the Palmén-Newton model, however there are significant discrepancies (Figure 3.5). In July, when the southern hemisphere experiences winter, the Hadley belt of high pressure is evident at 30–35° S centred over the eastern Pacific Ocean, the southern Atlantic and southeast Indian Ocean. Mid-latitude lows along the polar front are not evident on this map. In contrast, the northern hemisphere is experiencing summer in July, and Hadley cells are well developed off the coast of California and over the central Atlantic (the Bermuda or Azores High). High pressure is supplanted by low pressure over the Indian subcontinent, because of intense convective heating over this landmass in summer. In addition, the jet stream swings north of the Tibetan Plateau and draws in moist unstable air over India from the adjacent oceans. This forms part of the Indian monsoon. Weak low pressure associated with the polar front appears only over Siberia and northern Canada.

In the southern hemisphere (austral) summer, the solar equator shifts south of the equator, and the low pressure heating centre over the Indian subcontinent switches to the Indonesian–Australian 'maritime' continent. The Hadley belt also shifts southward by about 10°, although it remains stationary over the east Pacific. This latter aspect is crucial to ocean feedback on atmospheric circulation in this region, a feature that is examined in detail in Chapter 4. In addition, a small cell of high

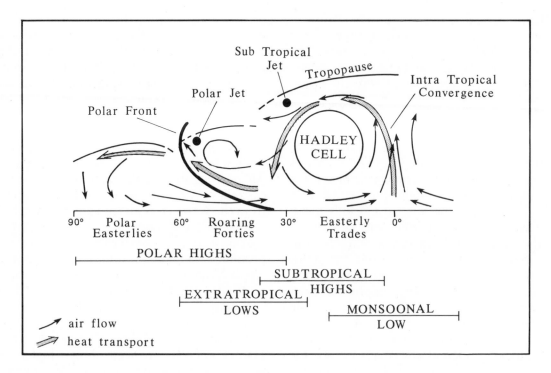

Figure 3.4 Palmén–Newton schematic model of general air circulation between the pole and the equator.

pressure develops over the Tasman Sea between Australia and New Zealand. A belt of low pressure develops around the Antarctic continent at the location of the polar front. In the northern hemisphere, the Hadley cell is weakened and appears only over the Azores in the Atlantic Ocean. The low pressure associated with the polar front is focused at two locations, over the Aleutian Islands in the North Pacific, and over Iceland. Lows are supplanted over the continental landmasses by two extensive high pressure cells. Snow cover enhances longwave emission and solar reflectance, permitting the centre of these continents to cool excessively, with attendant formation of surface high pressure. The Siberian high is so enhanced by this effect that it deflects the jet stream south of the Himalayas. Hence, while winds are northerly over the Indian subcontinent in summer, they flow seaward in winter. This fulfils the requirement of a monsoon climate: it is any location where winds reverse direction by 180° seasonally, leading to a defined wet and dry season.

Overall in the Palmén-Newton model, there is a continual flow of heat from the equator towards the poles. The pathway of this air is also located in Figure 3.4. Heat is transferred to the upper troposphere from the Earth's surface at the equator, moves poleward in the upper atmosphere above the Hadley cell, descends to the Earth's surface at mid-latitudes, is forced upwards along the polar front, and moves to the poles in the upper atmosphere. In return, cold air moves under the influence of polar highs towards the equator, and is heated in transit by increasing solar radiation. In the upper troposphere, air continually loses heat to space through longwave emission as part of the global radiation budget. The model diverges from reality over the northern hemisphere landmasses, especially the Indian subcontinent, and remains locked in position seasonally over the southern Pacific Ocean. Thus in terms of atmospheric circulation, the northern hemisphere can be considered a 'land' hemisphere, with the jet stream topographically controlled by mountains. In contrast, the southern hemisphere can be considered an 'ocean' hemisphere.

The Palmén–Newton model is restricted by two inherent facets. First, the location of pressure cells within the model is based upon averages over time. Second, because the model averages conditions, it tends to be static, whereas the Earth's atmosphere is very dynamic. Additionally, the concept of Hadley circulation is overly simplistic. In actual fact, rising air in the tropics is not uniform, but confined almost totally to narrow updrafts within thunderstorms.

Part A

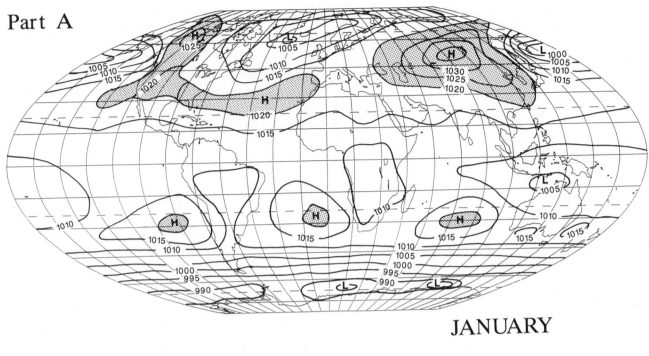

JANUARY

Part B

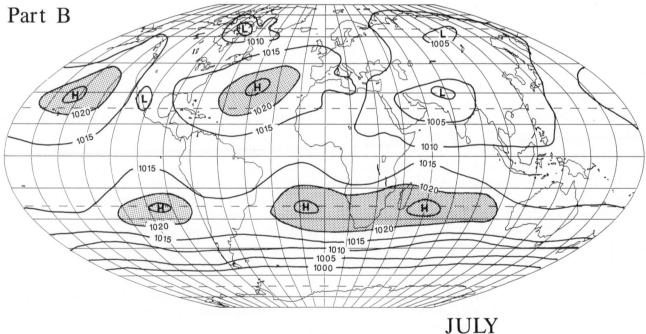

JULY

Figure 3.5 Pressure patterns for the globe in (A) January and (B) July (modified from Linacre and Hobbs, 1977). Areas with pressure above 1,020 hectopascals are shaded.

At higher latitudes, large scale circulation is distorted by relatively small eddies. Other factors also control winds. For example, over the Greenland or Antarctic icecaps, enhanced radiative cooling forms large pools of cold air, which can accelerate downslope under the effect of gravity because of the low frictional coefficient of ice. Alternative models that overcome some of these limitations are presented later in this chapter.

SECONDARY FEATURES OF ATMOSPHERIC CIRCULATION

Fronts
(Moran et al., 1991)

A front is a relatively narrow zone between two adjoining air masses where temperature, humidity, barometric pressure, and wind speed and direction change rapidly. This change is usually enhanced at the Earth's surface, although fronts have a vertical component as well. Winds usually converge towards the front such that warm, less dense and more buoyant air rises above cooler, more dense, and less buoyant air. A front is considered stationary if there is no movement of one air mass at the expense of the other (Figure 3.6). If the warm air mass begins to displace the colder air mass, then the front is termed a warm front. Similarly a cold front represents the displacement of a warmer air mass by a cooler one. Because warm air is more buoyant, air along a warm front tends to rise over colder air at an angle of less than 0.4°. Condensation and resulting cloud formation usually occurs up to 1,000 kilometres in front of this moving warm front. In contrast, cold air rapidly displaces warm air, forming a front angle of 0.5–1.0°. This leads to condensation over a shorter distance, and the formation of thunderstorms, with the possibility of attendant hazards such as hail, heavy rain, lightning and tornadoes. Cold fronts typically move between 30 and 45 kilometres per hour, while warm fronts move more slowly. Thus, where warm and cold fronts lie relatively close to each other, it is possible for the cold front to catch up with the warm front and lift it off the Earth's surface. In this case, the front is occluded (Figure 3.6). Fronts can form spontaneously, simply because the density contrast between two air masses increases due to convergence or subsidence of cold air. This process is termed frontogenesis. If the density contrast between air masses decreases, then the front becomes undefinable. Fronts are usually not marked on a weather chart unless the temperature difference between air masses is greater than 1°C, and the air masses have regional continuity lasting at least a day. For example, sea breezes producing a temperature drop greater than 1°C are not marked on synoptic weather charts, because they are too ephemeral and localised. In addition, fronts do not appear at extremes in latitude because air conditions, both in the tropics and at the poles, are relatively uniform.

Tropical cyclones
(Anthes, 1982; Nalivkin 1983; Gray, 1984; Bryant, 1991; Gray et al., 1994)

Tropical cyclones (known as hurricanes in North America, or typhoons in Japan and southeast Asia) are intense low pressure vortices that develop over warm tropical oceans or seas. They are associated with a range of hazards including strong winds, storm surges, high waves and torrential rainfall. On average, a tropical cyclone can dump 100 millimetres per day of rain within two hundred kilometres of the eye, and 30–40 millimetres per day between two hundred and four hundred kilometres distance. In the largest cyclones, winds can exceed 300 kilometres an hour. Climatically, tropical cyclones are an important mechanism moving latent heat from the tropics to mid-latitudes. Tropical cyclones derive their energy from the evaporation of water over oceans, particularly the western parts where surface temperatures are warmest. As the sun's apparent motion reaches the summer equinox in each hemisphere, surface ocean waters begin to warm over a period of two or three months, to temperatures in excess of 26°C. Tropical cyclones thus develop between December and May in the southern hemisphere, and between June and October in the northern hemisphere.

At least seven conditions are required to develop a tropical cyclone. First, efficient convective instability is required, causing near vertical uplift of air throughout the troposphere. Cyclonic depressions begin to develop when ocean temperatures exceed 26–27°C, because the lower atmosphere becomes uniformly unstable at these values. With further temperature increases, the magnitude of a tropical cyclone intensifies. Tropical cyclones tend not to form above 30°C because smaller scale turbulence begins to dominate, with the breakdown of near-vertical air movement. Once developed, cyclones rapidly decay below temperatures of 24°C. These relationships have implications for global warming. While the area of ocean favourable for cyclone formation may increase, there is an upper temperature limit constraining growth. Second, a trigger mechanism is required to begin rotation of surface air converging into the uplift zone. This rotation most likely occurs in waves developed within the trade winds. On surface charts these waves are evident as undulations in isobars parallel to degrees of latitude. Third, while easterly waves may begin rotation, Coriolis force has to be sufficient to establish a vortex. Cyclones cannot

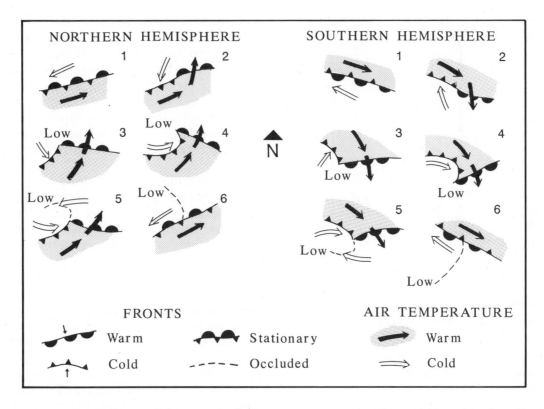

Figure 3.6 Stages in the development of extratropical depressions as viewed in the northern and southern hemisphere. The diagram shows (1 and 2) initial development of wave form; (3 and 4) mature front development and full extent of low; (5 and 6) occlusion (Bryant, 1991).

form within 5° of the equator for this reason. Nor can cyclones cross the equator, because cyclones rotate in opposite directions in each hemisphere. Fourth, cyclones cannot sustain themselves over land (unless the land is already flooded with water at temperatures above 24°C), because they are dependent upon transfer of heat, from the surface to the upper atmosphere, through the process of evaporation and release of latent heat of condensation. Fifth, tropical cyclones must form an 'eye' structure (Figure 3.7). If the convective zone increases to around 10–100 kilometres' distance, then subsidence of air takes place in the centre of the zone of convection, as well as to the sides. Subsidence at the centre of uplift abruptly terminates convection, forming a 'wall'. Upwardly spiralling convection intensifies towards this wall, where surface winds become strongest. Once the convective wall forms with subsidence in the core, the cyclone develops its characteristic eye structure. Subsidence causes stability in the eye, cloud evaporates, and calm winds result. The eye is also the area of lowest pressure. Around the eye the release

of latent heat of condensation can raise air temperatures up to 20°C above those at equivalent levels. Sixth, cyclones cannot develop if there are substantial winds in the upper part of the troposphere. If horizontal wind speed aloft increases by more than 10 metres per second over surface values, then the top of the convective column is displaced laterally, and the structure of the eye cannot be maintained. Cyclones tend to develop equatorward of, rather than under, the direct influence of strong westerly winds. Finally, the central pressure of a tropical cyclone must be below 990 hectopascals. If pressures are higher than this, then uplift in the vortex is not sufficient to maintain the eye structure for any length of time. The record low pressure measured in a tropical cyclone was 870 hectopascals in Typhoon Tip, northeast of the Philippines, in October 1979.

The actual number of cyclones each year is highly variable. In the Australian region, there were on average ten cyclones per year between 1909 and 1980. The number has been as low as one per year and as high as nineteen. In the Caribbean region,

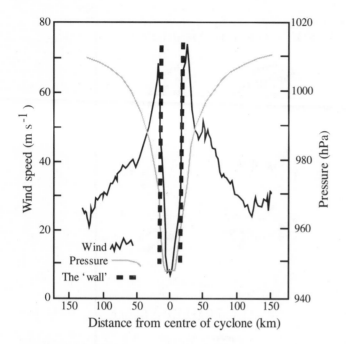

Figure 3.7 Wind and pressure cross-sections through Hurricane Anita, 2 September 1977 (based upon Sheets, 1980 and taken from Bryant, 1991).

the aggregated number of days of hurricanes has varied from none to fifty-seven. These variations are not random, but are linked to the state of trade wind circulation across the Pacific Ocean and Caribbean (the Southern Oscillation to be discussed in the following chapter); the direction of equatorial winds in the stratosphere (the Quasi-Biennial Oscillation); and rainfall, pressure and temperature patterns in West Africa. Using thirteen predictors based upon these conditions, it is possible to forecast over 70% of the tropical cyclone activity and potential cyclone destruction in the Atlantic Ocean region at the beginning of the hurricane season.

Extratropical cyclones
(Linacre and Hobbs, 1977; Whipple, 1982; Bryant, 1991)

Extratropical cyclones (also called depressions or mid-latitude lows) are usually considered to be low pressure cells that develop along the polar front. These lows have cold cores and an effective diameter that is much larger than that of a tropical cyclone (2000 kilometres versus 400 kilometres respectively). Their intensity depends upon the difference in temperature between colliding cold and warm air masses. They always travel in an easterly direction, entrapped within the westerlies. Their location depends upon the location of Rossby waves at the polar front, around 40° north and

south of the equator. In the southern hemisphere, this location occurs south of landmasses, while in the northern hemisphere, it coincides with landmasses that are heavily populated. Lows can intensify over bodies of water because they gain added energy through latent heat of evaporation from the water surface. However, the rapid uplift of very warm, humid air over landmasses can minimise any difference in storm intensity over land and water.

The theory for the formation of polar front lows was formulated by Vilhelm Bjerknes and his colleagues between 1918 and 1922. Outbreaks of cold polar air flow rapidly towards the equator, and clash with much warmer air masses in midlatitudes. Coriolis force dictates that this polar air has an easterly component, while the movement of warmer air on the equatorial side of the cold front has a westerly component (Figure 3.6). Warm air will be forced upwards by the advancing cold air because it is more buoyant. Winds on each side of the developing front travel in opposite directions and spiral upwards in a counterclockwise direction. This rotation establishes a wave or indentation along the front. Warm air begins to advance eastwards and poleward into the cold air mass, while cold air begins to encircle the warm air from the west. A v-shaped frontal system develops, with warm and cold fronts at the leading and trailing edges respectively. The lifting of warm, moist air generates low pressure and convective instability, as latent heat of condensation is released in the upper atmosphere. This uplift enhances the rotational circulation, increasing wind flow, and producing precipitation over a wide area. The low pressure cell advances northeastwards, steered by the jet stream attached to the polar front at the top of the tropopause. In this regard, these lows are an important mechanism for transporting heat poleward. The trailing cold front causes rapid uplift of warm air, leading to thunderstorm formation and, under extreme instability, tornado generation. Occlusion occurs as the cold front outruns the warm front. In the final stage, the polar front is reestablished at the ground, with decreasing rotational air movement stranded aloft. Without continued convection, cloud formation and precipitation slowly decrease, and the storm cell dissipates. Under some circumstances, the advancing cold front may entrap another lobe of warm air, generating a secondary low to the southeast of the original one. It is even possible for the rotating air aloft to circle back to the origin site of cyclogenesis, and trigger the formation of a new low pressure wave along the polar front.

These extratropical lows can produce some of the highest winds and rainfalls recorded at mid-latitudes. The intensity of such storms has been variable historically. Comparatively, our present climate, especially over eastern North America and western Europe, is benign in terms of the occurrence of these storms. For example, in the Middle Ages, horrendous storms were produced by these lows over Europe. The All Saints Day Flood of 1–6 November 1570 was probably the worst; an estimated 400,000 people were killed. These extreme storm events appear to be a signature of a changing climate at the mid-latitudes. It is noteworthy that, at a time when the climate of Europe is perceived as warming, these super-storms have begun to increase in frequency. The Fastnet Race storm of August 1979, and the 15 October 1987 gale which swept through the English Channel with a central pressure of 958 hectopascals (the lowest ever recorded in England) are two recent examples supporting this tendency.

East coast lows
(Sanders and Gyakum, 1980; Bryant, 1991)

East coast lows are storm depressions that develop, without attendant frontal systems, over warm bodies of water usually off the east coast of continents between 25 and 40° latitude. The development of such lows exhibits the forcing of both topography and sea surface temperature patterns upon atmospheric instability. The lows gain latent heat from warm ocean currents which tend to form on the western sides of oceans. In some regions, where easterly-moving low pressure systems pass over a mountainous coastline and then out to sea over this warm poleward flowing current, there may develop intense cyclonic depressions which have structural features and intensities similar to tropical cyclones. This development is shown schematically in Figure 3.8. These regions include: the east coasts of Japan, characterised by the Japanese Alps and the Kuroshio current; the United States, characterised by the Appalachians and the Gulf Stream; and southeastern Australia, characterised by the Great Dividing Range and the East Australian current. Lows may intensify also off the east coast of South Africa dominated by the Drakensburg range and the Agulhas current. The lows are not associated with any frontal structure, but tend to develop under, or downstream of, a cold-core low or depression that has formed in the upper atmosphere. The depressions are associated on the surface with a strong high pressure system that may stall or block over the adjacent ocean. The initial sign that a low is developing is a poleward dip in surface isobars on the eastward side of these highs.

The lows intensify and reach the ground near the coastline, due to convection that is enhanced by onshore winds crossing cold upwelling water inshore, and impinging upon steep topography at the coast. Topography deflects low level easterlies towards the equator, causing convergence and convection that can explode the formation of the cyclone. Some lows develop rapidly within a few hours to become 'bombs'. This explosive development occurs when pressure gradients reach 4 hectopascals per 100 kilometres. Convection of moisture-laden air is enhanced by cold inshore water that maximises condensation and the release of latent heat. Along the east Australian coastline, the magnitude of the sea surface temperature anomaly does not appear to be important in triggering a bomb. Rather, the temperature gradient perpendicular to the coast (the zonal gradient) is the crucial factor. This gradient must be greater than 4°C per 0.5° of longitude, within 50 kilometres of the coastline, for a low to intensify.

East coast lows develop preferentially at night, at times when the maritime boundary layer is most unstable. East coast lows tend to form in late autumn or winter when steep sea surface temperature gradients are most likely. However, they are not necessarily restricted to these times, and can occur in any month of the year. In Australia, a 4.5-year cycle is apparent in records of these storms over the last forty years, with a preference for such lows to develop in transition years between El Niño–Southern Oscillation (ENSO) and La Niña events. In extreme cases, the lows can develop the structure of a tropical cyclone. They may become warm-cored, obtain central pressures below 990 hectopascals, develop an eye that appears on satellite images (Figure 3.9), and generate wind speeds in excess of 200 kilometres an hour. These eye structures have been detected off the east coast of both Australia and the United States. Once formed, the lows travel poleward along a coastline, locking into the location of warmest water. Such systems can persist off the coastline for up to one week, directing continual heavy rain onto the coast, producing a high storm surge, and generating waves up to ten metres high. Heaviest rainfalls tend to occur towards the tropics, decreasing with increasing latitude. This rainfall pattern shifts

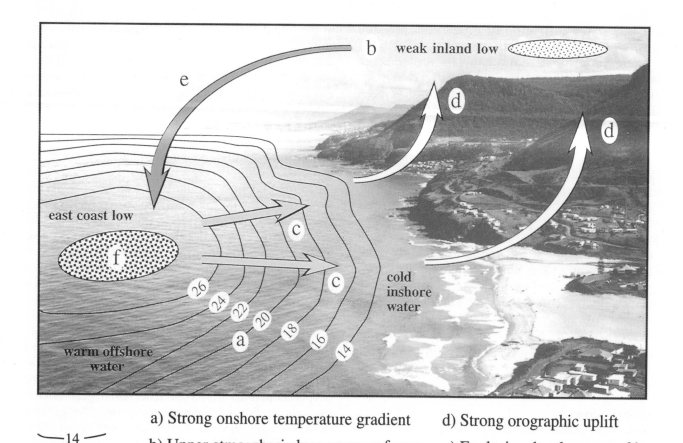

a) Strong onshore temperature gradient
b) Upper atmospheric low pressure forms
c) Strong surface air flow
d) Strong orographic uplift
e) Explosive development of low
f) Intense surface low forms

—14— isotherm °C

Figure 3.8 Schematic representation of the development of an east coast low or 'bomb'.

towards the equator from summer to winter, reflecting the intensification and seasonal movement of high pressure systems. The Ash Wednesday storm of 7 March 1962 in the United States, and the 25 May 1974 storm in Australia, are classic examples of these storms. The latest storm to occur in the United States was on 21–22 December 1994; it turned on explosively, reaching a central pressure of 970 hectopascals, developed a warm core with attendant eye structure, and generated winds of 160 kilometres per hour and waves 11.9 metres high.

Monsoons
(Yoshino, 1971; Hamilton, 1979; Suppiah, 1992)

The term monsoon is derived from the Arabic word 'mausim', referring to any climate where there is a distinct 180° change in wind direction between summer and winter, with seasonal alternation between copious rain and aridity. This shift in wind is most prominent in tropical regions affected by the seasonal passage of the intertropical convergence. In its simplest form, monsoonal air circulation behaves as a regional sea breeze in summer and a land breeze in winter. Heating of landmasses in summer, with the generation of thermal lows, causes uplift and instability in the lower atmosphere. Oceans adjacent to these landmasses are relatively cooler, and moist air flows from the ocean into centres of convection over the land. Latent heat evaporated from the oceans is released into mid-troposphere levels, intensifying the low pressure cell, and strengthening the monsoon circulation. This process is particularly strong over the Tibetan Plateau and Himalayan Mountains. In winter, landmasses cool by longwave emission, while the oceans tend to remain warmer because of their larger thermal capacity. Cool, stable, subsiding air then flows from the landmasses toward the oceans.

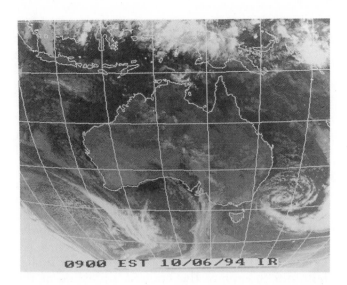

Figure 3.9 Infrared satellite image of an east coast low off the southeast coast of Australia, 10 June 1994. Note the visible 'eye' structure that this hybrid storm has developed. Synoptic conditions for this storm appear in Figure 3.16.

While a weak monsoon exists over southwestern United States, the most noteworthy monsoons occur over western Africa and the Indian subcontinent. The Indian monsoon in the northern hemisphere summer affects the Arabian and Indian peninsulas and southeast Asia. Climatically, the centre of heating that develops here is contiguous throughout the year, simply shifting to the Indonesian–Australian 'maritime' continent in the austral winter. Because of continuity of the monsoonal low between the two hemispheres, atmospheric circulation over most of Asia and the southwest Pacific is interlinked. Climatic changes affecting one section tend to be manifested over the whole region. For instance, cold surges, from the Siberian high pressure cell in the northern hemisphere winter, pass around the Tibetan Plateau over east Asia, and across the South China Sea. These surges increase convection over southeast Asia and northern Australia, generating active cloud formation over the 'maritime' continent of Indonesia–Australia. The onset of the Australian summer monsoon influences the formation and intensity of tropical cyclones in the southern hemisphere.

The Indian–Southeast Asian monsoon is controlled to a large extent by atmospheric processes over the Tibetan Plateau. This plateau is over 2,000 kilometres long and 600–1,000 kilometres wide. Much of it lies above five thousand metres elevation. Hence, the atmosphere over the plateau is much thinner than it would be at sea-level (500–700 hectopascals, dropping to as low as 300 hectopascals over the highest mountain peaks). The Tibetan Plateau acts as a mechanical barrier to the jet stream, and as an upper atmospheric heat source. Shortwave radiation in summer is more effective at heating the plateau surface because less of this radiation is attenuated in the atmosphere. In July, the radiative heating of air above the plateau may be five times that occurring in the adjacent Ganges Valley. The strength of the monsoon relies heavily upon formation of an upper-air, thermally driven anticyclone in the upper third of the atmosphere (100–300 hectopascals) over the plateau. This circulation forms an easterly upper jet stream over southern India, and deflects the upper-air westerly jet stream from its winter position south of the plateau, to its summer position north of it. The movement of the westerly jet over a distance of one thousand kilometres is trigger-like, such that the onset of the monsoon tends to recur each year at the same time, and progresses across the Indian subcontinent predictably in the space of a few weeks. The first appearance of the monsoon occurs in the last week of May in Sri Lanka; by mid-June rains have progressed northwest across India; and by mid-July the whole subcontinent is affected by monsoonal circulation. The demise of the monsoon follows a reversed sequence between September and January.

The Indian monsoon is strongly linked to sea surface temperatures in the southern hemisphere, and to the degree of insolation around 30° S in the Indian Ocean. The latter is the preferred location of the Hadley subtropical belt of high pressure in the austral winter. Clear skies at this time of year permit increased solar radiation to reach the ocean surface. Warm winds, generated under high pressure, enhance evaporative cooling of the ocean. Over annual and interannual time scales, colder, southern hemisphere, sea surface temperatures in the Indian Ocean are associated with: increased evaporation, enhanced transfer of latent heat from the ocean to the atmosphere, increased southern hemisphere trade winds, and increased transfer of heat northwards across the equator. This leads to a stronger monsoon.

There are variations to this pattern. Both the sub-Sahara and Indian monsoons are notorious for their failure, resulting in drought and calamitous famines. The Indian monsoon and its attendant circulation are also intricately linked to the

Southern Oscillation, a quasi-periodic phenomenon synonymous with alternating regional flooding and regional drought. These relationships are examined more fully in Chapter 4. As well, the monsoon is not steady in time, but tends to go through active and dormant phases linked to a 40–50-day cycle of longwave radiation, cloudiness and wind present in the tropics, and to regional feedback processes. For example, as land is flooded by monsoonal rain, solar radiation is used in evaporation, but not for heating of soil and the air above it. This tends to cool recently wetted areas, suppressing convection until the soil dries out enough to resume the sensible heat flux. In consequence, monsoonal rains tend to undergo a 15–20-day period of alternating activity and dormancy.

CONCEPT OF MOBILE POLAR HIGHS
(Leroux, 1993)

The monsoonal circulation described above does not fit well within the Palmén–Newton general circulation model. In actual fact, the positioning of Hadley cells, and the semi-permanency of features such as the Icelandic or Aleutian Lows, is a statistical artifact. There is not an Icelandic Low which is locked into position over Iceland. Nor are there consistent trades or polar easterlies. Air circulation across the surface of the Earth is dynamic, as is the formation and demise of pressure cells. In reality, the Icelandic Low may exist as an intense cell of low pressure for several days, and then move eastward towards Europe within the westerly air stream. Climate change in the Palmén–Newton model implies the movement, or change in magnitude, of these centres of activity. For example, weakening of the Icelandic Low conveys the view that winter circulation is less severe, while expansion of Hadley cells towards higher latitudes suggests that droughts should dominate mid-latitudes.

There are other problems with terminology in the Palmén–Newton model which have a historical basis. The depression, extratropical cyclone or polar low was initially explained as a thermal phenomenon, but later linked to frontal uplift along the polar front, to upper air disturbances and, recently, to planetary waves or Rossby waves which produce undulations in the path of the westerly jet stream at high altitudes. The jet stream is supposedly tied to the polar front, and related to cyclogenesis at mid-latitudes. However, the connection between the jet stream and the polar front is approximate, and there has never been a clear relationship established between the jet stream in the upper troposphere and the formation of low pressure at the Earth's surface. Indeed, no single theory adequately explains the initial formation of lows at mid-latitudes. The Palmén–Newton general circulation model is a good teaching model, but it is not an ideal model for explaining the causes of climate change.

Conceptually, the Palmén–Newton model in its simplest form has two areas of forcing: upward air movement at the equator from heating of the Earth's surface, and sinking of air at the poles because of intense cooling caused by longwave emission. At the equator, air moves from higher latitudes to replace the uplifted air, while at the poles subsiding air moves as a cold dense mass hugging the Earth's surface towards the equator. The lateral air movement at the equator is slow and weak, while that at the poles is strong and rapid. Polar surging can reach within 10° of the equator, such that some of the lifting of air in this region can be explained by the magnitude and location of polar outbursts. Hence, the dominant influence on global air circulation lies with outbursts of cold air within polar high pressure cells. Each of these outbursts forms an event termed a mobile polar high. There is no separate belt of high pressure or Hadley cells in the subtropics. These cells, as they appear on synoptic maps (Figure 3.5), are simply statistical averages over time of the preferred pathways of polar air movement towards the equator.

Mobile polar highs, developing in polar regions, are initially maintained in position by surface cooling, air subsidence and advection of warm air at higher altitudes, as shown in Figure 3.4. When enough cold air accumulates, it suddenly moves away from the poles, forming a 1,500 metre-thick lens of cold air. In both hemispheres, polar high outbreaks tend to move from west to east to conserve vorticity. Additionally, in the southern hemisphere, pathways and the rate of movement are aided by the formation of katabatic (gravity) winds off the Antarctic icecap (Figure 3.10). In the northern hemisphere outbreak pathways are controlled by topography; there is a preference for mobile polar highs to form over the Hudson Bay lowland, Scandinavia and the Bering Sea (Figure 3.10). The distribution of oceans and continents, with their attendant mountains, explains why the mean trajectories followed by these highs are always the same. For instance, over Australia, a polar outbreak

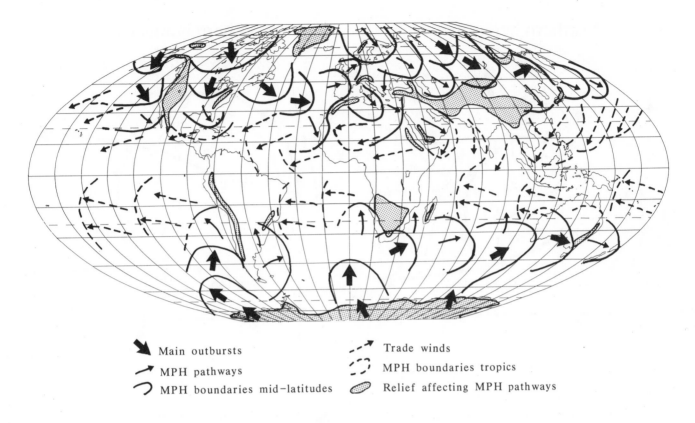

Figure 3.10 Pathways for mobile polar highs (MPH) and resulting trade wind circulation in the tropics (based on Leroux, 1993).

always approaches the continent from the south-west, then loops across the continent towards the equator, and drifts in the Hadley belt out into the Tasman Sea. Highs never move directly northward across the continent, or sweep up from the Tasman Sea. In North America, mobile polar highs regularly surge southward across the Great Plains for the same reason.

Because mobile polar highs consist of dense air, they deflect less dense, warm air upwards and to the side. The deflection is greatest in the direction they are moving. Hence the polar high develops an extensive bulbous, high pressure vortex, surrounded downdrift by a cyclonic branch or low pressure cell (Figure 3.11). Typically, the high pressure cell is bounded by an arching, polar cold front with a low pressure cell attached to the leading edge. However, in the northern hemisphere, individual outbreaks tend to overlap so that the low pressure vortex becomes contained between two highs. This forms the classic v-shaped, wedged frontal system associated with extratropical lows (Figure 3.6). Thus lows, and upper westerly jet streams, are a product of the displacement and divergence of a mobile polar

high. The intensity of the low pressure cell becomes dependent upon the strength of the polar high, and upon its ability to displace the surrounding air. Strong polar highs produce deep lows; weak highs generate weak lows. In a conceptual sense, a deeper Icelandic or Aleutian low must be associated with a stronger mobile polar high.

The repetitive pathways of highs, when averaged over time, produce the illusion of two tropical high pressure belts on each side of the equator. These belts form Hadley cells. Mobile polar highs can propagate into the tropics, especially in winter. Here, their arrival tends to intensify the easterly trade winds. More importantly, yet little realised, is the fact that strong polar highs produce stronger monsoons, although in a more restricted tropical belt. It has already been pointed out that the Indian monsoon is strongest when southern hemisphere subtropical highs are intensified around 30° S. During Ice Ages, mobile polar highs off the Antarctic icecap should be more intense. As a result, so should the Indian monsoon.

In summary, cold polar air is forced by mobile polar highs towards the equator, while at the same time sensible and latent heat from the tropics are

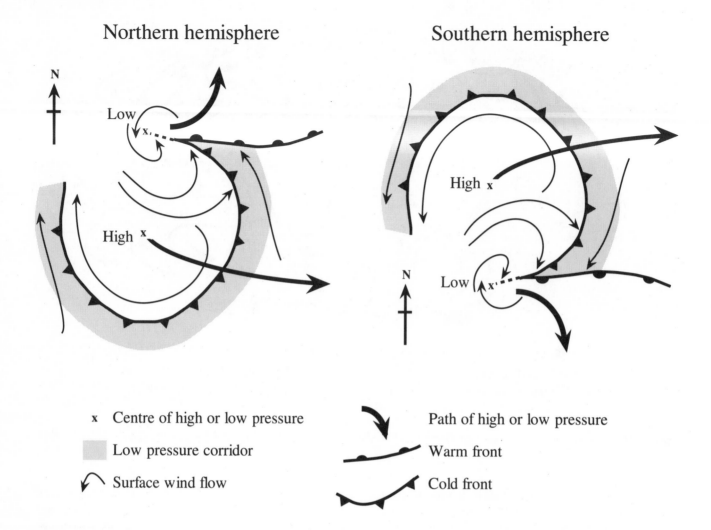

Figure 3.11 Schematic view of the dynamic structure of a mobile polar high and its associated mid-latitude low (based on Leroux, 1993).

forced poleward. This drives global air circulation and satisfies the need in the atmosphere for heat exchange between the equator and poles. The concept of mobile polar highs implies that the climate of high latitudes is the key to world climate and global climate change. These aspects are examined in more detail in Chapters 6 and 7, which treat climates subjected to glaciation and enhanced 'greenhouse' warming, respectively.

COMPUTER GENERATED CIRCULATION MODELS
(Cubasch and Cess, 1990; Gates et al., 1990; Trenberth, 1992)

Over the past 20 years, schematic models have been supplanted by computer simulation models (called General Circulation Models or GCMs), initially developed for short-term weather forecasting. The models are based upon the physical laws of conservation which describe the spatial redistribution of momentum, heat and water vapour in the atmosphere. All of these processes are simulated using simplistic, non-linear equations that describe the behaviour of a fluid (air or water) on a rotating body (the Earth), under the influence of a temperature contrast between the equator and the poles. The equations are solved using numerical methods that subdivide the atmosphere vertically into discrete layers (Figure 3.12). The spatial resolution of GCMs is constrained by the speed and memory of existing computers. Typical first generation models have a horizontal resolution of 300–1,000 kilometres, and use between two and nineteen vertical layers through the atmosphere.

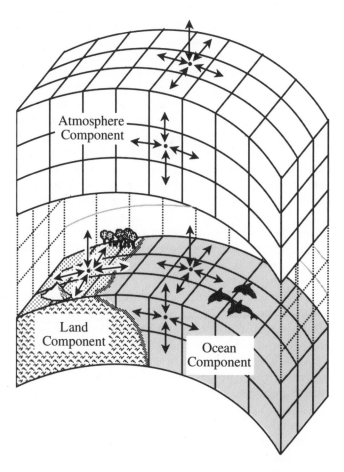

Figure 3.12 Schematic representation of a computer general circulation model, coupling the Earth's surface and atmosphere to the ocean. The arrows represent the fluxes that must be calculated at each step. The dimensions and number of layers vary depending upon the model and computer used.

The spatial resolution has to be increased to at least 25 kilometres before regional predictions can be made. Second generation models, incorporating the ocean, have a resolution of 200–1000 kilometres, with time scales of hours to years, and a vertical stratification of between two and twenty levels. About a dozen different simulation models exist with slightly different algorithms for handling climatic processes, varying scales of spatial resolution, differing numbers of segments through the atmosphere; and differing approaches to the simulation or parametrisation of clouds, convection, sea-ice, snow, and, importantly, the thermal inertia and interaction of the oceans.

All GCMs make assumptions about the Earth's land and ocean surfaces. For instance, the models deal poorly with the fact that oceans, which cover 70% of the globe, control sea surface temperatures

dynamically over time. Over half of the solar radiation reaching the Earth's surface is absorbed by oceans, where it is stored, used to drive ocean currents, or released back to the atmosphere as latent heat of evaporation. Feedbacks exist between currents and the atmosphere in terms of the exchange of momentum, heat and water vapour. Ocean flows consist of eddies with dimensions of approximately 50 kilometres, larger vortices and currents, and large scale gyres covering a hemisphere. Oceans develop stratification between warm, upper and cold, deep water separated by the thermocline, with overturning of surface and deep water because of salinity and density differences. Finally, sharp changes in sea surface temperature exist in the ocean especially associated with equatorial upwelling. Few of these aspects are captured in GCMs.

GCMs, especially the earlier ones, also suffer from some serious fundamental flaws. In the models, the coupled ocean and atmosphere often behave as a chaotic system, such that seemingly small errors feedback on each other, producing unrealistic projections. It may be impossible to overcome these limitations. GCMs also fail to account adequately for the dampening effect of the ocean and sea-ice upon air temperature fluctuations, and to consider adequately the negative feedback mechanisms produced by added cloud and moisture in the atmosphere under a warmer climate. For instance, recent satellite measurements have shown that middle level cloud has a forcing upon incoming solar radiation that is three to seven times greater than the forcing by 'greenhouse' gases upon outgoing longwave emissions. The models do not dynamically model smaller scale transfers in the atmosphere generated each day by 44,000 thunderstorms, and each year by 60 or 70 tropical cyclones. However, the most serious flaw is the fact that most models have to incorporate a flux adjustment for energy transfers through the atmosphere, or between the ocean and atmosphere. If left to run in a steady state situation, the models tend to drift upwards towards warmer temperature. This basically means that the models create energy over time. The magnitude of this drift in some cases is as high as 10 watts per square metre, or 4% of the total solar radiation flux available for heating of the Earth–atmosphere system.

GCMs produce considerable variability in simulating sea-level pressure, surface air temperature, precipitation, and zonal wind at 200 hectopascals, although the gross elements of each of these parameters can be replicated zonally. GCMs can simulate regionally most of the air pressure aspects of general atmospheric circulation shown in Figure 3.5

(Figure 3.13A). The Hadley belts of high pressure are depicted for both hemispheres throughout the year, and the models replicate the Siberian high and Indian monsoon which are features of continentality. The models also simulate some of the low pressure tracks associated with the polar front, however air pressure in the southern hemisphere, particularly towards the Antarctic, is poorly mimicked. Winds, especially in the upper atmosphere, where the frictional effects of the Earth's surface are negligible and jet streams operate, are well simulated by GCMs (Figure 3.13D). GCMs are exceptionally good at replicating global surface air temperature patterns (Figure 3.13B). Except in the Antarctic, simulated air temperatures agree with those predicted by the Earth's net radiation budget (Figure 3.13B versus Figure 2.12). However temperatures may be underestimated by as much as 3°C overall, and in the Antarctic, by as much as 10°C. Precipitation simulations are very inaccurate (Figure 3.13C). While broad and regional features can be simulated, significant errors occur for the southeast Asian summer monsoon, and for rainfall across the tropical Pacific Ocean controlled by the trade winds. The seasonal cycle in precipitation is in error by 20–50%, with overestimation in winter. The accuracy of GCMs in simulating soil moisture, snowfall and albedo is even poorer, particularly at high latitudes in the northern hemisphere.

The commonest use of GCMs has been in the modelling of temperature and precipitation changes forecast under enhanced 'greenhouse' warming. The need for reliable forecasts is driving the development of very sophisticated models that not only incorporate 'greenhouse' gases, but also aerosols. The outcomes of these models are discussed in more detail in Chapter 7.

REGIONAL CIRCULATION PATTERNS

North America
(Bryson and Hare, 1974)

The North American continent spans all the latitudes where heat transfers take place to balance the Earth's radiation balance. It is also the only continent bounded by mountain ranges: the Rocky Mountains to the west form a barrier to the easterly passage of air flow in the lower troposphere, and the Appalachians to the east steer weather systems passing over the eastern seaboard. For these reasons, the continent is dominated by the interplay of tropical and polar air masses, undergoes extreme seasonality, and is influenced by continentality. In winter, air masses over the Yukon are subject to extreme cooling through longwave radiation emission. Very cold, intense high pressure cells move from this region southward between the corridor of the Rocky Mountains and the Appalachians (Figure 3.14). These polar outbursts often trigger cyclogenesis where they meet warm, moist subtropical air masses near the Gulf of Mexico. Winter mid-latitude lows then track northeast across the east coast, steered topographically by the Appalachians. These storms can also intensify under the influence of warm water in the Gulf Stream, forming east coast lows which can reach tropical cyclone strength. While strong winds and waves generated by these events are a major coastal hazard, the occurrence of precipitation in the form of snowfall can effectively shut down most transport, commerce and industrial activity over the eastern half of the United States.

Mobile polar highs in summer tend to track north of the Appalachians. However, in the transition between summer and winter, extreme temperature contrasts of 20°C or more can be obtained across the Great Plains of the United States between polar air, flowing unimpeded southward, and very moist air forcing its way northward from the Gulf of Mexico. Severe thunderstorms with tornado outbreaks tend to develop along the cold front separating these two air masses in spring. Summers are dominated by the intrusion of warm, humid air northward up the Mississippi River valley, leading to heat wave conditions lasting up to two weeks over the central and eastern half of the continent. Deadly heat waves, penetrating into Canada, have plagued the continent repeatedly over the past century.

Hurricanes are also a major feature of the southeastern United States, and have been known to penetrate inland beyond the Appalachian Mountains. While the intensity of these storms appears to have increased in absolute terms in recent times (for instance, in the cases of Hurricanes Hugo and Andrew), records show that the strength of hurricanes in fact is returning to the more severe levels that prevailed in the 1950s. The southward penetration, and waning of high pressure across the Great Plains, also induces a weak monsoonal circulation over the southwest of the United States, alternating between dry winters and wetter summers.

Mobile polar highs also affect the west coast of the United States. Here outbreaks from the Bering

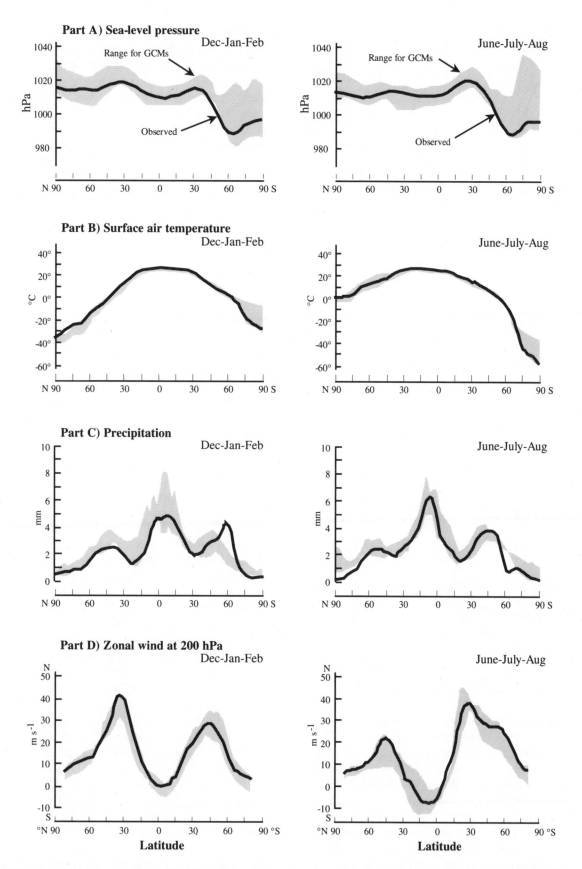

Figure 3.13 Actual versus computer simulated results for (A) sea-level pressure, (B) surface air temperature, (C) precipitation and (D) upper tropospheric wind (based on Gates et al., 1990).

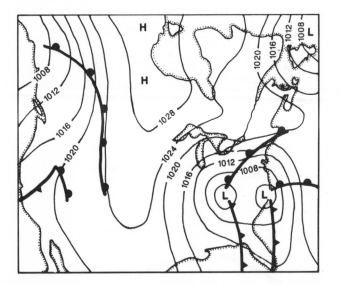

Figure 3.14 Synoptic map of North America, February, 1984 (Bryant, 1991). Mobile polar highs are channelled south across the centre of the continent by the Rocky Mountains to the west, and Appalachians to the east. Cyclogenesis occurs in front of the high over the Gulf of Mexico, with lows tracking northeast along the eastern seaboard. These storms move substantial quantities of latent heat from the tropics to high latitudes.

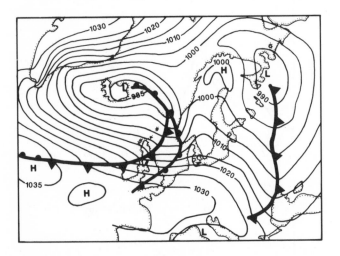

Figure 3.15 Synoptic map of northern Europe for 26 February 1949 (modified from Wallén, 1970). The intense low over Iceland is typical of winter and will track eastwards across Scandinavia in westerly air circulation. The low has been triggered by a strong mobile polar high over the North Atlantic, and eventually will track across France.

Sea can redirect jet streams southward over the Rocky Mountains, leading to cold conditions further eastward across the continent. More significant is the effect of warmer sea surface temperatures

in the Pacific Ocean. These temperature variations are related to the Southern Oscillation, and can alter the location of the Aleutian Low, or force the jet stream either northward around the Rockies, or directly across the southern part of the continent. Heavy rainfall in California is related to these sea surface temperature fluctuations, while the occurrence of storms is linked to the southward displacement of low pressure over the North Pacific. These changes also have prognostic teleconnections (long distance coherence) with short-term climate change over southeastern Asia and Australia. The signatures of these changes are described in the following chapter.

Europe
(Wallén, 1970; Boucher, 1987; Bryant, 1991)

Europe is the most maritime of continents, lying exposed to westerly winds crossing the North Atlantic Ocean. A belt of water that includes the Mediterranean, Black and Caspian Seas, encircles Europe to the south, while the North and Baltic Seas fragment the northern part of the continent. Westerly winds crossing the Atlantic Ocean are deflected by two east–west mountain ranges, the Alps in southern Europe, and the Atlas Mountains of North Africa. Rather than forming barriers to air movement, these two mountain ranges steer mobile polar highs across the continent. Those arriving from North America are deflected, either northward across the British Isles and Scandinavia, or through the Mediterranean Sea into the Middle East. Outbreaks of cold air over Scandinavia tend to be forced by this circulation eastward over Russia.

Europe has a climate that is moderated by the North Atlantic Ocean, not so much because of the warm Gulf Stream, but because strong winds evaporate moisture from the ocean in general. This latent heat is then released into air masses travelling across Europe. Warmer surface water is drawn into the region from the subtropics in this process. There is thus a strong temperature decline from west to east across Europe, especially in winter. The strong winds are produced by the movement of mobile polar highs from eastern North America across the North Atlantic. In the process, strong mid-latitude lows are generated in the region of Iceland (Figure 3.15). These lows are more vigorous in winter, and track eastward across northern Europe. In summer, when mobile polar highs are weaker, the continent comes under the influence of high pressure centred over the

Azores Islands. This produces summer dry conditions in southern Europe.

Climatic variability in Europe depends upon interplay between zonal versus meridional air circulation. The former brings heavier rainfall, warmer temperatures and storm conditions to northern Europe, and stable drier weather in the south. Meridional flow is conditioned by blocking of high pressure cells over the eastern Atlantic or northwestern Europe. In this situation, low pressure cells tend to track north–south across Europe. Beneath the high pressure, climate is dry and stable; however, deflection of low pressure tracks brings aseasonal storm conditions beneath its path. Whenever mobile polar highs are intense, storminess is exacerbated in northern Europe. This picture is complicated by regional cyclogenesis generated around the British Isles or over the eastern Mediterranean in the lee of the Alps. Historically, North Sea storms have been of enormous intensity, and have led to an appalling death toll with substantial effects on society and the course of history. The variability in the magnitude of such storms appears to be a clear signature of changing climate globally in response to changes in evaporation over the North Atlantic. Many of these aspects are dealt with in detail in other chapters.

Australia
(Gentilli, 1971; Bridgman, 1987; Colls and Whitaker, 1990)

Australia is an island continent with 40% of its landmass lying within the tropics, and 75% lying below 500 metres above sea-level. The effect of topography upon temperature and air mass movement is negligible. Topographic effects occur along the eastern Great Divide, on the island of Tasmania, and in the centre of the continent in the vicinity of the Musgrave and Macdonnell Ranges. The continent is subject to oceanic influences from warm currents on both its east and west coasts (the East Australian and Leeuwin Currents respectively). Australia lies at the equatorial end of mobile polar high penetration into the mid-latitudes from the Antarctic. When pressure is averaged seasonally, then the continent is dominated by the Hadley cell. Changes seasonally simply reflect a 5–8° latitudinal shift in the movement of these averaged high pressure belts in conjunction with the seasonal movement of the solar equator. The change between summer and winter in Australia is ambiguous. The passage of highs tends to stall in summer over the Indian Ocean southwest of Perth, in the Australian Bight and over the

south Tasman Sea. This reflects the tendency for mobile polar high pressure cells to move more slowly over locations where atmospheric subsidence is enhanced by surface cooling of air overlying cold water. In winter, highs linger over the continent because cooling is more effective over land.

The northern half of Australia is influenced by easterly trade winds induced by the passage of mobile polar highs. In winter, these trade winds affect only the northern extremity of the continent; but in summer, moist trade winds passing over the warm Coral Sea bring rainfall to the northern half of the continent. Northern Australia also forms part of the Indonesian–Australian 'maritime' continent, and thus comes under the influence of the Asian monsoon. Heat lows over the Pilbara region of Western Australia and the Gulf of Carpentaria enhance monsoonal circulation. Sea temperatures lag this peak in warming by several months. As a result, the timing of tropical cyclones lies preferentially between February and April.

Anomalies in Australian climate relate to the aseasonal pathways of mobile high pressure cells, or intensification of centres of instability, within the trade winds and over the Australian continent in summer and autumn. For instance, the occurrence of east coast lows provides some of the highest rainfalls in eastern Australia (Figure 3.16). The abnormal positioning of blocking high pressure to the south of Australia exacerbates these storms. Climate change in the Australian region is very much controlled by the behaviour of the Asian monsoon, the strength of trade winds, and the intensity of mobile polar highs. These aspects are also described in other chapters.

URBAN CLIMATES
(Oke, 1978, 1979; Landsberg, 1981a,b)

Smaller scale circulations are beyond the scope of this text; however, urban climates are important for three reasons. First, as of 1990, over 45% of the world's population lived in cities, and that figure is forecast to increase to over 50% by the year 2000. Table 3.1 presents the growth in the number of cities with over one million people since 1950. Such cities often are as much as 5°C warmer in their centres than their surrounding hinterlands. For Paris and Montreal, this 'heat island' effect is as much as 10–14°C. The number of cities affected by temperature increases due to heat islands will double by the year 2025. Second, the human

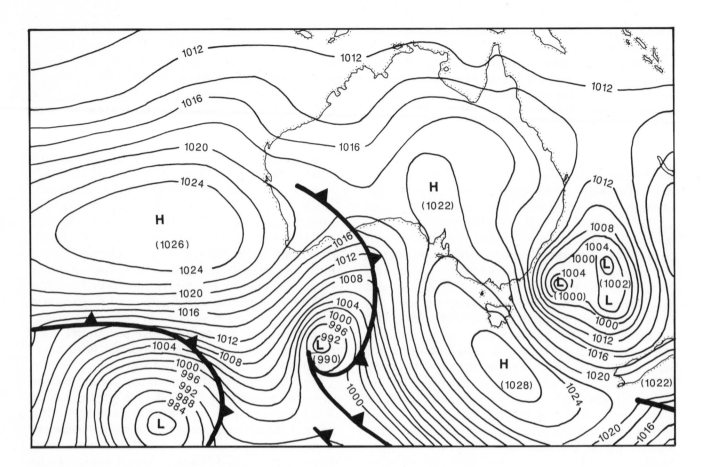

Figure 3.16 Synoptic map of the Australian region, 8 June 1994. Mobile polar highs tend to cross the Australian continent around 30°–35° S. Cyclogenesis often occurs to the southwest of Australia, but rarely with the classic V-shaped frontal pattern shown here. East coast lows develop over warm water in the southwestern Tasman Sea preferentially in late autumn. Note the similarities between the mobile polar highs and lows shown here, and the model in Figure 3.11.

impact of climate change, especially global warming, will be felt most within these heated cities. Third, the supposed temperature increases of the past century have been recorded mainly in cities, and those increases may not necessarily reflect global warming as much as they do the growth of those cities. This latter aspect is discussed in detail in other chapters.

The physical nature of a city is considerably different from its surroundings or, given the present growth of megalopolises, from what the landscape used to look like. For historical reasons most cities have developed in valleys or depressions, close to water. Their climates thus are shaped by mountain, land and sea breezes, and by temperature inversions. The non-urban landscape usually is vegetated to varying degrees (by forest, grassland or agricultural crops), has a significant exchange of water between the ground and the atmosphere (infiltration of rainfall versus evaporation), and is

relatively smooth over distances of 100–200 metres. Vegetated areas are cooler because evapotranspiration incorporates latent heat. Vegetated surfaces also have a higher reflectance, thus decreasing the total absorption of solar radiation at the ground. In addition, the underlying soil contains air-filled voids having lower thermal capacity.

The urban landscape is different. The portion of vegetated land is less, having been replaced by roads, parking lots, buildings, paved yards, forecourts and playgrounds. This sealing of the city surface prevents infiltration of rainfall into the ground and minimises evaporation. A higher proportion of rainfall goes into runoff, and the store of groundwater is lessened unless artificial watering of lawns and parks is maintained. Concreted and paved surfaces have lower albedos, greater heat conductivity and greater thermal conductivity. These surfaces accumulate large stores of heat energy which can be reradiated to the atmosphere quickly at night. These stores are supplemented by artificial heat emanating from artificial lighting, vehicle exhaust and internal heating of buildings. In the United States, this accounts for 10–15% of incoming solar radiation in cities with more than one million inhabitants. High buildings can also shade the city, and because they protrude irregularly into the sky, trap more solar radiation. Because heat is released from buildings higher in the atmosphere, lapse rates are altered and instability is favoured, with increased convection, cloudiness and precipitation over the urban area. This is aided by more dust and particulate pollutants in the air, providing more hygroscopic condensation nuclei. In general, cities tend to have 5–30% more rain than their hinterlands. These conditions reduce air temperatures in the city during daytime. Thus the urban surface tends to be wet and warm, absorbing solar radiation during the day and being less affected by cooling at night.

The greatest difference in heating between a city and its surrounds tends to occur from three to five hours after sunset, with the difference diminishing towards morning. The effect is most noticeable in winter and at high latitudes. Temperatures do not increase smoothly towards the city centre, but rise abruptly at the edge of the built-up area. The intensity of this warming is enhanced at low wind speeds. For instance, in London, England, the heat island all but disappears at wind speeds above 11 metres per second. This relationship is dependent upon the size of a city. For Hamilton in Canada, a city of 300,000, the heat island disappears at wind speeds above 6–8 metres per second; but for Palo Alto, United States, with a population of 33,000, the heat island disappears above wind speeds of only 3 metres per second. The intensity of warming is also highly correlated with the size of a city, and it follows that, as a city grows, so does the size of its heat island (Figure 3.17).

Climatically, the most unnatural aspect of a city is its structure. Cities, because of their irregular building heights, form very rough landscapes. This is most apparent in their central business districts where skyscrapers, rising hundreds of metres, alternate with roads. This effect increases the frictional coefficient of the ground surface. As a result, winds generally decrease over the city by as much as 20–25%. This has the effect of lifting in height equivalent wind speeds by 100–200 metres, and creating a large pool of slow-moving air above the city that accumulates water vapour, urban dust, pollutants and gases. Many of these gases are

Table 3.1 Number of cities with more than 1 million and 4 million inhabitants, 1950–2020 (based on the United Nations, 1985).

Year	Number of cities	
	>1 million inhabitants	>4 million inhabitants
1950	78	13
1960	114	19
1970	160	23
1980	222	35
1990	276	48
2000	408	66
2010	511	90
2025	639	135

'greenhouse' gases; however the urban atmosphere also contains large quantities of sulphate aerosols that block out incoming solar radiation, negating any enhanced 'greenhouse' warming. Air tends to spiral (anticlockwise in the northern hemisphere, clockwise in the southern hemisphere) into the city centre, notably at night because of urban heating. Some cities, such as Chicago and Winnipeg, with rectangularly gridded road systems orientated with at least one axis parallel to prevailing winds, may be windier and gustier because air can freely flow down the canyon-like gaps between buildings. When higher wind speeds exist, buildings can increase turbulence.

Cities are subject to enhanced climatic hazards. For example, regional heat waves become more

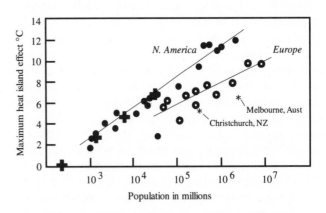

Figure 3.17 Relationship between urban population size and maximum heat island effect in North America and Europe. The crosses refer to the development of a new city, Columbia, Maryland, from a population of less than 1,000 to 25,000 inhabitants (based on Landsberg, 1981b and Oke, 1979).

severe in the city, and moderate rainfalls can become catastrophic because of reduced infiltration and greater runoff. Stagnant air allows moisture to accumulate in the urban atmosphere. If regional humidities are high to begin with, and the atmosphere is conditionally unstable, then convective cells can be drawn into the city, and undergo dramatic intensification into major thunderstorms, with attendant flash flooding, hail and near-tornadic winds. For example, two of the three most costly disasters in Australian history have been hail and windstorms over the Sydney metropolitan area. The heaviest rainfalls associated with thunderstorms in the Sydney region occur over the central business district. Similar anomalies have been observed over Berlin, London and Washington. In Paris, while summer thunderstorms have become more violent over the past decade, the city receives less rainfall in winter than the surrounding countryside because the heat island deflects the passage of rain-bearing weather systems.

CONCLUDING COMMENTS

Atmospheric circulation, the movement of pressure cells and their model representation are dynamic. The Palmén-Newton model, like any model, attempts to represent the real world in simplistic terms. It is an excellent teaching model, and allows students to understand the weather patterns they observe in their environment. However, the world's weather patterns do not follow this model. For instance, the Hadley cell is difficult to locate simply by examining sequences of weather charts. In Australia, high pressure cells are just as likely to appear south of the continent in winter as they are in summer, although when averaged they tend to appear where predicted. The polar front is just as elusive. Unfortunately, computer simulations accept as realistic the static features of this model. The concept of mobile polar highs offers a dynamic alternative explanation of atmospheric circulation. Students can grasp its underlying concepts just as easily. However, its main benefit is its ability to explain atmospheric circulation patterns under different climate regimes, regardless of whether the climate is colder than present, or undergoing 'greenhouse' warming. These views are expanded in other chapters.

Local climates do not figure prominently in discussions of climate change. However the greatest climate change of the past century may not be global warming, but the development and agglomeration of urban climates. Some cities are now so big that, if they do not already dominate their region, they will have the potential to do so in the twenty-first century. The idea that growing cities are responsible for recent temperature increases is not novel. It has been discussed in many significant documents including the reports of the Intergovernment Panel on Climate Change. Other aspects of climate change attributed to enhanced 'greenhouse' warming also may be logically related to urban growth. For instance, increased cloudiness has also occurred over recent decades. Most cloud measurements are taken at the same urban locations as temperature, and may be showing nothing more than increased cloudiness caused by urban growth. Certainly clouds cast a wider shadow regionally than simple temperature increases from confined point sources. These aspects are outlined further in Chapter 7.

Finally, atmospheric circulation shows teleconnections between hemispheres. Similarly, there are key regions that can affect large expanses of the globe if they undergo climate change. For example, there are links between the Siberian high and the southeast Asian monsoon. As well, mobile polar highs originating from the Antarctic, and crossing the Indian Ocean, can impact upon the Indian monsoon. Other significant teleconnections and key centres for climate forcing exist and are described in Chapter 4. It is worth keeping these in mind when searching for mechanisms to explain rapid climate change, and the contemporaneous occurrence of that change across the globe.

REFERENCES AND FURTHER READING

Anthes, R.A. 1982. 'Tropical cyclones: their evolution, structure and effects'. *American Meteorological Society Meteorological Monograph* v. 19 no. 41 208p.

Boucher, K.R. 1987. 'Climate of Europe'. In Oliver, J.E. and Fairbridge, R.W. (eds) *Encyclopedia of Climatology*. Van Nostrand Reinhold, New York, pp. 428–445.

Bridgman, H.A. 1987. 'Climate of Australia'. In Oliver, J.E. and Fairbridge, R.W. (eds) *Encyclopedia of Climatology*. Van Nostrand Reinhold, New York, pp. 144–160.

Bryant, E.A. 1991. *Natural Hazards*. Cambridge University Press, Cambridge, 294p.

Bryson, R.A. and Hare, F.K. (eds) 1974. *Climates*

of North America: World Survey of Climatology. v. 11. Elsevier, Amsterdam, pp. 1–47.

Colls, K. and Whitaker, R. 1990. *The Australian Weather Book.* Child & Associates, Sydney, 175p.

Cubasch, U. and Cess, R.D. 1990. 'Processes and modelling'. In Houghton, J.T., Jenkins, G.J. and Ephraums, J.J. (eds) *Climate Change: The IPCC Scientific Assessment.* Cambridge University Press, Cambridge, pp. 69–91.

Gates, W.L., Rowntree, P.R. and Zeng, Q.-C. 1990. 'Validation of climate models'. In Houghton, J.T., Jenkins, G.J. and Ephraums, J.J. (eds) *Climate Change: The IPCC Scientific Assessment.* Cambridge University Press, Cambridge, pp. 93–130.

Gentilli, J. 1971. *Climates of Australia and New Zealand.* Elsevier, New York, 405p.

Gray, W.M. 1984. 'Atlantic seasonal hurricane frequency Part 1: El Niño and 30 mb Quasi-biennial oscillation influences'. *Monthly Weather Review* v. 112 pp. 1649–1667.

Gray, W.M., Landsea, C.W., Mielke, P.W. and Berry, K.J. 1994. 'Predicting Atlantic Basin seasonal tropical cyclone activity by 1 June'. *Weather and Forecasting* v. 9 p. 103–115.

Griffiths, J.F. 1976. *Climate and the Environment.: The Atmospheric Impact on Man.* Paul Elek, London, 148p.

Hamilton, M.G. 1979 *The South Asian Summer Monsoon.* Arnold, Melbourne, 72p.

Harrison, E.F., Minnis, P., Barkstrom, B.R. and Gibson, G.G. 1993. 'Radiation budget at the top of the atmosphere'. In Gurney, R.J., Foster, J.L. and Parkinson, C.L. (eds) *Atlas of Satellite Observations related to Global Change.* Cambridge University Press, Cambridge, pp. 19–38.

Landsberg, H.E. 1981a. 'City Climate'. In Landsberg, H.E. (ed.) *World Survey of Climatology: General Climatology.* v. 3 Elsevier, Amsterdam, pp. 299–334.

Landsberg, H.E. 1981b. *The Urban Climate.* Academic, New York, 275p.

Leroux, M. 1993. 'The Mobile Polar High: a new concept explaining present mechanisms of meridional air-mass and energy exchanges and global propagation of palaeoclimatic changes'. *Global and Planetary Change* v. 7 pp. 69–93.

Linacre, E. and Hobbs, J. 1977. *The Australian Climatic environment.* Wiley, Sydney, 354p.

Miller, A. 1971. *Meteorology.* 2nd ed. Merrill, Columbus, 154p.

Moran, J.M., Morgan, M.D. and Pauley, P.M. 1991. *Meteorology.* 3rd ed. Macmillan, New York, 586p.

Nalivkin, D.V. 1983. *Hurricanes, Storms and Tornadoes.* Balkema, Rotterdam, 597p.

Oke, T.R. 1978. *Boundary Layer Climates.* Methuen, London, 372p.

Oke, T.R. 1979. *Review of urban climatology.* World Meteorological Organisation Technical Note No. 169, 100p.

Sanders, F. and Gyakum, J.R. 1980. 'Synoptic-dynamic climatology of the "Bomb"'. *Monthly Weather Review* v. 108 pp. 1589–1606.

Sheets, R.C. 1980. 'Some aspects of tropical cyclone modification'. *Australian Meteorological Magazine* v. 27 pp. 259–86.

Suppiah, R. 1992. 'The Australian summer monsoon: a review'. *Progress in Physical Geography* v. 16 pp. 283–318.

Trenberth, K.E. (ed.) 1992. *Climate System Modeling.* Cambridge University Press, Cambridge, 788p.

United Nations 1985. *Estimates and Projections of Urban, Rural and City Populations 1950–2025: The 1982 Assessment.* United Nations Department of International Economic and Social Affairs, New York.

Wallén, C.C. (ed.) 1970. *Climates of Northern and Western Europe: World survey of Climatology.* v. 5 Elsevier, Amsterdam.

Whipple, A.B.C. 1982. *Storm.* Time–Life Books, Amsterdam, 176p.

Yoshino, M.M. (ed.) 1971. *Water Balance of Monsoon Asia.* University of Hawaii Press, Honolulu, 308p.

4

The Role of Oceans

OCEAN CLIMATE PROCESSES
(Linacre and Hobbs, 1977; Pickard, 1979; Beer, 1983; Tolmazin, 1985)

Ocean circulation is conditioned by the properties of seawater under the effects of changing salinity, density and temperature, and by forcing of atmospheric circulation. In combination, these effects are responsible for the present state of the ocean under the Earth's existing climate. More importantly, these changes in ocean dynamics are responsible for much of the short- and long-term climate change normally attributed to the atmosphere.

Thermal expansion

The density of water is 1,000 kilograms per cubic metre at 4°C. Below this temperature, water becomes less dense and tends to rise to the surface. When it freezes, ice floats. The addition of salt increases the density of water if the temperature is constant. Seawater's density varies between 1,034 and 1,035.4 kilograms per cubic metre over 90% of the ocean (by convention 1,000 is subtracted from this value and these densities are reported as 34–35.4‰). Salt lowers the freezing point of water such that ocean water freezes at a temperature of -1.9°C. The maximum temperature that can occur in the open ocean is 30–31°C. Above these values, the loss of heat through latent heat of evaporation balances the addition of heat through absorption of solar radiation. In addition, thick, high-level clouds form as a result of intense convection at these temperatures. This limits incoming solar radiation. Surface temperatures in shallow seas can reach higher values if solar radiation can penetrate to the seabed, and heat the water from the bottom. These processes need to be considered when making predictions about sea surface temperature rises under global warming.

Without the influence of wind, the vertical and horizontal flow characteristics of water in oceans is dependent upon interrelationships amongst salinity, density and temperature. Warming seawater expands at the ocean surface. This warmed water is mixed by wind, producing a layer 50–200 metres deep that is isothermal. Between 500 and 1000-metre depths, temperature decreases rapidly, and then more slowly through the cold bottom water. The zone of sharply changing temperature is called the thermocline. Its depth can be increased by downwelling, and by the accumulation of warm surface water driven by wind against a coastline.

Warming water expands in volume by one centimetre for each 1°C increase in temperature. This raises sea-level locally, providing a hydraulic head that drives water downslope towards cooler water. Because water warms more in the tropics, there is a general tendency in the oceans for water to flow from the equator towards the poles. For example, along the east coast of the United States, the gradient in sea-level can be as much as one metre between the Caribbean Sea and the Labrador Sea. This supra-elevation drives the Gulf Stream. A similar warming effect in the Coral Sea initiates the East Australian Current southward along the east coast of Australia.

Climatic controls and salinity sinks

Rainfall over the ocean initially floats on the surface because it is less dense. Rain can dilute seawater, slowly lowering its density. In contrast, evaporation and the formation of sea-ice enhance salinity. New sea-ice has a salinity of 5 to 15‰, decreasing towards 0‰ as the ice ages and displaces brine. High density water, being more dense, sinks. Thus, wherever evaporation or the formation of sea-ice is greatest, salinity sinks develop. Generally, precipitation and the influx of freshwater discharged from rivers dominate the northern hemisphere, while evaporation dominates the southern hemisphere. This pattern causes a flow of ocean water from the northern to the southern hemisphere. Additionally, precipitation causes sinking in the ocean because it adds weight to existing mean sea-level, while evaporation lowers sea-level and generates surface inflow. Thus, as a first approximation, both vertical and horizontal flows in the ocean can be accounted for by precipitation and evaporation differences.

Two regions of increased salinity generated by these climatic effects tend to be crucial, not only as sinks, but also as controlling mechanisms for climate change. The area in the North Atlantic, under the influence of strong winds generated by mobile polar highs, has high winds blowing across relatively warm waters. Wind aids evaporation leaving behind a more dense brine that sinks to the ocean bottom. As a consequence, surface water moves in to replace the evaporated water and the water that has sunk. Most of this replacement water comes from the tropics. The release of latent heat of evaporation through this process, in addition to the presence of the Gulf Stream, is responsible for the enhanced temperatures of northern Europe. However, if the North Atlantic salinity sink were to stop operating, Europe would become colder.

The North Atlantic salinity sink forms a global conveyor belt circulating saline bottom water between the southern and northern hemispheres, and across the major oceans of the world. The saline water travels along the bottom of the Atlantic Ocean and eastward into the Indian Ocean adjacent to the Antarctic continent (Figure 4.1). Here the North Atlantic Deep Water upwells in plumes, providing an important source of heat for the colder Antarctic circumpolar current, and contributing to oceanic heating of the atmosphere between 60° and 75° S. In the process, this large circulation belt also aids the equalisation of the latitudinal temperature difference created by the inequality of the Earth's radiation budget (Figure 2.12). Around the Antarctic, sea-ice formation causes brine to sink to the seabed, enhancing the bottom flow. The saline deep current flows between Australia and New Zealand, threads its way through the South Pacific islands and upwells in the North Pacific because of shallowing and constricting bathymetry along the west coast of North America. Upwelling is aided by the removal of surface water by strong westerlies generated by the Aleutian Lows. The rate of flow involved is equal to one hundred times the discharge of the Amazon River. The complete circuit of this bottom flow takes around 500 to 1,000 years. Surface water drifts back from the Pacific Ocean into the Indian Ocean through the Philippine–Indonesian archipelago (the Indonesian Throughflow), around the south of Africa, and then northwards back to the North Atlantic Ocean. The switching-off of the North Atlantic salinity sink has major ramifications for the global spread of Ice Ages, a factor discussed in more detail in Chapter 6.

Interaction between the ocean and the atmosphere can exist within three equilibrium states: thermal, saline and latitudinal. A change from the dominance of one state to another is not gradual, but sudden. In this sense, ocean–atmosphere interaction forms a chaotic system. Within the thermal state, cold water tends to sink at each pole because of cooling of ocean water. The saline state has high evaporation around the equator, mainly in the southern hemisphere, which increases the density of seawater and causes it to sink in the tropics. In the latitudinal state, seawater dominantly sinks around the Antarctic because of the large scale freezing of seawater. The North Atlantic salinity sink forms part of this latter regime, which presently dominates climate. Climate change involves swings between these states, but at different time scales. For instance, at the decade level, the latitudinal mode dominates. The present warming in northern Europe may be a manifestation of change in this mode. It is an important change, because it affects the strength of the westerlies and the location of storm tracks in the northern hemisphere. Much of the short-term noise in climate in the northern hemisphere is linked to changes in this mode. At the century time scale, the salinity sink in the tropics dominates. Most historical climate records are just beginning to detect changes in this mode.

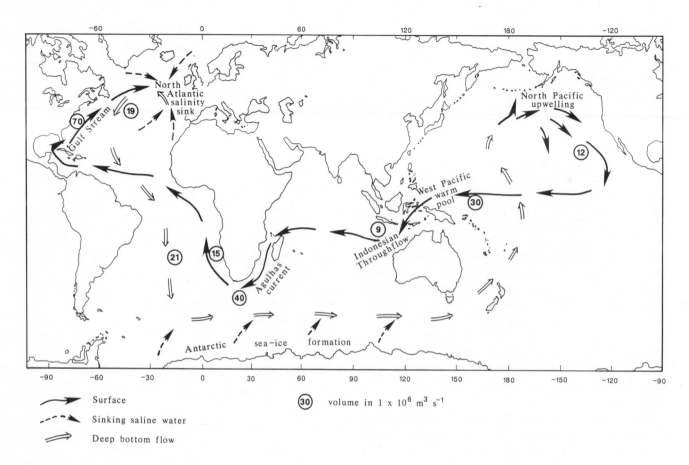

Figure 4.1 The North Atlantic salinity sink and deep ocean current system (transverse Mercator projection).

At longer time scales, which include Ice Ages, the thermal equilibrium state dominates.

Ekman transport

One of the greatest controls upon ocean currents is surface wind. When wind blows steadily across a body of water, it tends to drag surface water in the same direction. However, the moving water is affected by Coriolis force, and tends to be deflected to the right of wind flow in the northern hemisphere, and to the left in the southern hemisphere. The degree of this deflection increases with water depth (Figure 4.2). Surface water interacts viscously with the water underneath, and begins to drag it in the same direction but at a slower speed. This water is also deflected by Coriolis force, increasing the angle between the current at depth and the surface wind. This combination of viscous interaction and Coriolis deflection continues downwards in the water column, until either the seabed is reached or the current velocity becomes non-existent. In cross section, water flow

with depth traces out a logarithmic spiral. At a great enough depth water can actually flow opposite the wind direction. The phenomenon is termed Ekman transport.

Ekman transport has three main consequences for water circulation in the world's oceans. First, in ocean gyres which operate over a hemisphere, water tends to migrate towards the centre of the gyre, raising sea-level locally by about two metres. The added weight of water causes subsidence or downwelling at the centre of the gyre, while deeper water upwells towards the boundary to replace the water moving inwards. Logically, water at the centre of a gyre tends to flow under gravity downslope to zones of upwelling. However, Coriolis force acts upon this flow and deflects water movement perpendicular to this gradient. As a result of this interaction, elevated centres of gyres, and bordering troughs in the ocean, tend to be self-sustaining. Second, winds blowing parallel to a coastline cause water to move away from the coastline at the surface and towards the landmass at depth (Figure 4.3). Upwelling of colder water

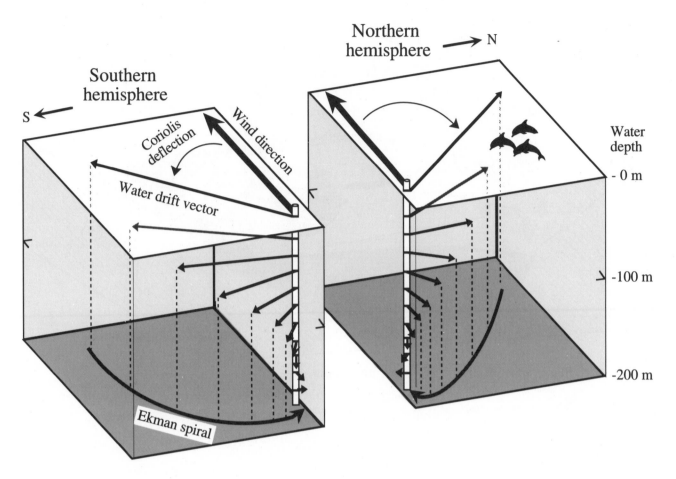

Figure 4.2 Schema of Ekman transport in a water column for both hemispheres (based on Beer, 1983).

occurs near the shore. High pressure cells located adjacent to a coast exacerbate this process. The stronger the wind, the shorter the time to initiate upwelling, and the stronger the event. Upwelling of cold water can produce the type of temperature gradient in coastal sea surface temperature required to develop east coast lows. Where high pressure semi-permanently stays in position over time, such as in the southeastern Pacific Ocean off the coast of Peru, then upwelling of cold water because of Ekman transport becomes semi-permanent. High pressure cells with preferred pathways over the ocean generate cold currents flowing towards the equator along the west coasts of continents. Such upwelling regions are most prominent off the west coasts of South America (Humboldt current), California (California current) and Africa (Benguela current). Finally, if pairs of counter-rotating gyres exist, then an upwelling zone of cold, deeper water will develop between them. The largest ocean gyres in the world exist in each hemisphere of the Pacific

Ocean under the influence of subtropical high pressure. These highs are responsible for easterly trade winds; however, along the equator, surface water tends to move poleward because of Ekman transport even though Coriolis force is very small. Sea-level elevations along the equator as a result are depressed, but more importantly; temperatures here are up to 5°C cooler because of upwelling.

Atmosphere–ocean feedbacks

Ocean surface temperatures partially control atmospheric temperature and climate. Both the ocean and the atmosphere are locked into positive feedback that tends to give the Earth's climate system stability, and resistance to sudden perturbations and climate change. Alternatively, changes in ocean temperature will affect atmospheric temperature structure and rainfall distribution. Normally, currents elevate sea-level where they impinge upon the side of an ocean. The process is reinforced by the winds that drive the currents. For

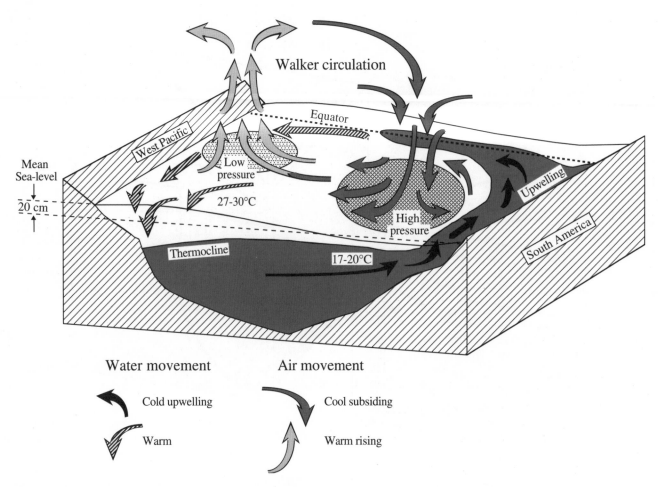

Figure 4.3 Positive feedback amongst atmospheric pressure, ocean circulation, ocean temperature and sea-level. The figure schematically mimics normal ocean–atmospheric conditions in the south Pacific Ocean.

example, the easterly trades in the southern Pacific Ocean pile up water by about 20 centimetres along the western boundary (Figure 4.3). Because warm surface water is being moved, sea surface temperatures are higher in the west, reaching values of 27–30°C. These high temperatures enhance evaporation, and permit the transfer of latent heat of evaporation from the ocean to the atmosphere. Subsequent condensation creates vertical instability leading to the formation of low pressure. Air moving into this low pressure reinforces the original wind pattern, and the complete process becomes self-sustaining as long as warm water is maintained at the western side of the Pacific. A similar positive feedback process operates on the eastern side of the Pacific. Cold water between 17 and 20°C upwells along the South American coast, because of Ekman transport under the influence of a quasi-permanent high pressure cell (Figure 4.3). The cold water cools the air above it, which begins

to subside because it is more dense. Subsiding air prohibits condensation, and leads to conditions of aridity and stability. It also reinforces high pressure at the Earth's surface. High pressure under Coriolis force causes wind to flow westward along the equator and northward along the South American coast. This airflow removes warm surface water westward, and reinforces the upwelling of cold water at the coastline. Thus, there is a tendency for the oceans to stabilise the climatic state of the atmosphere. When this state is disturbed, then climate in the short-term in the subtropics can undergo dramatic change.

GLOBAL CURRENTS
(Beer, 1983; Tolmazin, 1985)

The largest control upon the main ocean currents is the recurrent location of high and low pressure cells over the ocean and their associated wind patterns. Both lows and highs, in their transit over oceans, pause over warmer and cooler waters respectively. When averaged over time, as in the Palmén-Newton model, highs tend to centre over the oceans (Figure 3.5). Whereas the apparent position of the sun migrates seasonally through nearly 47° of latitude, the average position of lows and highs rarely shifts by more than 5°. As wind blows across the ocean surface, some of its momentum is transferred to the water surface through wind stress. Wind stress is related to the square of wind velocity, such that stronger winds have a disproportionate effect upon the generation of ocean currents. The surface layer of the ocean moves considerably faster than layers of water further down at depth. Wind effects are transmitted downwards through turbulence, convection and viscous drag directed by Ekman transport. Because of this downward flux in momentum, surface currents in the ocean rarely reach velocities above 2 metres per second. For this reason, most wind-driven currents are called drifts.

The main surface currents of the world's ocean are outlined in Figure 4.4. The actual volumes of flows flowing in the currents are summarised in Table 4.1. Meridional flows tend to form confined currents in contrast to zonal flows which drift over broad areas. The west sides of oceans are dominated by warm currents flowing poleward, and suddenly swinging eastward at mid-latitudes under the influence of Coriolis force and the sloping boundary with cold water below. The largest of these is the Gulf Stream, with an average flow of 100×10^6 m^3 s^{-1} located off the east coast of North America.

The east sides of oceans are dominated by cold currents that are the product of upwelling caused by the dominance of high pressure cells. These currents are weaker than their western counterparts. The largest such current is the Humboldt current with an average flow of 18×10^6 m^3 s^{-1}, off the west coast of Peru. There is one main exception to this pattern, and that occurs off the west coast of Australia where the warm Leeuwin current flows poleward. The Leeuwin current is the product of throughflow between the west Pacific and Indian Oceans via the Indonesian Archipelago, due either

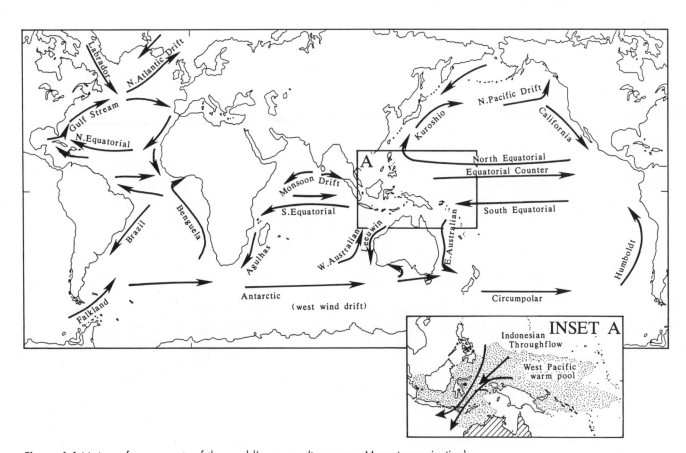

Figure 4.4 Main surface currents of the world's oceans (transverse Mercator projection).

Table 4.1 Flow rates of major ocean currents.

Current	Flow rate (× 10^6 m³ s⁻¹)
Antarctic Circumpolar	185
Gulf Stream	55–75
Kuroshio	65
Agulhas	40
Pacific North Equatorial	30
Equatorial countercurrent	25
East Australian	20
Humboldt	18
Benguela	15
California	12
Indonesian Throughflow	5–18
West Australian	10
Brazil	10
South Pacific Equatorial	10

to the piling up of water in the west Pacific Ocean by the trade winds (Figure 4.4), or to the thermal forcing of the west Pacific warm pool. Flow velocities through the Lombok and Makassar Straits in Indonesia can reach values of 0.3–0.6 m³ s⁻¹. This water flows along the northern coast of Australia, and then turns south under Coriolis force down the west coast. The volume of water involved is 5–18 × 10^6 m³ s⁻¹. Water takes thirty months to travel from the centre of the west Pacific warm pool to about 20° S along the west Australian coast. The expected northward flowing current for the east side of the Indian Ocean still exists in the form of the West Australian current, but is forced further offshore because of the Leeuwin current.

Some of the largest volumes of water occur as zonal flow. The cyclonic winds of the roaring forties generate a strong easterly drift around the Antarctic continent that carries 185 × 10^6 m³ s⁻¹ of water. Smaller, but still substantial flow is also carried westward in each hemisphere adjacent to the equator by the North and South Equatorial Pacific currents. Squeezed between these is a narrow countercurrent that flows eastward in the northern hemisphere. This countercurrent is a product of Ekman transport which raises sea-level on the non-poleward side of each equatorial current. Lower sea-level is sandwiched in between. However, because the mean centre of the solar equator lies in the southern hemisphere in the Pacific Ocean, part of the South Equatorial current actually resides in the northern hemisphere. The trough between the North and South Equatorial currents lies in the northern hemisphere, and water flowing into it is deflected to the east by Coriolis force to form a single narrow countercurrent north of the equator. Weaker equatorial currents also exist in the Atlantic and Indian Oceans. A single countercurrent also exists in both these oceans for the same reason as outlined above; however in the case of the Indian Ocean, the trough between the equatorial currents lies completely within the southern hemisphere.

GYRES, EDDIES, POOLS AND RINGS
(Beer, 1983; Tolmazin, 1985)

Zonal drifts form five large, elliptical gyres between 4,000 and 8,000 kilometres in diameter in the world's oceans. Currents are fastest towards the west sides of gyres in the northern hemisphere. The largest two gyres exist in each hemisphere of the Pacific Ocean (Kuroshio–California–North Equatorial, and Humboldt–South Equatorial–East Australian–Antarctic Circumpolar). The Atlantic Ocean contains another two gyres with the stronger one lying in the North Atlantic (Gulf Stream–Canary Island, and Brazil–Antarctic Circumpolar–Benguela). The fifth gyre exists in the Indian Ocean (Agulhas–Antarctic Circumpolar–West Australian). The intensity of these gyres is interrelated through global atmospheric–oceanic teleconnections. The gyres also inhibit meridional flow of water between hemispheres. Similar isolation occurs at convergence zones. For example, the Antarctic Convergence near the Antarctic Circumpolar current prevents icebergs drifting further north towards the subtropics.

Strong current velocities can lead to turbulence, instability in flow or meandering. The Gulf Stream and the East Australian current are not linear streams of water, but consist of meanders with attached ring and pool structures. The meanders are not well constrained laterally and, in moderately flowing systems such as the East Australian or Leeuwin currents, can pinch off to form an eddy. As the eddies become more isolated from the main current, they form isolated pools. Pools can be up to 150 kilometres in diameter, and extend 500 metres or more below sea-level, with rotational velocities of 0.7–1.5 metres per second (Figure 4.5). Warm eddies and pools tend to form on the poleward side of a current and move erratically in the general direction of the original

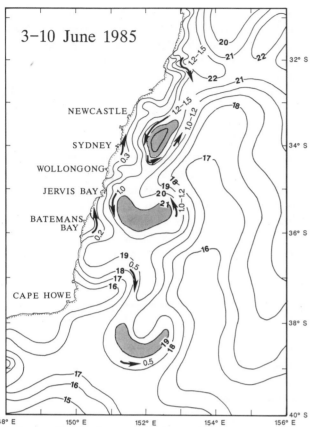

Figure 4.5 The formation of eddies and pools within the East Australian current (based upon Naval Weather Centre, Royal Australian Navy maps). Surface velocities and current directions are approximate.

current. Cold eddies and pools form on the equatorial side and move against the current. Sometimes a current may disintegrate completely into simply a series of eddies and pools. Warm pools tend to ride higher on the ocean surface, while cool pools form depressions. Water tends to flow downslope in each, but is deflected by Coriolis force parallel to isotherms. Water within a pool thus can be isolated from the surrounding ocean for periods lasting as long as one or two years. Because warm pools are more buoyant than the surrounding seawater, they can bob up and down below the ocean surface, disappearing for a period of time, only to reappear at the surface weeks or months later. Eventually the pools dissipate enough heat to the surrounding atmosphere, or mix sufficiently with the surrounding ocean, that they lose their identity.

Rings are the product of more rapidly meandering flow such as that found in the Gulf Stream, Kuroshio and the Antarctic Circumpolar currents. Rings can be 150–300 kilometres in diameter and

extend 2,500–3,500 metres downstream. Whereas an eddy or pool usually has a uniform temperature, rings have a narrow band of water circulating tightly around a core with a different temperature. The ring represents the cut-off current while the core represents water of different temperature adjacent to the main current, which has been trapped by the pinched-off meander. In the northern hemisphere, smaller warm-core rings rotate clockwise parallel to the Gulf Stream on its north side (Figure 4.6). Larger, more slowly westward moving cool-core rings with anticlockwise rotation are produced on the southern side. About three or four warm-core rings form compared to between eight and fourteen cold-core ones. Rings rotate with a surface speed that can reach 1.5 metres per second, drift at a speed of 3–5 kilometres per day, and can persist from 1 to 3 years. The rings on the south side of the Gulf Stream form part of a slow moving volume of water called the Sargasso Sea that drifts towards North America. Cold-core rings from this sea may be reincorporated back into the main Gulf Stream after a few months.

Both pools and rings carry the physical and biological material contained within them at the time of their formation. They are also important mechanisms for transporting heat energy in the ocean. If pollution discharged at the coastline, or dumped at sea, becomes trapped in a ring, then it does not disperse easily in the ocean. Pools and rings can impinge upon the shelf, where they generate secondary flows, upwelling, and local sea-level changes across the shelf and at shore. Along the east coast of Australia, pools are important in establishing the sea surface temperature gradient perpendicular to shore that intensifies an ordinary low pressure cell into an east coast 'bomb'. The pool provides moisture and latent heat for the sustained generation of heavy rainfall along the adjacent coast.

SCALE AND RESPONSE TIME OF THE ATMOSPHERE AND OCEAN
(Pickard, 1979; Calder, 1991)

The ocean and atmosphere have very different scales of activity and response times. The oceans are three hundred times more massive than the atmosphere, and thus tend to have circulation systems that are smaller. Eddies are the oceanic equivalent of atmospheric vortices. While vortices

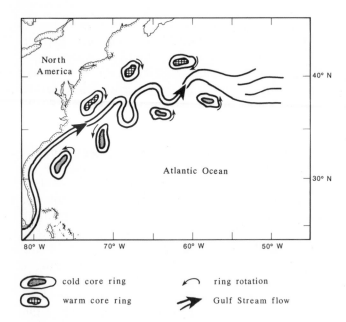

| cold core ring | ring rotation |
| warm core ring | Gulf Stream flow |

Figure 4.6 Schematic representation of ring structures associated with the Gulf Stream (based upon Tolmazin, 1985).

can be 1,000 kilometres in diameter in the atmosphere, they are only 10–100 kilometres in diameter in the ocean. Yet oceans can have gyres up to 7,000 kilometres in diameter, whereas the largest equivalent in the atmosphere are high pressure cells with diameters of a few thousand kilometres. The ocean is also shallower than the atmosphere, four kilometres in depth versus the seven- to twelve-kilometre thickness of the troposphere. Height differences in sea-level across the Pacific Ocean are normally only 20 centimetres, reaching 60 centimetres under extreme conditions. Major ocean currents can also elevate sea-level locally by one metre. In the atmosphere, high and low pressure systems can change the weight of air locally by 20%, and the overall change in the height of the atmosphere varies by 40% between the poles and the equator.

Ocean temperatures are also less variable than in the atmosphere. This is due to the higher heat capacity of water in the upper layers of the ocean. However at depth, ocean temperature is fairly constant, varying by less than 5°C below 1,000-metre depths. The thermal characteristics of the ocean change much more slowly that those of the atmosphere, while ocean currents can persist longer than flows in the atmosphere. The turnover of water in the ocean is also much slower than that in the atmosphere. Nuclear testing and volcanic eruptions have shown that the mixing time of air between the hemispheres

is about two years, although vortices may prohibit interchange with the polar regions. Most volcanic dust settles from the troposphere and much of the stratosphere within two years. Within six or seven years of a major eruption, there is virtually no trace of volcanic dust in the atmosphere. In contrast, fallout from nuclear testing has indicated rapid mixing of surface waters in the oceans only down to depths of 1,000 metres within two to three years. Radiocarbon dating suggests that the residence time for deep water in the Atlantic is about 300–450 years, while that in the Pacific Ocean is about 550–950 years. The North Atlantic salinity (thermohaline) sink thus takes about 500–1,000 years to recirculate water in the world's oceans.

THE SOUTHERN OSCILLATION
(Philander, S.G. 1990; Bryant, 1991; Glantz et al., 1991; Jacobs et al., 1994; Allan et al., 1996; Glantz, 1996; Trenberth and Hoar, 1996)

Normal circulation

The Southern Oscillation and interrelated phenomena such as Walker circulation, El Niño, ENSO events and La Niña affect over 60% of the globe. The Southern Oscillation is an atmosphere–ocean feedback process that tends to oscillate spatially, making the phenomenon erratic and, over the short term, responsible for climate change leading to extreme climatic hazard events such as droughts and floods. Simplistically, in the northern hemisphere summer, heating shifts from equatorial regions to the Indian mainland with the onset of the Indian monsoon. Air is drawn into the Indian subcontinent from adjacent oceans and landmasses, and returns via upper air movement, either to southern Africa or the central Pacific. In the austral summer, this intense heating area shifts to the Indonesian–Australian 'maritime' continent (Figure 4.7A), with air then moving in the upper troposphere to the east Pacific. Convection is so intense that updrafts penetrate the tropopause to provide one of the main routes by which water vapour and tropospheric pollutants enter the stratosphere. These convective cells are labelled stratospheric 'fountains'.

Throughout the year, an area of high pressure persists over the equatorial ocean west of South America. The high pressure off the Peruvian coast

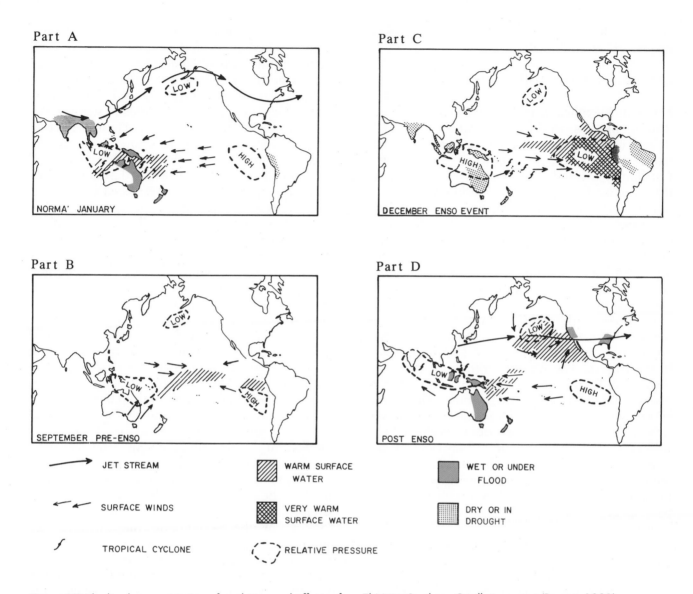

Part A

NORMAL JANUARY

Part B

SEPTEMBER PRE-ENSO

Part C

DECEMBER ENSO EVENT

Part D

POST ENSO

→ JET STREAM

←← SURFACE WINDS

∫ TROPICAL CYCLONE

WARM SURFACE WATER

VERY WARM SURFACE WATER

RELATIVE PRESSURE

WET OR UNDER FLOOD

DRY OR IN DROUGHT

Figure 4.7 Idealised representation of evolution and effects of an El Niño–Southern Oscillation event (Bryant, 1991).

is intense, and is locked into position because of cold ocean water and subsiding air feedback processes (Figure 4.3). The zonal position of this high pressure shifts seasonally less than 5° in latitude. Air flows from this region, across the Pacific, as an easterly trade wind back to the the region of heating in the western Pacific Ocean or Indian subcontinent. This circulation is zonal in contrast to the meridional circulation inherent within the Palmén–Newton general circulation model. The persistent easterlies blow warm surface water across the Pacific Ocean, piling it up in the west Pacific warm pool in the Philippine Sea. The supra-elevation of sea-level in the west Pacific Ocean is approximately one centimetre for each

1°C difference in temperature between the west Pacific and the South American coast. Normally this supra-elevation amounts to 13–20 centimetres (Figure 4.3). The easterly trade winds maintain this condition and, at the same time, drive surface currents westward adjacent to the equator in both hemispheres. Ekman transport operates on these currents and cold water upwells in a narrow zone at the equator (Figure 4.8A). This sharp temperature gradient at the equator cannot yet be modelled accurately by GCMs. These atmospheric and current systems are very stable, existing beyond the annual climatic cycle. The easterly trade wind flow is termed Walker circulation, after the Indian meteorologist, Gilbert

Walker, who described the phenomenon in detail in the 1920s and 1930s.

El Niño–Southern Oscillation (ENSO) events

For some inexplicable reason, this quasi-stationary heating process weakens in intensity or breaks down completely, every three to five years. This tendency to fluctuate is known as the Southern Oscillation. An index of its intensity and fluctuations can be derived simply by subtracting the barometric pressure for Darwin, Australia, from that of Tahiti. This value is presented for 1851 onwards in Figure 4.9. When the index is positive, Walker circulation prevails; when the index is negative, Walker circulation weakens or even reverses. At the latter times, higher pressure becomes established over the Indonesian–Australian area, while lower pressure develops over warm water off the South American coast. The easterly trade winds abate and can even be replaced by westerly flow in the tropics. The western rainfall area shifts to the central Pacific, and drought replaces normal or heavy rainfall over eastern Australia. This failure is known as an El Niño–Southern Oscillation (ENSO) event. Such conditions explained the Great World Drought of 1982–1983 which affected most of Australia, Indonesia, India and South Africa. Warming over significant parts of the globe is typical of ENSO events. Generally, the percentage of the Earth's surface that becomes cloudless exceeds the percentage that receives increased cloud during such events. The clear skies allow more solar radiation to reach the Earth's surface.

The exact cause of change in Walker circulation is not known; however the change most likely occurs during the periods when the low pressure centre in the West Pacific migrates seasonally between the Indian subcontinent and the Indonesian–Australian region, in either March–April or August–September. Five precursors to the failure of Walker circulation stand out. First, sea-ice in the Antarctic and snow cover in central Asia tend to be more extensive beforehand. Second, the strength of the Indian monsoon relies heavily upon formation of an upper air, thermally driven anticyclone between 100 and 300 hectopascals over the Tibetan Plateau, that leads to the formation of an easterly upper jet stream over southern India. Zonal easterly winds at 250 hectopascals over the Indian monsoon area also have been found to weaken up to two months before the onset of an ENSO event. Third, the Southern Oscillation appears to be a combination of the 2.2-year Quasi-Biennial Oscillation and a longer period cycle centred around a periodicity of five years. Fourth, changes in the behaviour of southern hemisphere, mobile polar highs are linked to ENSO events. Strong Walker circulation is related to strong westerlies between 35° and 55° S. These westerlies appear to lock developing heat lows in the austral spring over the Indonesian–Australian 'maritime' landmass. If the normally strong latitudinal flow in the mobile polar highs is replaced by longitudinally skewed circulation in the Australian region, then low pressure can be forced into the Pacific during the transition from northern to southern hemisphere summer. Increased southerly and southwesterly winds east of Australia preceded the 1982–1983 ENSO event, and flowed towards a South Pacific Convergence Zone that was shifted further eastwards. While highs were as intense as ever, they had a stronger meridional component, were displaced further south than normal, and were stalled over the east Australian continent. Fifth, ENSO events are more likely to occur one year after the south polar vortex is skewed towards the Australian–New Zealand sector. A significant correlation has been found amongst this eccentricity, the Southern Oscillation and the contemporaneous occurrence of rainfall over parts of southern hemisphere continents. Finally, the causes may be interlinked. For example, more snow in Asia the winter before collapse of Walker circulation may weaken the summer jet stream, leading to failure of the Indian monsoon. Or more sea-ice in the Antarctic may distort the shape and paths of mobile polar highs over the Australian continent, or displace the south polar vortex.

Whatever the cause, the movement of low pressure east, beyond the Australian continent, leads to westerly air flow at the western edge of the Pacific Ocean. Because there is no easterly wind holding supra-elevated water against the western boundary of the Pacific, warm water begins to move eastwards along the equator via an enhanced Equatorial Countercurrent (Figure 4.7B). Normally, the thermocline separating warm surface water from cooler water below is thicker (200 metres) in the west Pacific than in the east (100 metres). As warm water shifts eastwards, the thermocline rises in the west Pacific. One of the first indications of an ENSO event is the appearance of colder water north of Darwin. A zone of convective instability also moves from the western Pacific to the normally

Part A

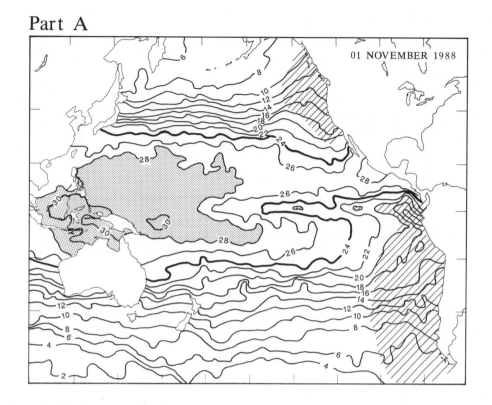

Part B

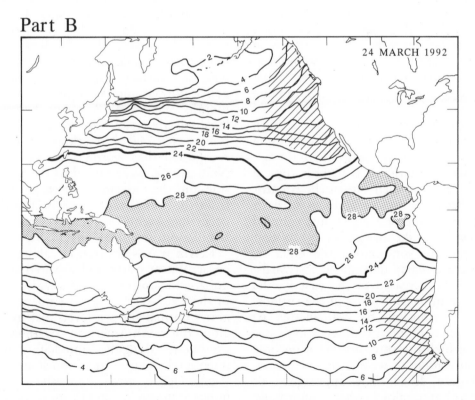

Figure 4.8 Sea surface temperature pattern in the Pacific Ocean (A) under normal Walker circulation; (B) at the peak of an ENSO event (based on NOAA maps). Areas with sea surface temperatures greater than 28°C are shaded with stippling. Those below 20°C adjacent to the coast of the Americas are shaded with slanted lines.

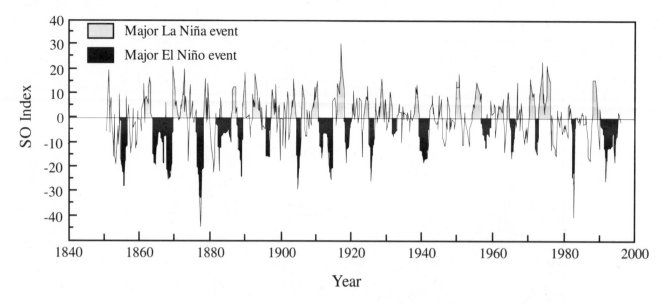

Figure 4.9 The Southern Oscillation Index 1851–1995. Based on Tahiti–Darwin pressure, normalised for the period 1951–1980, and amplified by a factor of ten.

drier, central Pacific islands. Cyclones which usually form in the west Pacific drift eastward towards Tahiti and Tonga, while the easterly trades weaken. This has two effects. First, because the warm water which has piled up in the west Pacific Ocean has nothing to hold it there, it now continues to surge across the Pacific as a succession of Kelvin waves (slow moving waves of low amplitude with lengths of thousands of kilometres), reaching the eastern side within a couple of months and swamping the cold water that normally resides there (Figure 4.8B). In unusual circumstances, paired tropical cyclones may develop on each side of the equator in November. Because tropical cyclones in each hemisphere spiral in opposite directions, coupled cyclones can combine (like an egg beater) to generate strong westerly wind along the equator, that may accelerate water movement eastwards. Second, sea-level is increased in the east Pacific as warm water swamps cold water at the surface, thickening the thermocline to 200 metres or more. The presence of warm water suppresses the formation of high pressure, making it become less intense, and leading to weaker or failed easterly trades (Figure 4.7C). The appearance of enhanced warm water, along what is normally an arid coastline, leads to abnormally heavy rainfall, causing floods in the arid Andes rainfall region, and drought in southern Peru, Bolivia and northeastern Brazil.

The appearance of warm water usually occurs in a weakened form each year along the Peruvian coast around Christmas. This annual warming is termed the El Niño, which is Spanish for 'Christ Child'. However when Walker circulation collapses, this annual warming becomes exaggerated, with sea surface temperatures increasing 4–6°C above normal and remaining that way for several months (Figure 4.8B versus Figure 4.8A). This localised above-average warming is termed an El Niño event. Jacob Bjerknes in the early 1960s recognised that warmer waters, during some El Niño events, were linked to the Southern Oscillation and the movement of warm water eastwards from the west Pacific Ocean. These more widely occurring events are termed El Niño–Southern Oscillation, or for brevity ENSO, events.

Kelvin waves are trapped along the east Pacific coast. Warm water and elevated sea-levels spread north and south reaching as far north as Canada. In the 1982–1983 event, temperatures were 10°C above normal and sea-levels rose by 60 centimetres along the South American coastline, and by 25–30 centimetres along the United States west coast. The intense release of heat by moist air over the central Pacific Ocean causes the westerly jet stream in the upper atmosphere, and the position of the Aleutian Low, to shift towards the equator (Figure 4.7D). Often a ridge of high pressure develops over the west coast of North America, deflecting westerly surface winds and the jet stream from their normal pathways across the continent. This can lead to dramatic changes in climate across the continent. During the 1982–1983 ENSO event, the jet stream passed through the middle of the continent, causing heavy rainfall in the southern United States, record-breaking mild temperatures along the American east coast, and heavy snowfalls in the southern Rocky Mountains. In contrast, the 1976–1977 event produced record-breaking low temperatures and heavy snowfall in the eastern half of North America, because the polar jet stream was diverted north of the Rocky Mountains into Arctic regions. The need to conserve vorticity sent the jet stream plunging south through eastern America, bringing with it cold Arctic air.

Finally, with the spreading of warm water in the northern part of the Pacific, easterly air circulation begins to reestablish itself in the tropics (Figure 4.7D). The depression of the thermocline in the east Pacific during an ENSO event causes a large wave, also called a Rossby wave, to propagate westwards along the thermocline boundary. This wave is reflected off the western boundary of the Pacific, and as it returns across the Pacific, it slowly raises the thermocline to its pre-ENSO position. However the return to Walker circulation can be anything but normal. The pool of warm water now moving back to the west Pacific drags with it exceptional atmospheric instability, and a sudden return to rainfall that breaks droughts in the western Pacific.

Where the Southern Oscillation is effective in controlling climate, it is responsible for about 30% of the variance in rainfall records. More importantly, it causes extreme, short-term, climate change over 60% of the globe. The climatic effects of an ENSO event are widespread, and its influence on both extreme climate hazards and short-term climate change is significant. The 1982–1983 event resulted in economic damage totalling $US8 billion worldwide. Southern Peru, western Bolivia, Venezuela and northeastern Brazil are affected most in South America by ENSO events; however, greatest devastation occurs in the west over southern Africa, India, Indonesia and Australia. For instance, in Indonesia, over 93% of monsoon droughts are associated with ENSO events, and 78% of ENSO events coincide with failure of the monsoon. In Australia 68% of strong or moderate ENSO events produce major droughts in eastern Australia. ENSO events have recently been shown to contribute to drought in the African Sahel region, especially in Ethiopia and Sudan. There is a strong coherence in the timing of both flood and drought on the Nile in Africa, the Krishna River in central India, and the Darling River in Australia.

Other climatic hazards such as tropical cyclones are affected by ENSO events. For instance, the duration and incidence of tropical cyclones in the equatorial Atlantic is profoundly depressed during ENSO years. Here, upper tropospheric winds between 0° and 15° N must be easterly, while those between 20° and 30° N must be westerly, for easterly wave depressions and disturbances to develop into tropical cyclones (hurricanes). During ENSO events, upper westerlies tend to dominate over the Caribbean and the western tropical Atlantic, giving conditions that suppress cyclone formation. Years leading up to an ENSO event have tended to produce the lowest number of tropical cyclone days in the Atlantic over the past century. Additionally, since 1955, the number of icebergs passing south of 48° N has been highly correlated to the occurrence of ENSO events. ENSO events may trigger a concomitant response in the much weaker North Atlantic Oscillation, involving the Icelandic Low and intensity of average high pressure over the Azores. A strengthened Icelandic Low enhances the number of icebergs. Alternatively ENSO events may change wind stress components along the North American east coast, thus exacerbating the production and southward movement of icebergs. Both the 1972–1973 and 1982–1983 ENSO events produced over 1,500 icebergs south of 48° N, a number that dramatically contrasts with non-ENSO years, when less than one hundred icebergs per year were recorded this far south.

Once an ENSO event is triggered, the cycle of climatic change operates over a minimum of two years. In historical records, the longest ENSO event lasted four years from early 1911 to mid-1915. However recent events have dispelled the belief that ENSO events are biannual. Following the 1982–1983 ENSO event, the warm water that travelled northwards along the western coast of North America slowly crossed the North Pacific, deflecting the Kuroshio current a decade later. As a result, northern Pacific sea surface temperatures increased abnormally, affecting general circulation across the North American continent. It is plausible that the drought on the Great Plains of the United States in the summer of 1988, and even the flooding of the Mississippi River basin in the summer of 1993 (attributed by most to the 1990–1995 ENSO event) are prolonged North American climatic responses to the 1982–1983 El Niño.

Warm water left over from the 1982–1983 ENSO event still persists in the North Pacific, and should be detectable there until the year 2000. The oceanographic effects of a major El Niño thus can have decade-long persistence outside the tropics.

The 1990–1995 ENSO event was even more anomalous. This event appeared to wane twice, but continued with warm central Pacific waters for five consecutive years, finally terminating in July 1995, a timespan that is unprecedented. Probability analysis indicates that an ENSO event of five years' duration should recur only once every 1,500 to 3,000 years. The changes in climate globally over these five years have been as dramatic as any observed in historical records. Two of the worst cyclones ever recorded in the United States, Hurricane Andrew in Florida and Iniki in Hawaii in August 1992, occurred during this event. Hurricane Andrew was unusual because Atlantic hurricanes should be suppressed during ENSO events. The Mississippi River system recorded its greatest flood ever in 1993 (and then again in 1995) surpassing the flood of 1973. Record floods devastated western Europe in 1994–1995. Eastern Australia and Indonesia suffered prolonged droughts that became the longest on record. Cold temperatures afflicted eastern North America in the winter of 1993–1994, together with record snowfall, while the western half of the continent registered its highest winter temperatures ever. Record high temperatures and drought also occurred in Japan, Pakistan and Europe in the summer of 1994.

Not all of the climatic extremes of 1990–1995 were consistent with that formulated for a composite ENSO event. For example, the Indian monsoon operated in 1994, while eastern Australia recorded its worst drought. Even locally within Australia, while most of the eastern half of the continent was in drought in 1992–1993, a 1,000 square kilometre region south of Sydney received its wettest summer on record. Whether or not all of these climatic responses can be attributed to the 1990–1995 ENSO is questionable, but they have occurred without doubt in regions where ENSO teleconnections operate. Additionally, the unprecedented nature of the 1982–1983 and 1990–1995 ENSO events may be linked to significant volcanic eruptions. While the El Chichon eruption of 1981 in Mexico did not trigger the 1982–1983 El Niño, it may have exacerbated its intensity. Similarly, the 1990–1995 event corresponded well with the eruption and subsequent global cooling generated by significant volcanic eruptions in 1991 of Mount Pinatubo, Philippines and Mount Hudson, Chile; and in 1992 of Mount Spurr, Alaska.

La Niña events

Exceptional Walker circulation is now being recognised as a phenomenon in its own right. With the return to easterly trades, the cool water that develops because of Ekman transport off the South American coastline, and in a 1–2° band around the equator, may become as cold as 20°C (Figure 4.8A). Exceptionally warm water may pool over the west Pacific, bringing with it increased convection and rainfall. This phenomenon is termed La Niña, Spanish for 'the girl child'. Because the percentage of the Earth's surface that receives increased cloud generally exceeds the percentage that becomes cloudless, La Niña events are globally associated with cooler temperatures. The 1987–1988 La Niña event was one of the strongest in forty years, and brought record breaking flooding in 1988 to the Sudan, Bangladesh and Thailand in the northern hemisphere summer. Regional flooding also occurred in China, Brazil and Indonesia. In Australia, this event led to dry winters sandwiched between exceedingly wet summers over the period 1988–1990. Record deluges and flooding of eastern rivers occurred on an extensive scale. During this period, Lake Eyre in the interior of Australia filled twice in consecutive years, the same number of times as over the previous century (other fillings in 1950 and 1974). La Niña thus may be as important as ENSO events in generating extremes in flood and drought over large sections of the globe.

CONCLUDING COMMENTS

The occurrence of ENSO and La Niña events is somewhat enigmatic, in that, what is described as normal often is based upon limited historical records that are used to form a composite picture of the behaviour of changes in the Pacific ocean–atmosphere system. It is possible that the behaviour of the Southern Oscillation changes from one decade to another. Indeed, from the 1870s to after the turn of the twentieth century, ENSO events appeared regularly and were strong. Except for the major event of 1940–1942, they weakened until the 1950s, and since 1972 have dominated over

La Niña events. This latter upswing in the frequency and strength of ENSO events is unusual, and has a recurrence interval of once in two thousand years. The recent changes have been attributed to global warming; however, ENSO events can produce warmer global temperatures because they are associated with clearer skies.

Computer simulations of Southern Oscillation behaviour over the twentieth century indicate that this phenomenon has only a small effect outside the tropics. While sea surface temperature anomalies dominate the patterns of floods and drought in the tropics, at mid-latitudes, climatic variations appear related to abnormal tracking of mobile polar highs and blocking of pressure cells. There may be too much emphasis placed upon ENSO and La Niña events as causes of short-term climate change globally. Finally, the prognostic characteristics of ENSO and La Niña events may be unwarranted. The Southern Oscillation may simply be a chaotic system with little predictability. The three- to five-year quasi-cyclic behaviour of the Southern Oscillation; the tendency to switch between two opposing states, one that is warm and dry (ENSO) and the other that is cold and wet (La Niña); the occurrence of record rainfall adjacent to areas of record drought; and the unusual aspects of the 1990–1995 ENSO event, are all characteristics suggesting that a chaotic process may be operating.

REFERENCES AND FURTHER READING

Allan, R., Lindesay, J. and Parker, D. 1996. *El Niño Southern Oscillation and Climate Variability*. CSIRO Publishing, Melbourne, 408p. IBM CD-ROM.

Beer, T. 1983. *Environmental Oceanography: An introduction to the Behaviour of Coastal Waters*. Pergamon, Oxford, 262p.

Bryant, E.A. 1991. *Natural Hazards*. Cambridge University Press, Cambridge, 294p.

Calder, N. 1991. *Spaceship Earth*. Penguin, London, 208pp.

Glantz, M.H. 1996. *Currents of change: El Niño's impact on climate and society*. Cambridge University Press, Cambridge, 194p.

Glantz, M.H., Katz, R.W. and Nicholls, N. 1991. *Teleconnections linking Worldwide Climate Anomalies: Scientific Basis and Societal Impact*. Cambridge University Press, Cambridge, 535p.

Jacobs, G.A., Hurlburt, H.E., Kindle, J.C., Metzger, E.J., Mitchell, J.L., Teague, W.J. and Wallcraft, A.J. 1994. 'Decade-scale trans-Pacific propagation and warming effects of an El Niño anomaly'. *Nature* v. 370 pp. 360–363.

Linacre, E. and Hobbs, J. 1977. *The Australian Climatic Environment*. Wiley, Sydney, 354p.

Philander, S.G. 1990. *El Niño, La Niña and the Southern Oscillation*. Academic, San Diego, 293p.

Pickard, G.L. 1979. *Descriptive Physical Oceanography*. 3rd ed. Pergamon, Oxford, 233p.

Tolmazin, D. 1985. *Elements of Dynamic Oceanography*. Allen and Unwin, Boston, 181p.

Trenberth, K.E. and Hoar, T.J. 1996. 'The 1990–1995 El Niño–Southern Oscillation event: longest on record'. *Geophysical Research Letters* v. 23 pp. 57–60.

—II—

CHANGE

5

Scales of Climate Change: Pleistocene to Modern

THE PLEISTOCENE ICE AGES
(Dawson, 1992; Williams et al., 1993; Andersen and Borns, 1994)

Ice Ages are not unique to the Earth's history. They also occurred in the Precambrian and Permian, although it is uncertain whether temperature fluctuated to the same degree as during the Pleistocene. The exact age when the Pleistocene began has been difficult to define because records of early glaciations are lacking. Additionally, the Earth's climate drifted into a colder regime throughout the Pliocene or Late Tertiary. Indeed, glacial ice was present over the Antarctic during most of the past sixty million years, as that continent drifted over the South Pole. Certainly, by 2.4 million years ago temperatures began alternating globally by 4–10°C every 40,000–100,000 years, to the extent that icesheets began to develop during cold phases, termed glacials. Eventually massive icesheets formed over the landmasses of the northern hemisphere during glacials, with significant alpine glaciers developing globally. Each time the Earth was locked in ice, the level of the oceans dropped from their present levels by 100–130 metres. This left broad plains fringing continents, instead of submerged continental shelves, and increased the size of continental landmasses by 23 million square kilometres. The latest Ice Age, termed the Last Glacial, peaked 22,000 years ago. Note that the Last Glacial Maximum is commonly reported in the literature at 18,000 radiocarbon years before the present era (BP), which is actually 22,000 calendar years BP. The present icecaps of Greenland and Baffin Island are now small vestiges of these former icesheets. The Antarctic icecap has remained a permanent fixture throughout both warm and cold periods, although its size waxed and waned in tune with global temperature fluctuations.

Our present-day temperatures occupy a warm period, termed an interglacial. The change from the Last Glacial was abrupt; temperatures rose permanently to their present values in as little as three to five years. Interglacials are unusual, because for 90% of the Pleistocene, global temperatures have been colder, and global shorelines have occupied positions near the edge of present continental shelves. Even today, alpine glaciers are present along the equator in Africa and Papua New Guinea. The Earth's glaciers and icesheets presently contain about 25% of the volume of ice that was present at the peak of the Last Glacial. The Pleistocene can thus be defined as a cold period with significant bursts of warm climate and elevated sea-levels.

EVIDENCE OF PLEISTOCENE CLIMATE FLUCTUATIONS

Dating techniques
(Aitken, 1990; Andersen and Borns, 1994)

The chronology of past climate has been unravelled since 1950 with the advent of a range of absolute and derivative dating techniques that includes radiocarbon or ^{14}C, thermoluminescence, electron spin resonance, uranium/thorium, and amino acid racemisation techniques. The term absolute does not necessarily imply total accuracy, because these techniques are subject to sampling, measurement and environmental errors. For example, radiocarbon years back to 18,000 BP are too young by 22–27%. Very accurate absolute dates can also be obtained by counting tree-rings (dendrochronology), or the annual variation in sediment accumulating in lakes (varve counting). However, these latter methods are environmentally limited. Derivative dating is based upon surrogate (proxy) evidence usually derived from stratigraphic sections. The latter depend upon some marker or reference date in the section to convert the sequence to an absolute time series. Magnetic susceptibility and oxygen isotope ratio ($\partial^{18}O$) sequences are examples of derivative dating techniques.

Oxygen isotope ratios are a very powerful tool in palaeoclimate reconstruction. Oxygen exists commonly as ^{16}O and the rarer isotope ^{18}O. At present the heavier isotope, ^{18}O, on average makes up about 0.19% of seawater and 0.15% of freshwater. Both isotopes behave the same chemically. However, in water, those molecules containing the heavier ^{18}O isotope tend to evaporate less readily than those containing the lighter ^{16}O isotope. Under present conditions, this is not a problem as most water evaporated from the ocean returns there within a short period of time. However, during a major period of glaciation, the oceans become 'soupier' with respect to ^{18}O at the expense of ^{16}O, which has evaporated and preferentially been stored in icecaps. If the ratio of ^{18}O to ^{16}O is measured at these times, it will be higher in the oceans than in icecaps. By convention, this ratio is abbreviated to $\partial^{18}O$, presented in tenths of a percentage, and referenced to a standard.

Small, single-cell organisms known as foraminifera incorporate oxygen isotopes as calcium carbonate in their shells. When they die, their skeletal remains settle to the ocean bottom, where they accumulate and preserve a continuous time series of climatic change. The absorption of ^{18}O is also temperature dependent; but in certain parts of the world's oceans, ocean temperatures have varied little between glacial and interglacial cycles. In very select regions, where there are not strong currents to disperse this biological fallout, where biological activity is profuse, and where disturbance by biota (bioturbation) is non-existent, a time series of $\partial^{18}O$‰ has built up throughout the Pleistocene (and beyond). Chronological control is provided by

radiocarbon dating, the presence of the Brunhes–Matuyama magnetic reversal at 740,000 years BP, and by marker horizons of ash from large volcanic eruptions such as Mount Toba, Indonesia 78,000 years ago.

The $\partial^{18}O$ deep sea record

As recently as the 1960s, it was believed that the Pleistocene consisted of only four major Ice Ages, as revealed by distinctly different glacial deposits in North America and Europe. The deep sea oxygen isotope record indicates that there have been at least twenty major Ice Ages throughout the Pleistocene. A composite record of deep ocean $\partial^{18}O$‰ values is shown in Figure 5.1 for the last 1.1 million years. Interglacials have smaller $\partial^{18}O$‰ values than glacials. The numbers at the top of the graph, by convention, identify each major climatic stage. Glacials are given even numbers and interglacials are given odd numbers. Significant departures in temperature can exist within each stage. These substages are given letters beginning with the most recent event. Various nomenclature also exists for the most recent stages. For instance, the Last Glacial is known as the Wisconsin or the Würm glaciation in North America and Europe respectively. The previous interglacial, between 80,000 and 130,000 years BP is known as the Last Interglacial or Eemian. It consisted of three warm peaks, the warmest of which occurred at 125,000 years BP. The latter is often referred to as the substage 5e warm peak. The magnitude of this warming was about 1–2°C above present-day temperatures. Our present climate, the Holocene interglacial, is shown at the extreme left of the time series. Comparable warm temperatures have occurred during only 10% of the Pleistocene.

The Earth's climate tends to drift slowly into Ice Ages which are rapidly terminated by sudden warming. There are also subtle changes in the characteristics of each stage with time. Since stage 11, interglacials have occurred every 100,000 years and generally lasted 30,000–40,000 years. The Holocene only began 11,500 years ago, and if it follows the trend for recent interglacials, should last for another 20,000 years. The 100,000 periodicity between interglacials becomes less dominant before stage 11. At the time of the Brunhes–Matuyama magnetic reversal, glaciations occurred every 40,000 years. The reason for this shift in cyclicity is still speculative. Finally, the time series presented in Figure 5.1 should not be considered definitive. The $\partial^{18}O$‰ values are

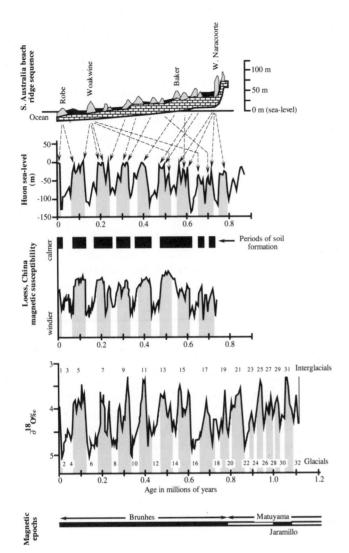

Figure 5.1 Composite record, for the last 750,000–1,100,000 years, of the raised beach ridge sequence of South Australia; global sea-level measured from raised coral terraces on the Huon Peninsula, in Papua New Guinea; the magnetic susceptibility record in loess in Xifeng, China; the $\partial^{18}O‰$ deep sea record; and major magnetic reversals (from Pirazzoli et al., 1993; Huntley et al., 1993; Andersen and Borns, 1994).

subject to considerable variation both in magnitude and time. For instance, a land-based record suggests that the Last Interglacial began about 10,000 years before the time shown in Figure 5.1. There is also strong evidence, from Europe, that the Earth's climate temporally warmed close to present-day values around 30,000 and 60,000 years ago. These small scale warmings during cooler glacial stages are termed interstadials.

The Pleistocene record inferred from geomorphic evidence
(Pirazzoli et al., 1993; Huntley et al., 1993; Andersen and Borns, 1994)

The oxygen isotope record is not the only chronological evidence of the Pleistocene climatic variations. Each time glaciation became widespread, water was withdrawn from the ocean and global sea-levels dropped by 100–130 metres. Whenever climate stabilised, coral reefs or coastal barriers were built at shorelines. Unfortunately, because present sea-level is higher than for 90% of the Pleistocene, most of this coastal evidence lies submerged on the continental shelves. However, there are key locations where tectonic uplift has been rapid enough that interglacial shorelines, and even some glacial ones, are preserved in a staircase fashion. The best coral reef evidence is preserved on the Huon Peninsula, Papua New Guinea and Sumba Island, Indonesia; while the best coastal barrier sequence lies at the mouth of the Murray River valley, in South Australia. In addition, major glacials generated substantial amounts of wind-blown dust or loess that accumulated at several unglaciated locations. Two of the best records exist in central Europe and on the Chinese Loess Plateau. In the latter location, high pressure over the Tibetan Plateau generated strong winds that repetitively deposited fine silt, to a depth of 150 metres, during glaciations covering the complete timespan of the Pleistocene. These land-based sequences can be dated at many points in time using many of the modern dating techniques described above. These sequences also preserve the Brunhes–Matuyama magnetic reversal as well as major geomagnetic fluctuations. Examples of these records are also presented in Figure 5.1.

The Huon sea-level record shows that sea-level has risen higher with successive interglacials over the last 800,000 years, until it globally stood three to five metres higher than present during the Last Interglacial substage 5e. During major glacials, sea-level dropped a minimum of 80–100 metres below present values. Such falls shut down the Indonesian Throughflow, and isolated the Arctic Ocean completely from the Pacific Ocean and partially from the Atlantic. The coastal barrier record in South Australia supports this sea-level record; however, the location of barriers is not sequential because varying sea-level elevations caused later barriers to overlap older ones. Evidence of earlier interglacials, when sea-levels did not reach present levels by tens of metres, are preserved sometimes more seaward in the sequence. The deposits of the last two interglacials, with their major substages, cover older deposits. Alternate glacials and interglacials in the Chinese loess record are defined by magnetic susceptibility variations. Magnetic susceptibility was higher during interglacials or interstadials, when soils formed under less windy conditions. The windiest periods were relatively brief, being restricted to glacial peaks.

LATE PLEISTOCENE CLIMATE

Together the four time series in Figure 5.1 offer the best evidence of generalised Pleistocene climate change. However, more detailed records exist over the last 250,000 years that reveal the nature and causes of very rapid temperature fluctuations. The records are derived from ice cores from the Greenland and Antarctic icecaps, and sediment sequences in the North Atlantic and Indian Oceans.

The recent $\partial^{18}O$ record
(Dansgaard et al., 1993)

The centre of the Greenland and Antarctic icecaps have built up in situ, without significant lateral displacement, over the last 150,000–250,000 years. Each icecap entraps air representative of the atmosphere's composition over this timespan, as well as a $\partial^{18}O‰$ record that should mirror Figure 5.1. Smoothed records of $\partial^{18}O‰$ from both polar icecaps are presented in Figure 5.2A. The age of the ice can be resolved to within two or three years. Annual ice layers can simply be counted back to about 14,000 years BP, in a similar fashion to tree-ring counting, while accelerator mass spectrometer ^{14}C dating can date the ice back to 70,000 years BP. Besides accumulating snow, polar icecaps also accumulated particle fallout from the atmosphere, mainly measured by changes in chloride, calcium, sulphates and nitrogen oxides. Chloride is derived from sea spray, and is more common during cold periods because of exceptionally vigorous wind circulation. Calcium levels are strongly dependent upon the source of dust, mainly in the form of loess. Wind and aridity over the continents enhance the formation of salt lakes and evaporites caused by excessive evaporation. The acid content of ice reflects the present of marine biota, volcanic sulphate fallout and nitrates. Because of its closeness to a non-glaciated landmass, the Greenland icecap preserves this information better than does the Antarctic (Figure 5.2B).

The detailed records reveal that interglacials tend to occur suddenly in the space of a few hundred years. On the other hand, climate tends to drift slowly into glacials that peak towards the end of the cold cycle. If $\partial^{18}O‰$ values can be related to temperature, then temperatures similar to the Holocene have only occurred during 10% of the last 250,000 years. The Penultimate Interglacial, about 200,000 years ago, was less pronounced than the Last Interglacial, which occurred as three warm peaks from 133,000 to 70,000 years BP. The latter interval is longer than that derived from the marine $\partial^{18}O$ record. The Last Interglacial thus was prolonged, and abruptly interrupted by two periods of glacial-like temperatures. In addition, the Antarctic record indicates that the main substage 5e may have also have been interrupted by cool spells lasting from two thousand to six thousand years. This has implications for climate change in the Holocene, which is often likened to stage 5e of the Last Interglacial. The Holocene is considered climatically stable, and has lasted 11,500 years without similar, major temperature fluctuations. The records shown in Figure 5.2 indicate that this may be atypical of interglacials. Holocene stability is more characteristic of the Penultimate Interglacial, but even here temperatures fluctuated more than they do today. Human civilisation appears to have developed during a period of abnormally warm and unchanging climate.

Noise, 'flickerings' and Dansgaard–Oeschger oscillations
(Dansgaard et al., 1971, 1989; Taylor et al., 1993; Bender et al., 1994)

Climatic time series naturally contain noise, extreme values and oscillations. The Pleistocene record is no different. In fact, the Last Ice Age contains more noise than that shown in the modern record (Figure 1.2). Much of this evidence is derived from detailed chemical analysis of icecores from the Greenland icecap, where a slight shift in the magnitude or location of the mid-latitude jet stream accentuated climatic variation in the northern hemisphere. For instance, under very cold conditions, mid-latitude winds were much stronger and dustier over North America. During the Last Ice Age, dust transport over Greenland was at least forty times greater than it is at present. When climate warmed, these winds decreased in intensity and less dust blew off the North American continent across Greenland.

Since the Last Interglacial, there have been

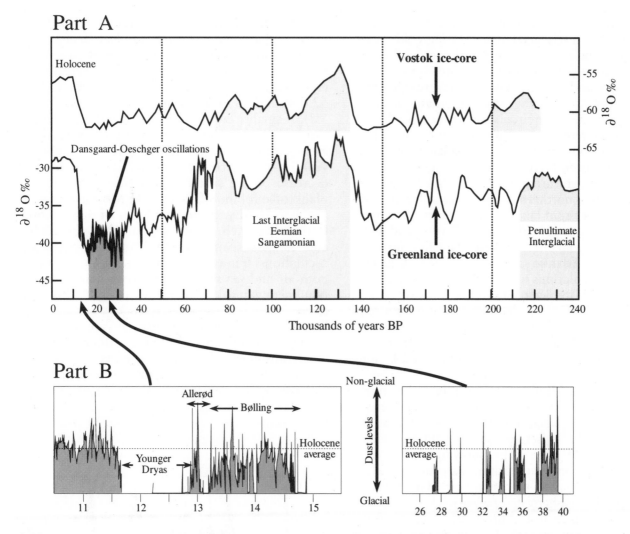

Figure 5.2 Late Pleistocene climatic records: (A) the $\partial^{18}O‰$ record from the Greenland and Antarctic icecaps (based on Dansgaard et al., 1971, 1993). The Dansgaard–Oeschger oscillations are marked. (B) detailed records of electrical conductivity measurements from the Greenland icecap (based on Taylor et al., 1993). These latter measurements are a proxy for the continental dust content falling on this icecap.

twenty-two interstadial warm events with temperature fluctuations of 6–7°C. Changes have occurred within the space of ten to thirty years. Unless a warm event lasted more than 2,000 years, it was restricted to the northern hemisphere, and not recorded in Antarctica. Longer warm interstadial events were distinctly different from the present Holocene. While major interstadials in some parts of the northern hemisphere witnessed substantial ice melting and warmer temperatures, shorter ones were characterised by short bursts of dustless circulation that never at any time mimicked that of an interglacial. At these times, oscillations from one extreme temperature state to another took place within the space of less than five to twenty years. The transitions at

the start and end of these interstadials occurred within a decade. These 'flickerings' required extremely rapid reorganisation of atmospheric circulation, coupled with the ocean surface.

The fluctuations between 17,000 and 34,000 years ago were extremely rapid and are known as the Dansgaard–Oeschger oscillations, after the researchers who first detected them (Figure 5.2). Approximately six oscillations, with as many minor fluctuations, took place. Each major oscillation lasted about one thousand years, with temperature changing in under a year. During these oscillations, temperature over Greenland changed by 7°C, snowfall by a factor of two, and atmospheric dustiness by a factor of a hundred. Some warmings lasted only a few hundred years. At no time did temperatures lock into an interglacial state. During cold cycles, stronger wind circulation blew voluminous dust out of the interior of North America, northwards over the Greenland icecap. The Dansgaard–Oeschger oscillations represent major reorganisation of northern hemisphere circulation patterns, with large changes in temperature and precipitation. These oscillations, and many of the smaller 'flickerings', could have been forced only by changes to the latitudinal temperature gradient, at high to mid-latitudes in the northern hemisphere. Ocean cores indicate that rapid changes in sea surface temperature also accompanied these oscillations and 'flickerings', with cold phases linked to the shutting down of the North Atlantic salinity sink. The Dansgaard–Oeschger oscillations support the notion that glacial climate was bistable; it behaved as a chaotic system, alternating between at least two equilibrium temperature states that were very different from each other. When a change in state occurred, climate fluctuated repetitively between the two states, locking predominantly into a glacial one.

The Allerød–Bøling Interstadial, beginning around 14,700 BP, represents a false start to the Holocene Interglacial. For 1,500 years climate circulation oscillated between glacial and interglacial states, without ever really showing any sign of remaining in one equilibrium state or the other. Then about 12,900 BP climate circulation in the northern hemisphere descended back into glacial conditions which lasted for another 1,200 years. This temporary setback, known as the Younger Dryas, is discussed in more detail later in the chapter. The Last Glacial finally ended very abruptly around 11,500 BP, with climatic conditions that dominated the Last Ice Age disappearing completely within thirty years. Average temperatures over the Greenland icesheet rose by 7°C. Since then, Holocene climate has been characterised by small temperature fluctuations, none of which have approximated those of the Last Glacial. Holocene atmospheric circulation became less vigorous than during the preceding Glacial period in the northern hemisphere.

Heinrich iceberg events
(Heinrich, 1988; Bond et al., 1992; Broecker, 1994; Clark, 1994)

The Last Glacial was also characterised by profuse dumping of chaotic mixtures of sediment in the North Atlantic, transported by icebergs from icecaps in Greenland and North America. Debris

can be traced 3,000 kilometres westwards, in increasing amounts, towards Hudson Strait. Since the Last Interglacial, there have been six major events, termed Heinrich events after their discoverer, centred at 14,300, 21,000, 28,000, 41,000, 52,000 and 69,000 radiocarbon years BP (Figure 5.3). A smaller event occurred 11,500 years BP and can be linked to the end of the Last Glacial. Each event lasted from one thousand to two thousand years, and occurred concomitantly with marked decreases in sea surface temperature, salinity and planktonic foraminifera growth. Most events were not associated with major phases of deglaciation. Rather, Heinrich events are linked to the coldest and longest spans of the Dansgaard–Oeschger oscillations. Immediately after each iceberg swarm, both air and sea temperatures rose sharply before plummeting into another cold period. This bundling of Dansgaard–Oeschger oscillations, bounded by Heinrich events at approximately 10,000 year intervals, is termed a Bond cycle.

Ice rafting events are related mainly to changes in the dynamics of the Laurentide icecap. Icecap collapse occurs during the coldest stages of an Ice Age. At these times, icesheets are thickest, and have begun to trap geothermal heat at their base.

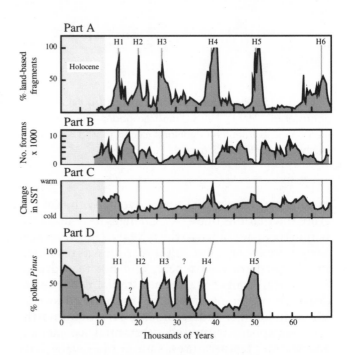

Figure 5.3 Record of Heinrich ice rafting events in the North Atlantic: (A) ice-rafted sediment derived from land sources; (B) number of foraminifera; (C) change in sea surface temperature (SST); (D) pollen record of *Pinus* from Florida. (Based on Bond et al., 1992 and Grimm et al., 1993.)

If basal melting occurs, the icesheet rests on a layer of water, rather than being frozen to its bed. When this happens, the ice can rapidly surge forward. If the icesheet overlies deformable sediment, such as occurs in Hudson Strait, rapid collapse of the ice dome is ensured as the ice margin advances at a rate of hundreds to thousands of metres each year. Collapse shifts the mass of the icesheet laterally into marginal oceans, where iceberg calving takes place. The complete process of icecap thinning can take hundreds of years because of the volume of ice involved. Computer simulation has shown that the 10,000 year quasi-cyclicity of Heinrich events is illusionary, and simply reflects the time required for icesheets to thicken sufficiently to trap geothermal heat, and begin basal melting.

The sharp drop in ocean salinity during each event probably 'turned off' the North Atlantic thermohaline sink, ensuring that colder temperatures in the northern hemisphere were transmitted to northern Europe, if not globally, following each event. Heinrich events have global teleconnections. For instance, they appear in records of pine pollen in Florida (Figure 5.3), an area far removed from the Laurentide icesheet. They have also been linked to surges in alpine glaciers in Chile and New Zealand, vegetation changes across the North American continent, lake levels in Africa, and ice core changes in Antarctica.

The Ice Age temperature paradox of the tropics
(Webster and Streten, 1978; CLIMAP, 1981; Prell 1985; Anderson et al., 1989)

It is indisputable that the Earth's landmasses were ubiquitously cooler than present during the Last Glacial Maximum. However, the ocean record is less clear-cut. The tropical oceans appear to have been warmer than they are now by more than 2°C (Figure 5.4). This warming was concentrated in the Pacific Ocean, particularly in the southern hemisphere, and in belts presently dominated by high pressure. The results are so dichotomous that they have been termed a paradox.

Detailed studies are confirming the land–sea temperature paradox of the tropics originally mapped in 1981. For instance, the Coral Sea lies adjacent to the tropical mountains of Papua New Guinea, in the west Pacific. Here, snowlines were 1,000 metres lower at the peak of the Last Glacial, and temperatures were at least 6°C colder than at present. Even allowing for some modulation near the ocean, temperatures along the coast of the Coral Sea had to be about 3°C cooler than at present. Detailed $\partial^{18}O‰$ temperature determinations, from micro-fauna (foraminifera) in sea bottom cores, confirm the fact that the Coral Sea, north of 15° S, had temperatures within 1°C or slightly higher than their present values during the Last Ice Age. Sea surface temperatures only became colder by 3–4°C poleward of about 20° S.

Thus, in the Pacific Ocean, low-latitude sea surface temperatures, dominated by large subtropical gyres, appear to be a stable component of global climate even during Ice Ages. The pattern shown in Figure 5.4 has ramifications for the operation of the Southern Oscillation during the Last Ice Age. The enhanced warming has two nodes, in the Coral Sea and the east Pacific, which are the current centres of amplification of sea surface temperatures during La Niña and ENSO events respectively. However, there is no evidence that the present-day west Pacific warm pool (Figure 4.4) was warmer at this time. It is uncertain whether or not the Southern Oscillation was operating, but certainly these pockets of warm ocean water in the tropics and subtropics would have provided sufficient moisture for enhanced rainfall over these, and adjacent regions.

Ice Age general atmospheric circulation
(Clemens and Prell, 1991; Leroux, 1993)

The nature of atmospheric circulation during the Last Glacial has been subjected to much speculation. The concept of mobile polar highs, described in Chapter 3, also can be applied to a cooler globe dominated by icesheets. These highs were highly mobile with pathways that are affected by topography. Under colder conditions, mobile polar highs produced rapid general circulation, a situation that is presently approximated to a certain extent each winter. During glacial times, the prime forcing of air circulation was due to an enhanced temperature gradient between the poles and the tropics. The occurrence of an icesheet at high elevations greatly maximised longwave emission towards the poles, while the tropics maintained nearly the same balance between incoming solar radiation and longwave emission as exists at present. Mobile polar highs also responded to the presence of icesheets (Figure 5.5). Highs formed because of intense cooling at the centres of domed icesheets over North America and Scandinavia. Cooling was also aided by the fact that the icesheets were three to five kilometres high, a situation analogous to Antarctica today. Mobile polar highs flowed rapidly into the Atlantic across the

Part A

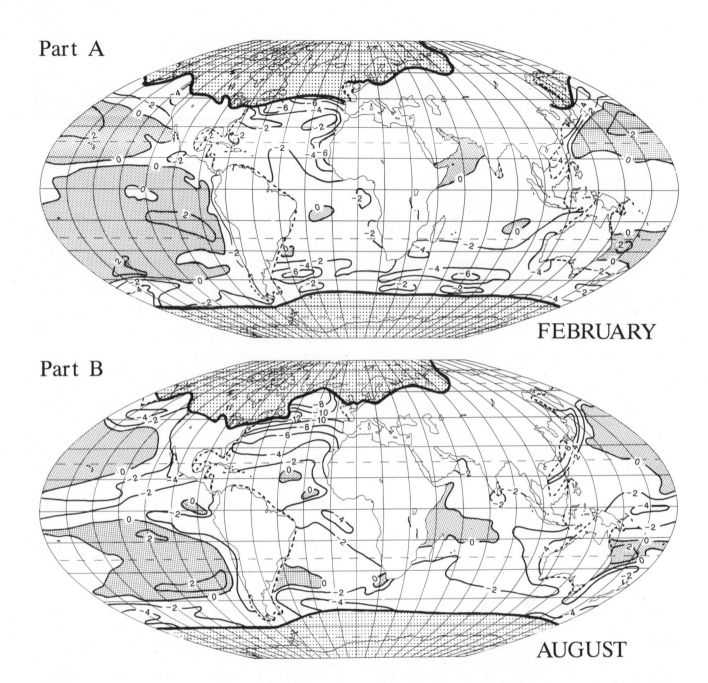

FEBRUARY

Part B

AUGUST

Figure 5.4 Differences in ocean temperature between the present and the peak of the Last Glacial Maximum, for the months of February and August (based upon CLIMAP, 1981). Icecaps are lightly stippled. Sea surface temperatures which were warmer than now are heavily stippled.

eastern seaboard of the United States. The strengthened highs forced rapid and voluminous movement of tropical air northwards in the Atlantic Ocean. Because sea surface temperatures between 40° and 60° N were still relatively warm at the beginning of icesheet growth, this air movement generated more precipitation that fell as snow at high and mid-latitudes. Over a 5,000- to 10,000-year period, 25×10^6 cubic kilometres of continental ice formed through this mechanism. The lack of insolation in summer prevented the previous winter's accumulation of snow from melting. As icesheets grew, the mobile polar highs were intensified by katabatic winds, reaching speeds of 200 kilometres per hour, as they flowed off the icesheets. While it has been proposed that

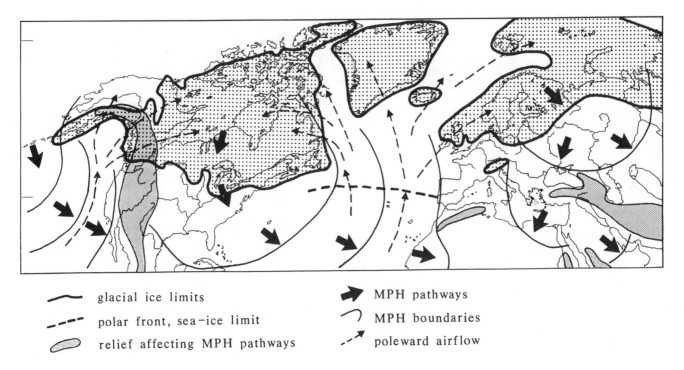

⌇ glacial ice limits	➤ MPH pathways
- - - polar front, sea–ice limit	⌒ MPH boundaries
⬭ relief affecting MPH pathways	⤏ poleward airflow

Figure 5.5 Mobile polar highs in the northern hemisphere at the peak of the Last Glacial Maximum (based on Leroux, 1993).

the Icelandic Low was very strong during the Last Ice Age, in reality only mobile polar highs were strong. Their tendency to enhance northward airflow from the tropics supplanted the need for cyclonic storms as a mechanism for increased snowfall over icesheets.

Ice Age global circulation was globally windier as shown by the increased accumulation of dust in ocean sediments, and within the Antarctic and Greenland icecaps. The intensified circulation produced by mobile polar highs caused the zone of tropical trades to contract. Concomitantly, wind speeds in the tropics increased. This situation led to greater deflation, and the intrusion of arid conditions into the Congo and Amazon rainforests. Temperature, evaporation and rainfall remained high in the tropics, but within a restricted band around the equator. At first, rainfall increased in the tropics; but, as mobile polar highs increased in intensity, atmospheric water was captured within their circulation and redirected towards higher latitudes.

There is debate about the nature of monsoon circulation during Ice Ages. The prevailing view is that monsoons were weaker at these times. Similarly, computer modelling suggests that as the Laurentide icesheet over North America expanded, the Tibetan Plateau cooled and the

Southeast Asian monsoon decreased in strength. This cooling was exacerbated by higher snow and ice cover, leading to a decreased planetary albedo. However, sampling of sediment deposited in the Arabian Sea over the past 350,000 years presents a different picture. Whenever the Indian monsoon is in operation, offshore winds sweep dust off the Arabian Peninsula, and deposit it in the adjacent ocean. Up to 80% of the sediment deposited in the northwest Arabian Sea is deposited during the summer monsoon. The stronger the monsoon, the stronger these offshore winds, and the coarser the sediment that can be transported. A record of the size of sediment deposited in the Arabian Sea is plotted in Figure 5.6 for the last 360,000 years. Analysis indicates that the Indian monsoon is strongest when solar radiation is greatest at 30° S, and there is an increased north–south pressure gradient resulting from strengthened southern hemisphere subtropical highs. These conditions were met during Ice Ages. When insolation in the northern hemisphere was low during glacials, it must have been higher in the southern hemisphere, because there is no decrease in the annual amount of solar radiation reaching the top of the Earth's atmosphere. In addition, mobile polar highs originating from the Antarctic icecap during glaciations were

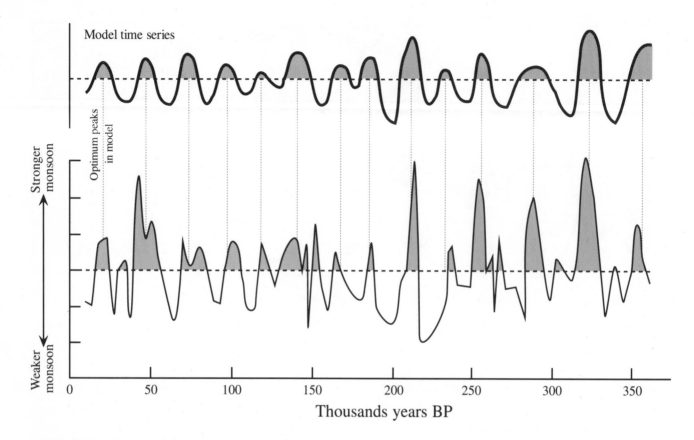

Figure 5.6 Proxy record of the intensity of the Indian monsoon over the last 360,000 years from sediments in the Arabian Sea (modified from Clemens and Prell, 1991).

also intensified. These should have increased high pressure over the Indian Ocean during glaciations. Thus, icesheets in the northern hemisphere, especially during Ice Ages, appear to have played a very small role in controlling the strength of the Indian monsoon.

Deglaciation
(Ruddiman and McIntyre, 1981; Lehman and Keigwin, 1992; Andersen and Borns, 1994; Blanchon and Shaw, 1995)

Glaciations end only when the copious amounts of precipitation that are directed northwards by mobile polar highs begin to fall as rain, rather than snow, under conditions of increased insolation in the northern hemisphere. Figure 5.7 presents a gradualistic view of icesheet melting ending the Last Glacial in the northern hemisphere. Also plotted is the location of the polar front, which was coincident with the southernmost location of sea-ice. In Europe, the continental icesheet began melting extensively over the

west coast of Europe by 15,000 BP, leaving only a remnant icecap in the highest parts of Ireland and Scotland. Despite a slight readvance during the Younger Dryas, the icesheet margin had retreated to Scandinavia by 10,500 BP. Two thousand years later the icesheet was gone. In North America, the continental icesheet consisted of three merged lobes: the Cordilleran, the Keewatin west of Hudson Bay and the Laurentian icesheet over northern Quebec. Glaciation extended northwards to cover all but the islands of the northwest High Arctic. Southernmost lobes covered the Dakotas, surged south of the Great Lakes, and blanketed New England. By 15,000 BP, these icesheets had contracted to expose the proto-Great Lakes, and had opened up a corridor between the Keewatin and Cordilleran icesheets. Extensive ice-marginal lakes, trapped by isostatically dipping topography, formed south of the icesheets. Major spillways drained southwards along the proto-channels of the present-day Missouri, Mississippi and Ohio Rivers, carrying between four and ten times the

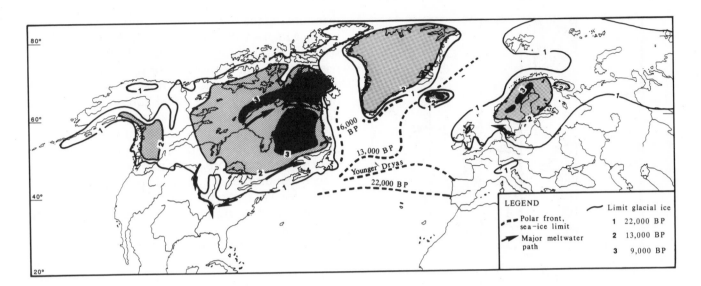

Figure 5.7 Location of icesheet fronts in North America and Europe during the most recent deglaciation (based upon Dawson 1992; Andersen and Borns, 1994), and of polar front and sea-ice limits in the North Atlantic (from Ruddiman and McIntyre, 1981).

discharge of present rivers flowing into the Gulf of Mexico. About 13,000 years BP, the ice front had retreated so much that the Cordilleran icesheet had separated from the main icecap. At this time, the Great Lakes–Saint Lawrence River passageway opened, allowing substantial meltwater to flow directly into the North Atlantic Ocean. The latter event triggered the Younger Dryas. Around 8,500 BP, the Keewatin–Labrador icesheets were separated from each other by ice calving through Hudson Strait. At this time, ice-marginal lakes catastrophically drained northwards into the Arctic Ocean, raising global sea-levels by 0.2–0.4 metres in the space of a few days. By 7,000 BP the stagnant Laurentian icesheet had completely melted, leaving only remnant glaciation on the east Arctic Islands.

More significant, climatically, was the retreat of sea-ice in the North Atlantic. At the height of the Last Ice Age, the polar front overlapped the most southward position of sea-ice offshore of Portugal. However by the Younger Dryas, sea-ice and the polar front lay between Iceland and the Grand Banks of Newfoundland. The Younger Dryas witnessed a significant extension of this front to 45°–50° N with an associated 7°–10°C cooling of surface waters. Following this interruption in deglaciation, the sea-ice front rapidly retreated into Davis Strait between Greenland and Baffin Island, with the North Atlantic thermohaline sink switching into its modern day mode. Today, the

sea-ice limit lies close to Greenland, and the polar front hugs the coast of North America, Greenland and Iceland.

Climate changed more rapidly than indicated by the retreating ice-margins plotted in Figure 5.7. Glacial climates switched to interglacial ones over timespans as short as three years. While insolation was a controlling factor on both icecap buildup and melting, it fails to explain the rapidity of these changes. The gradualist approach to deglaciation invokes progressive melting of icecaps that raised sea-levels enough to impinge upon continental icesheets, causing them to float and to calve. This led to rapid icesheet collapse, but still at timescales much slower than now indicated in marine records. Evidence from coral reefs in the Caribbean Sea indicates that there were three phases of rapid sea-level rise exceeding 45 millimetres per year. These events, each lasting at most between 140 and 290 years, raised sea-level by 6.5, 7.5 and 13.5 metres respectively at 7,600, 11,500 and 14,200 years BP (Figure 5.8). Each event occurred because of voluminous meltwater discharge to the oceans. The catastrophic events at 11,500 and 14,200 years BP also correlate with Heinrich events in the North Atlantic, and the concomitant collapse, within as little as one hundred years, of the Laurentide icesheet over North America. The catastrophic sea-level rise event around 7,600 years BP was due to wastage of the Antarctic icesheet, because the Northern Hemisphere icecaps were so

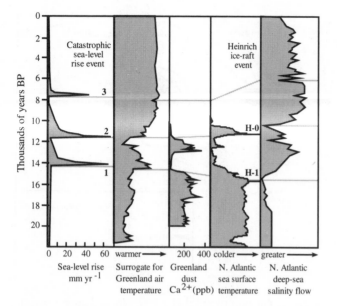

Figure 5.8 Atmosphere–ocean–icecap relationships for the northern hemisphere during the last deglaciation (from Blanchon and Shaw, 1995). Rate of sea-level rise is based upon coral sequences in the Caribbean Sea. Greenland temperature is a combination of palaeo-$\partial^{18}O$‰, snow and ice volume determinations. Dust is measured by the parts per billion of calcium (Ca^{2+} ppb) in ice cores. Sea surface temperature and deep sea salinity are inferred from foraminifera and ∂C^{13} ratios in seabed cores. All panels do not align in time because of errors in dating.

small by then that they could not have generated a 6.5-metre rise in sea-level.

The timing of these events relates to other ice-sheet and ocean records (Figure 5.8). Around 14,200 years ago, glacial air masses over Greenland were replaced within a ten-year timespan by warmer, moister and less dusty conditions. The North Atlantic was invaded by warm water from the subtropics, and the North Atlantic thermohaline sink was turned on. Around 12,900 years ago, ocean–atmosphere circulation returned to glacial conditions at the beginning of the Younger Dryas. This lasted until 11,500 years BP, when glacial conditions finally terminated abruptly within a three-year period, and the present Holocene Interglacial began its uninterrupted phase. The catastrophic sea-level events were not instigated by seasonal meltwater. It appears that as insolation increased in the northern hemisphere, meltwater accumulated within icecaps and then was suddenly released as catastrophic floods.

The largest mega-flood occurred within the Laurentide icesheet, producing regional-scale fields of drumlins, giant flutings and scoured bedrock features. Much of this water flowed down the Mississippi River into the Gulf of Mexico. This triggering event caused subsidiary mega-floods elsewhere, as the rapid rise in sea-level floated ice margins globally. The Laurentide icesheet decreased dramatically in elevation, removing this icesheet as a topographic control upon the path of glacial mobile polar highs, and permitting their return to routes more similar to present-day interglacial conditions. Interglacial atmospheric circulation forced the retreat of sea-ice in the North Atlantic, and reestablished the Gulf Stream. The retreat of sea-ice exposed the North Atlantic to increased westerly winds that enhanced evaporation, and set up the North Atlantic bottom saline conveyor belt. Warming occurred in northern Europe through the release of latent heat of evaporation, linking Scandinavian icesheet retreat to that occurring in North America. Global climate returned to its modern interglacial state as the altered pathways of mobile polar highs reorganised tropical and inter-hemispheric circulation.

The Younger Dryas
(Dansgaard et al., 1989; Dawson, 1992; Blanchon and Shaw, 1995)

Following the first icesheet collapse 14,200 years ago, enough of the icesheets remained that precipitation still fell as snow. The icesheets grew again until, around 12,900 BP, they had reestablished glacial atmospheric circulation. This period of renewed glaciation is termed The Younger Dryas (named after a small Arctic flower, *Dryas octopetala*, that reappeared in England). During the Younger Dryas, North Atlantic sea surface temperatures dropped 10°C, the Laurentide icesheet advanced again, icebergs and sea-ice spread as far south as 45° N in the Atlantic ocean, and Europe returned to periglacial conditions for a 1,400-year period. (Note that the age of the Younger Dryas is commonly reported in the literature as 11,000–10,000 radiocarbon years BP). The return to glacial conditions was also aided by the reorganisation of drainage from the Laurentide icecap. The catastrophic discharge of the 14,200 BP flood event caused the icecap margin to retreat enough to open up the Saint Lawrence spillway. This diverted cold meltwater, from the Mississippi River and the Gulf of Mexico, directly into the North Atlantic. The meltwaters had little time to warm up before flooding most of the surface of the North Atlantic. This cooled the atmosphere, reduced convection, and decreased the intensity of westerly winds over the North Atlantic. More importantly, because the ocean surface was covered by freshwater, evaporation did not affect surface salinities, and the thermohaline sink that existed in the North Atlantic was turned off. Once this sink stopped, warm waters from the tropics no longer flowed into the North Atlantic to moderate air temperatures. The temperature of northern Europe returned to glacial conditions, the Scandinavian icecap expanded, and air circulation throughout the northern hemisphere reverted to a glacial state. Other feedback mechanisms, such as an already thickening Laurentide icecap, may have stabilised the pattern of air circulation; but they did not cause the change directly.

The Younger Dryas terminated with a major meltwater discharge event from the Laurentide icesheet around 11,500 years ago. In less than twenty years, the climate of the North Atlantic became milder and less stormy, as sea-ice rapidly retreated northwards. Precipitation amounts increased by 50%, and temperatures over adjacent landmasses warmed by 7°C within fifty years. Dust amounts in Greenland ice-cores decreased by a factor of three, indicating a major readjustment of northern hemisphere general circulation towards present-day interglacial conditions. Further reductions in dust that occurred over the next two centuries reflected the rapid reestablishment of vegetation on Northern American loess areas, south of the limit of Laurentide glaciation. By 11,000 BP, a Holocene Interglacial climate was firmly established, and insolation had increased, to such an extent that snow was not able to accumulate sufficiently to reestablish glacial conditions. The remaining continental icesheets melted, ending the Last Ice Age.

HOLOCENE CLIMATE
(Lamb, 1985; Leroux, 1993; Wright et al., 1993)

Holocene temperatures can be considered stable relative to glacial ones; however, there have been significant regional temperature changes on the order of 1–2°C over the past 9,000 years. In the Arctic, sudden coolings of 4°C have been noted lasting two hundred years, before abruptly terminating. More significantly, dramatic changes in precipitation have occurred that have affected the development and continuation of human society.

Temperatures, particularly in the northern hemisphere, were controlled by the amount of insolation received in the summer, and the proximity to remaining icecaps. The amount of insolation, relative to present values, is presented in Figure 5.9 for the northern hemisphere. It shows an insolation peak of 8% above present summer values about 10,000 years ago. Temperatures did not parallel this peak everywhere, but instead tended to reach a maximum between 4,500 and 6,000 years BP. This peak is known as the Holocene Climatic Optimum. It occurred slightly later at higher latitudes in the northern hemisphere. This peak is set within a broader interval of warmer temperatures between 2,500 and 9,000 years BP, known as the Hypsithermal Climatic Interval. Proxy indicators of temperature, such as pollen records and inferred tree-line limits on mountains, have been used to outline the timing and magnitude of these changes (Figure 5.9). The tree-line limit in the Alps reached its greatest altitude around 7,000 years

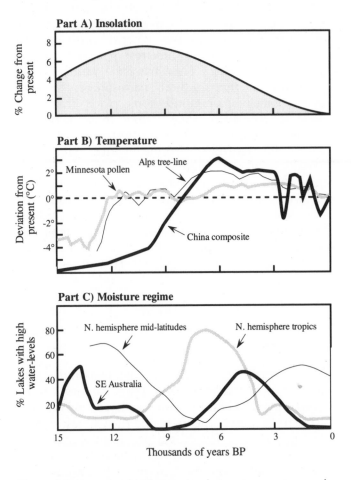

Figure 5.9 Records of Holocene insolation, temperature and moisture (modified from Lamb, 1985; Williams et al., 1993; Wright et al., 1993).

ago; but it has retreated two hundred metres in altitude since then in response to a temperature fall of about 2°C. This is mirrored to a certain degree by temperatures in China east of Beijing, where temperatures stayed 2°C above current values until 1,000 years ago. In contrast, in Minnesota, temperatures rapidly rose to their present levels about 10,000 years ago, and have fluctuated by only 0.5°C since then.

When temperatures were up to 2°C above present values, the effect of mobile polar highs was greatly restricted, with warming favouring the occurrence of deep thermal lows over the tropics. This situation led to amplification of the Southeast Asian and African monsoons. Under these conditions, rainforests reestablished and extended beyond their present limits. Lake sediments record the balance between thermally driven evaporation and precipitation at these times. Generally, the wettest and driest periods occurred slightly before the Holocene Climatic Optimum. In the northern hemisphere, lowest lake levels occurred about 7,000 years ago at mid-latitudes. However, in the northern tropics, this was a period with precipitation about 300 millimetres per year above present amounts. This tropical wetness also occurred in the southern hemisphere. In both areas, catastrophic floods took place, that were four to ten times greater than those witnessed today. Between 2,000 and 7,000 years ago, the temperate latitudes became wetter, while the tropics dried out. There is some indication that the temperate latitudes of the southern hemisphere behaved in a similar fashion. The changeover to a wetter climate at mid-latitudes probably reflects reduced evaporation because of cooler temperatures. Over the past 2,000 years hydrological regimes globally have tended to remain constant.

THE HISTORIC RECORD
(Lamb, 1982, 1985)

With written records, more detailed and direct indicators of temperature and precipitation can be determined. These records evolved during the Roman Empire in the west, and the Chinese dynasties in the east. The feudal period in Europe was characterised by detailed reporting of harvests and agriculturally important events. Of course these written records are restricted to the northern hemisphere, and then mainly to western Europe. Over the last 2,000 years, temperature fluctuations of

1.0–1.5°C have occurred with dramatic repercussions upon society. For instance, northern Europe was cold until the seventh century, after which temperatures warmed to a peak in the eleventh and twelfth centuries (Figure 5.10). This latter warming is known as the Mediaeval Warm Epoch or Little Climatic Optimum. During this period, the Vikings colonised Greenland and Iceland, and established regular trade across the North Atlantic as sea-ice receded northwards.

Temperatures collapsed dramatically, first after 1250, and then again after 1600, leading to the Little Ice Age. The changes in temperature were very rapid and punctuated by extremes. For example, in the 1250s climate suddenly cooled in northern Europe in the space of a decade. The 1420s witnessed six of the coldest years recorded in a millennium in Burgundy, France. Embedded within this decade is the warmest summer for the last one thousand years. The effects of these temperature changes were widespread across the

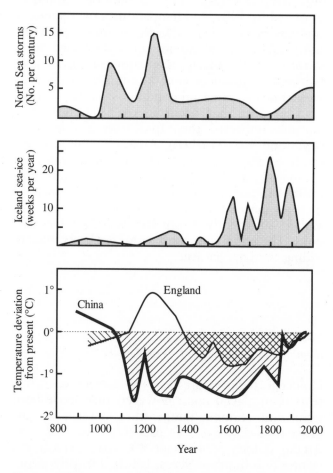

Figure 5.10 The record of storms, sea-ice change and temperature over the past 1,200 years in northern Europe and the North Atlantic Ocean (based on Lamb, 1982).

northern hemisphere. Sea-ice encroached upon Greenland and Iceland, first for five weeks of the year in the thirteenth and fourteenth centuries, and then for up to twenty weeks of the year, at the peak of the Little Ice Age, at the end of the seventeenth century (Figure 5.10). The temperature changes of the Little Climatic Optimum brought a dramatic increase in severe northern European storms. The impact on society was catastrophic. Four storms in the earlier period had a death toll exceeding 100,000. The 1–2 November 1570 storm killed 400,000 people. With initial cooling, the Viking colony in Greenland was abandoned in 1369, because sea-ice and storms prevented its resupply from Norway. The stranded colonists' progeny became stunted in growth, were picked off by marauding Eskimos, and finally starved to death. The last colony survived until about 1500. In some villages in England and France, crop failures and food shortages led to incidences of children being sold in exchange for food, and to cannibalism. Wheat was no longer viable, so farmers switched to rye. The wetter climate and improper storage facilities led to the growth of ergot fungus in rye. When ingested, the resulting condition, known as ergotism or Saint Anthony's fire, produced an effect similar to the drug LSD, and led to limbs turning black and dropping off. People hallucinated; whole villages went mad. The climatic conditions also aided the spread of the Black Plague, and led to increased religious persecution and warfare. These health impacts are described in more detail in Chapter 8.

THE MEASURED RECORD
(Linacre, 1992; Hulme, 1994, 1995; Jones et al., 1994; Schlesinger and Ramankutty, 1994; Nicholls et al., 1996)

The beginning of the modern record, based upon accurate measurements of climate, is difficult to define. Wind vanes for observing wind direction were developed around 850. In 1500, Leonardo da Vinci invented the hygrometer and designed a pressure plate anemometer. Galilei Galileo, or his pupil Santorio, invented the air thermometer in 1593, while Castelli invented a reliable rain gauge and began systematic measurements shortly afterwards. In 1643, Evangelista Torricelli invented the barometer, and in 1653, Ferdinand II of Tuscany established a network of meteorological observing stations. Christopher Wren invented a self-recording, tipping bucket rain gauge in 1662, and the

longest continuous systematic weather observations began in Paris in 1664. In England, regular rainfall measurements were begun by Richard Townley in 1677. In 1683, Edmund Halley published his map of global winds. This was followed in 1686 by his milestone theory on trade winds and monsoon circulation. An accurate network of global climate existed by 1850. The Southern Oscillation index (plotted in Figure 4.9) incorporates the first data from this network.

Temperatures

There are many recent compilations of global temperature. Many of these data sets are now available on the Internet. Figure 5.11 presents a compilation of corrected land and marine temperatures for the globe from 1854 to 1993. Temperatures are plotted as deviations from the average for the period 1950–1979. As well, the global temperature effect of the Southern Oscillation is shown. ENSO events generally lead to global warming because of the higher proportion of cloud-free areas over drought affected landmasses. In contrast La Niña events produce global cooling because of increased cloudiness over the western Pacific Ocean and over southern Asia. Upwelling of cold water along the South American coastline and equator (Figure 4.8A) leads to cooling of the Americas. The coldest December recorded in eastern North America coincided with the 1989 La Niña event. By inference a 'cloudy-sky' globe tends to be cooler at the surface than a 'clear-sky' one. No attempt has been made to incorporate satellite measurements in Figure 5.11. This aspect, and some of the biases in the record, are discussed in Chapter 7.

The time series in Figure 5.11 shows many aspects of a chaotic system. The global and hemispheric records show little trend during the nineteenth century, marked warming until 1940, slight cooling or erratic fluctuations to the mid-1970s, and rapid warming during the 1980s that appears to be continuing unabated into the 1990s. The rate of temperature change is between 0.40–0.45°C per century, being lowest in the northern hemisphere. Overall, temperatures have risen approximately 0.5°C since accurate measurements of temperature were initiated in the mid-1800s. The warmest seven years in descending order were 1995, 1990, 1991, 1988, 1987, 1983 and 1981. Between 1978 and 1991, global temperatures rose 0.34°C. However, the June 1991 Mount Pinatubo eruption lowered

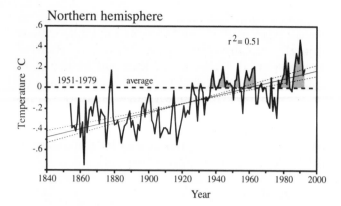

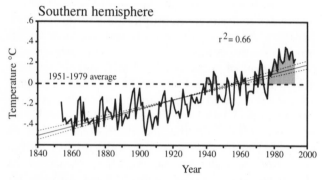

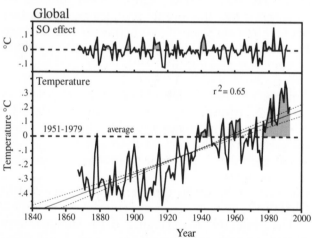

Figure 5.11 Combined land and marine temperature records, 1854–1993, for each hemisphere and the globe. Values are referenced to the 1951–1979 average. The SO effect is the global temperature change resulting from variations in the Southern Oscillation. (All data from Jones et al., 1994.)

same degree of fluctuation over time as the northern hemisphere. The southern hemisphere also supports global warming better over the period of record than does the northern hemisphere. This is probably due to the fact that the southern hemisphere is dominated by oceans, which moderate temperature fluctuations because of their large heat capacity and slower response time. As well, the northern hemisphere oceans have often responded quite differently from southern hemisphere ones. Temperatures can vary by 0.4–0.6°C contemporaneously between hemispheres, and from one year to another in the same hemisphere. For the period from 1920 to 1960, the southern oceans were up to 0.2°C cooler. Beginning in 1960, the southern oceans warmed while the northern oceans cooled. Only since 1980 has this trend changed, with oceans in both hemispheres warming simultaneously. These types of changes can have a strong influence on temperature as well as rainfall over adjacent landmasses. The difference in ocean temperature between the two hemispheres, certainly in the Atlantic Ocean, has been correlated with the presence of drought in northern Africa over the past thirty years.

The northern hemisphere temperature record also displays a long period of cyclicity of approximately 65–70 years, with a peak-to-peak magnitude of 0.2°C. Detailed analysis indicates that this long cyclicity does not occur in the southern hemisphere, where there are shorter nine- to twenty-year cycles with magnitudes of 0.1–0.2°C. The long-term cycle for the northern hemisphere is the result of regional oscillations on the order of 50 to 88 years. For example, there is a seventy-six-year cycle in the North Atlantic region with a peak-to-peak variation in temperature of 0.75°C. This cycle accounts for 55% of the noise in the detrended North Atlantic temperature record. The North Atlantic Ocean appears to be a crucial region where temperatures can be cyclically amplified, positively or negatively, beyond any hemispheric or global trend. As shown in Chapter 6, changes in temperature in the North Atlantic region have global consequences.

Precipitation

Precipitation records (including rain, snow, hail and sleet) have not commonly been presented as evidence of recent climate change. Such records are often discontinuous and subject to error. For instance, underestimation by gauges is probably as high as 11%. Today a total of 200,000 rain gauges

temperatures by 0.3–0.7°C over a two-year period. Since 1994, temperatures have risen again.

Ocean- and land-based air temperature records are remarkably coherent. Sea surface temperatures have fluctuated from year to year as much as land-based air temperatures (0.6°C). However, the southern hemisphere record does not show the cover most landmasses; however, there are no measurements from the world's oceans. Calibrated records, those that have been culled for inaccuracies, do not extend far back beyond 1900. Figure 5.12 presents a global average of precipitation based upon 5,899 stations, for the period from 1900 to 1993. Values are expressed as departures from the mean. While annual precipitation amongst stations can vary between 0 and 11,900 millimetres, the range averaged for the globe rarely exceeds ±50 (910–1135 millimetres). This year-to-year variation represents a significant component of the annual global hydrological cycle, and, when averaged over the world's oceans, represents a ±15 millimetre variation in global sea-level. Many of the fluctuations in Figure 5.12 show a three- to five-year cyclicity that is strongly coherent with fluctuations in sea-level on the order of 20–100 millimetres, at various locations around the globe. This cyclicity and the amplitude of sea-level fluctuations can be related to the Southern Oscillation.

The global record also displays many of the characteristics of a chaotic system. For instance, the variance in rainfall has changed continually over the past century. Rainfall fluctuations from year to year were greatest at the beginning of the twentieth century, and smallest during the warming period of the 1930s and 1940s. Many of the fluctuations, especially at the three- to five-year frequency, are also linked to significant ENSO and La Niña events. The 50-millimetre dip in global

rainfall between 1980 and 1983 corresponds with the strong ENSO event of 1982–1983. One of the crucial years in the record is 1952, when rainfall rose globally by 60 millimetres in the space of three years. This jump coincides with the large La Niña event of 1950–1951. In countries such as Russia, Canada and Australia, long-term precipitation increased respectively by 15%, 20% and 30% after this date. Since the 1950s, global rainfall has tended to decrease steadily.

In more detail, the global trend in rainfall fluctuations is most prominent over northern hemisphere landmasses, where the largest land surface is located. There has been minimal change in rainfall over the past century at mid-latitudes in the northern hemisphere, and a 5% increase at high latitudes, north of 50°. Much of the recent decrease in rainfall can be attributed to the drying out of the southern Sahara, the Sahel. This record has been superimposed upon the global time series in Figure 5.12. Between 1954 and 1983, there was a 90-millimetre decrease in rainfall in this region. At some locations, rainfall decreased by 75% over a twenty-year period, from 400 to less than 100 millimetres per year. The nature of this decrease is related to a cooling of the North Atlantic Ocean relative to the oceans of the southern hemisphere, and to a decrease in the intensity of the African monsoon. This drying out of the Sahel triggered repetitive calamitous droughts, followed by famine in many countries. The nature of precipitation changes in the northern hemisphere supports some

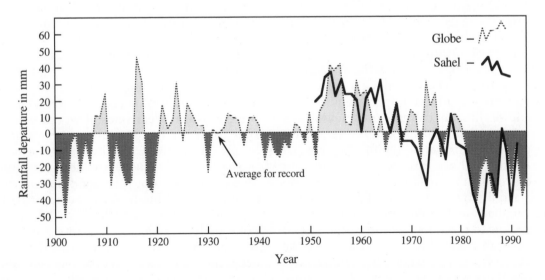

Figure 5.12 Annual deviations in global precipitation, 1900–1993 (data from NOAA, 1995). Values are referenced to the average for this timespan. Superimposed are scaled deviations for the Sahel region of Africa, 1950–1991 (data from Hulme, 1994).

regional computer simulations which predict increased precipitation for a warming globe due to enhanced 'greenhouse' gases. However, the cross-correlation between Figures 5.11 and 5.12, even when detrended, is poor at the global scale. At most, the historical records indicate a global increase of 3.2% in precipitation for each 1°C increase in temperature.

CONCLUDING COMMENTS

The 'flickerings' between glacial and interglacial climates during the Last Glacial were on the order of 5–7°C over the timespans of a decade in the Greenland region. A similar change in temperature occurred over a space of fifty years between the Younger Dryas and the Holocene in the North Atlantic region. The Holocene is supposedly immune from these temperature swings. Certainly, the rates of increase in temperature globally over the past century of 0.40–0.45°C are small compared to any these oscillations. And yet, dramatic local temperature swings do appear in the historic record. Along the east coast of Greenland temperature has increased by 8°C over the past century. A similar temperature change probably occurred there after the Little Climatic Optimum. How long these modern 'flickerings' can continue without triggering a more widespread change in our present atmospheric circulation remains conjectural. We should not be putting as much emphasis upon global climate change as upon climate change in crucial trigger regions. The North Atlantic Ocean is certainly one of these trigger regions that bears closer monitoring.

When the 'flickerings' take on the magnitude and frequency of the Dansgaard–Oeschger oscillations, there are significant implications for the development of human culture and civilisation. There is no doubt that humans occupied the northern hemisphere landmasses during the Last Glacial period. Neolithic culture was well developed, but it was one of nomadic subsistence. There was no attempt to acquire agriculture, not because it was too cold (one had only to move far enough south from the icecaps to reach temperatures that would sustain most domesticated plants), but because the temperature regime fluctuated too frequently. Temperature shifts which were only 20% of those that occurred during the Dansgaard–Oeschger oscillations had a major impact upon the type of grain that could be grown in northern Europe over the last one thousand years. Neolithic culture during the Last Glacial simply did not have the climatic stability of the Holocene to permit the domestication of plants, and the subsequent organisation of cities, that represents to us the development of civilisation.

The lesson of climate change in the Middle Ages should not be lost on today's society, given the apparent consensus about present global warming. Society in the thirteenth century had progressed to such an extent in Europe that the Industrial Revolution almost took place. However this breakthrough was interrupted by the climate deterioration that began about 1250. Not only did climate changes of only 1°C require a major agricultural adaptation, but they also had a major impact upon social fabric, human health and survival. Often, it was not so much the warming or the cooling that affected society, but the inability of society to adapt. Fortunately, our present global warming is not occurring at the same rate as the temperature fluctuations of the Middle Ages.

Dichotomous climatic states can also produce similar outcomes. The best example occurs in the tropics, where both the peak of the Last Glacial and the Holocene Climatic Optimum produced catastrophic flooding, greater than anything witnessed during the recent historic record. The concept of mobile polar highs explains this paradox best. During glacial periods, intense and rapidly moving polar high pressure cells gave more power to the trades and monsoons. This caused greater rainfall near the equator in a region that was more restricted than present. Contrary to logic, the equatorial oceans were also warmer, by 2–4°C at the peak of the Last Glacial Maximum, than they are at present. Even though the globe was in an Ice Age, the Earth still received the same average amount of solar radiation as it receives at present. Under clear skies, it would have been possible for ocean temperatures to warm up to higher levels. Thus strong trades, over warmer tropical oceans, would have provided the atmospheric instability and moisture conditions conducive to catastrophic rainfall over adjacent landmasses. During the Holocene Climatic Optimum, mobile polar highs were in a slower mode and were much weaker. Warming favoured the development of deep thermal lows over tropical continents, an aspect that amplified the resulting monsoonal circulation more than what occurs at present. This enhancement led to increased rainfalls, and greater and more frequent flood events. The field evidence supports both of these hypotheses. While different climatic circulation dominated the tropics between glaciations and warm epochs, the resulting climate, at least with regard to rainfall, was similar, with bigger and more frequent catastrophic floods. In predicting the impact of our present climate change, we should bear in mind this fact, that different climatic regimes can produce the same consequence.

REFERENCES AND FURTHER READING

Aitken, M.J. 1990. *Science-based Dating in Archaeology.* Longman, London, 274p.

Andersen, B.G. and Borns, H.W. Jr. 1994. *The Ice Age world.* Scandinavian University Press, Oslo, 208p.

Anderson, D.M., Prell, W.L. and Barratt, N.J. 1989. 'Estimates of sea surface temperature in the Coral Sea at the Last Glacial maximum'. *Paleoceanography* v. 4 pp. 615–627.

Bender, M., Sowers, T., Dickson, M., Orchardo, J., Grootes, P., Mayewski, P.A. and Meese, D.A. 1994. 'Climate correlations between Greenland and Antarctica during the past 100,000 years'. *Nature* v. 372 pp. 663–666.

Blanchon, P. and Shaw, J. 1995. 'Reef drowning during the last deglaciation: evidence for catastrophic sea-level rise and ice-sheet collapse'. *Geology* v. 23 pp. 4–8.

Bond, G., Heinrich, H., Broecker, W., Labeyrie, L., McManus, J., Andrews, J., Huon, S., Jantschik, R., Clasen, S., Simet, D., Tedesco, K., Klas, M., Bonani, G. and Ivy, S. 1992. 'Evidence for massive discharges of icebergs into the North Atlantic ocean during the last glacial period'. *Nature* v. 360 pp. 245–249.

Broecker, W.S. 1994. 'Massive iceberg discharges as triggers for global climate change'. *Nature* v. 372 pp. 421–424.

Clark, P.U. 1994. 'Unstable behavior of the Laurentide ice sheet over deforming sediment and its implications for climate change'. *Quaternary Research* v. 41 pp. 19–25.

Clemens, S.C. and Prell, W.L. 1991. 'Late Quaternary forcing of Indian ocean summer-monsoon winds: a comparison of Fourier models and general circulation model results'. *Journal Geophysical Research* v. 96 pp. 22683–22700.

CLIMAP Project Members 1981. 'Seasonal Reconstructions of the Earth's surface at the Last Glacial maximum'. *Geological Society of America, Map and Chart Series* MC-36 pp. 1–18.

Dansgaard, W., Johnsen, S.J., Clausen, H.B. and Langway, C.C. 1971. 'Climatic record revealed by the Camp Century ice core'. In Turekian, K.K. (ed.) *The Late Cenozoic Glacial Ages.* Yale University Press, New Haven, pp. 37–56.

Dansgaard, W., Johnsen, S.J., Clausen, H.B., Dahl-Jensen, D., Gundestrup, N.S., Hammer, C.U., Hvidberg, C.S., Steffensen, J.P., Sveinbjörnsdottir, A.E., Jouzel, J. and Bond, G. 1993. 'Evidence for general instability of past climate from a 250-kyr ice-core record'. *Nature* v. 364 pp. 218–220.

Dansgaard, W., White, J.W.C. and Johnson, S.J. 1989. 'The abrupt termination of the Younger Dryas climatic event'. *Nature* v. 339 pp. 532–533.

Dawson, A.G. 1992. *Ice Age Earth: Late Quaternary Geology and Climate.* Routledge, London, 293p.

Grimm, E.C., Jacobson, G.L., Watts, W.A., Hansen, B.C.S. and Maasch, K.A. 1993. 'A 50,000 year record of climate oscillations from Florida and its temporal correlation with Heinrich events'. *Science* v. 261 pp. 198–200.

Heinrich, H. 1988. 'Origin and consequences of cyclic ice rafting in the northeast Atlantic Ocean during the past 130 000 years'. *Quaternary Research* v. 29 pp. 143–152.

Hulme, M. 1994. 'Century-scale time series of regional rainfall anomalies in Africa'. In Boden, T.A., Kaiser, D.P., Sepanski, R.J. and Stoss, F.W. (eds) *Trends'93: A compendium of data on global change.* ORNL/CDIAC-65. Carbon Dioxide Information Analysis Center, Oak Ridge National Laboratory, Oak Ridge, Tennessee, pp. 964–973.

Hulme, M. 1995. 'Estimating global changes in precipitation'. *Weather* v. 50 pp. 34–42.

Huntley, D.J., Hutton, J.T. and Prescott, J.R. 1993. 'The stranded beach-dune sequence of south-east South Australia: a test of thermoluminescence dating, 0–800 ka'. *Quaternary Science Reviews* v. 12 pp. 1–20.

Jones, P.D. Wigley, T.M.L. and Briffa, K.R. 1994. 'Global and hemispheric temperature anomalies-land and marine instrumental records'. In Boden, T.A., Kaiser, D.P., Sepanski, R.J. and Stoss, F.W. (eds) *Trends'93: A compendium of data on global change.* ORNL/CDIAC-65. Carbon Dioxide Information

Analysis Center, Oak Ridge National Laboratory, Oak Ridge, Tennessee, pp. 603–608.

Lamb, H.H. 1982. *Climate, History and the Modern World*. Methuen, London, 387p.

Lamb, H.H. 1985. *Climatic History and the Future*. Princeton University Press, Princeton, 835p.

Lehman, S.J. and Keigwin, L.D. 1992. 'Sudden changes in North Atlantic circulation during the last deglaciation'. *Nature* v. 356, pp. 757–762.

Leroux, M. 1993. 'The Mobile Polar High: a new concept explaining present mechanisms of meridional air-mass and energy exchanges and global propagation of palaeoclimatic changes'. *Global and Planetary Change* v. 7 pp. 69–93.

Linacre, E. 1992. *Climate Data and Resources: A Reference and Guide*. London, Routledge, 366p.

Nicholls, N., Gruza, G.V., Jouzel, J., Karl, T.R., Ogallo, L.A. and Parker, D.E. 1996. 'Observed climate variability and change'. In Houghton, J.T., Meira Filho, L.G., Callander, B.A., Harris, N., Kattenberg, A. and Maskell, K. (eds) *Climate Change 1995: The Science of Climate Change*. Cambridge University Press, Cambridge, pp. 132–192.

NOAA 1995. '*Baseline climate datasets: global climate perspectives system*'. **http://www.ncdc.noaa.gov/onlineprod/prod.html**.

Pirazzoli, P.A., Radtke, U., Hantoro, W.S., Jouannie, C., Hoang, C.T., Causse, C. and Borel Best, M. 1993. 'A one million-year-long sequence of marine terraces on Sumba Island, Indonesia'. *Marine Geology* v. 109 pp. 221–236

Prell, W.L. 1985. 'The stability of low-latitude sea-surface temperatures: an evaluation of the CLIMAP reconstruction with emphasis on the positive SST anomalies'. *United States Department of Energy, Technical Report* No. TR025, 60p.

Ruddiman, W.F. and McIntyre, A. 1981. 'The North Atlantic Ocean during the last deglaciation'. *Palaeogeography Palaeoclimatology Palaeoecology* v. 35 pp. 145–214.

Schlesinger, M.E. and Ramankutty, N. 1994. 'An oscillation in the global climate system of period 65–70 years'. *Nature* v. 367 pp. 723–726.

Taylor, K.C., Lamorey, G.W., Doyle, G.A., Alley, R.B., Grootes, P.M., Mayewski, P.Q., White, J.W.C. and Barlow, L.K. 1993. 'The 'flickering switch' of late Pleistocene climate change'. *Nature* v. 361 pp. 432–436.

Webster, P.J. and Streten, N.A. 1978. 'Late Quaternary ice age climates of tropical Australasia: interpretations and reconstructions'. *Quaternary Research* v. 10 pp. 279–309.

Williams, M.A.J., Dunkerley, D.L., De Deckker, P., Kershaw, A.P. and Stokes, T. 1993. *Quaternary Environments*. Edward Arnold, London, 329p.

Wright, H.E., Kutzbach, J.E., Webb, T., Ruddiman, W.R., Street-Perrott, F.A. and Bartlein, P.J. (eds) 1993. *Global Climates since the Last Glacial Maximum*. University of Minnesota Press, Minneapolis, 569p.

6

Causes of Climate Change

INTRODUCTION

Many causes have been proposed over the past century for the onset and cyclicity of the Ice Ages. The causes of the Pleistocene Ice Ages can be grouped into four categories: factors external to the Earth–atmosphere system, factors with a strong geological expression, factors internal to the Earth–ocean–atmosphere system, and factors that have a significant randomness. External factors are mainly cyclic; however, by themselves, they cannot account for the temperature difference between glacial and interglacial periods. Some of these cycles, such as sunspots, have implications for recent climate change. Geological factors explain the overall cooling that has dominated the Pleistocene; but cannot explain the cyclicity of climatic fluctuations. Factors involving the Earth–ocean–atmosphere system are multi-faceted and, when linked to externally forced cycles, account best for the intensity and timing of glacial periods through positive and negative feedback mechanisms. The fourth factor acknowledges the fact that much climate change, even when associated with cycles, may be random. This chapter highlights the main causes that have been invoked under each of these general headings. The list is not meant to be comprehensive. Finally, while emphasis is placed upon large scale climate change, minor cycles and oscillations at the biennial-to-decade intervals, account for a higher degree of the noise in Pleistocene and recent climatic records.

EXTERNAL FACTORS

Changes in the sun's orbit within our galaxy
(Fairbridge, 1987b)

A fundamental cyclicity of 200–300 million years appears to underlie large scale geological processes, such as mountain building and plate movements. This cycle corresponds to the time that it takes for the solar system to orbit the Milky Way galaxy. A secondary consequence of this movement is the regular crossing by the sun, of the galactic plane, in a cycle of thirty-three million years. Around the time of each crossing, the solar system encounters more interstellar dust. This periodicity correlates with increased volcanism, meteor and comet impacts, mass extinctions, and reversals in the Earth's magnetic field. The Pleistocene also coincides with the passage of the solar system across the galactic disk. The amount of cosmic dust that the earth intercepts depends upon the tilting of the axis of the solar system. This tilting is not stable, but undergoes a 100,000 year periodicity with increases in cosmic dust correlating to interglacials.

Comets
(Gribbin, 1994)

Comet impacts with the Earth, or crossing the Earth's orbit, can also increase extraterrestrial dust fallout. Comets enter the inner solar system from the Kuiper belt lying outside the orbit of Neptune, or from the Oort belt beyond the outer boundary of the solar system. The gravitational attraction of Jupiter and Saturn can force a comet as large as 200 kilometres in diameter to traverse the inner part of the solar system once every 200,000 years. The best known comet impact occurred in the Yucatan Peninsula of Mexico at the Cretaceous-Tertiary boundary. This event threw large volumes of sediment up into the stratosphere, attenuating solar radiation significantly for months if not years, and leading to the extinction of the dinosaurs. Even objects smaller than the one kilometre in diameter can have a major impact on climate. The most recent event was the Tunguska comet explosion over Siberia in June 1908. Recent evidence suggests that the actual incidence of small objects hitting the Earth is more common than previously recognised. However, impact events are not regular enough to cause cycles of glaciation.

Comets do not have to crash into the Earth to affect the Earth's climate. Large comets orbiting the sun can be progressively broken into smaller pieces, in a similar fashion to the Shoemaker–Levy comet that was initially fragmented by passing close to Jupiter in 1992, and then crashed into that planet in July 1994. The Earth's orbit often intersects the tail of debris formed in the path of such a fragmenting comet orbiting the sun. At these times, the Earth can sweep up about 5,700 million tonnes of dust each year over a timescale of thousands of years. This steady cloud of debris can reduce the Earth's surface temperatures by 3–5°C over this timespan. Temperature changes of this magnitude are large enough to produce a major Ice Age. Some astronomers believe that the Taurid meteorite shower, which peaks in June and November each year, represents the aftermath of the breakup of a giant comet within the inner solar system, and that this comet has led to the glacial temperatures of the past 100,000 years. The Earth passes through the thickest part of this debris belt every 3,000 years. It also sweeps through at least ten other comet debris trails each year, each of which produces visible meteorite showers. However, little longer term cyclicity is possible, because any coherent comet debris is dispersed by the gravitational attraction of the planets within 10,000 years.

Changes in solar output (Simpson's theory)
(Fairbridge, 1987b; Magny, 1993; Schönwiese et al., 1994)

The sun does not emit radiation constantly over time. Exact sunspot records, while relatively long compared to measured temperature data, only

extend back to the fifteenth century (Figure 2.3), and thus cannot be correlated to any major changes in solar activity that may have affected Quaternary global temperatures. However, an important product of solar activity in the atmosphere is the formation of radioactive carbon-14 (^{14}C). When solar activity is high, and by inference sunspots numerous, cosmic rays are deflected from the Earth's stratosphere by the solar wind. This lessens the production of ^{14}C, and correlates to warmer temperatures. When solar activity is low, then ^{14}C productivity increases. The increase in ^{14}C during the Wolf (1280–1340), Spörer (1420–1530) and Maunder (1645–1714) sunspot minima substantiates this latter effect. These periods were also cooler in the northern hemisphere. The ^{14}C record for the past 7,000 years shows three other effects related to solar activity. First, there are varying short-term changes at 50 to 200-year intervals termed 'Suess wiggles', induced by varying solar activity. Second, there are changes related to variations in the Earth's geomagnetic field. Low geomagnetic field strength coincides with low solar activity, increased ^{14}C production, larger 'Suess wiggles', and higher climatic variability. In contrast, the Holocene Climatic Optimum around 6,000 BP, when temperatures were about 1–2°C warmer than present, was a period of stronger geomagnetic activity, increased solar activity, and decreased ^{14}C production. Finally, the record also contains a prevalent 2,300-year periodicity that can be linked to dramatic changes in the amount of solar radiation reaching various latitudes, and to cycles of tropical and subtropical aridity. The main times of climate deterioration during the Holocene occurred at 2,800, 5,500 and 7,100 years BP, and during the little Ice Age about 1500. These times coincide with major alpine ice advances in Europe.

While variable solar output can be linked to temperature fluctuations of 1–2°C during the Holocene, there is scant evidence that this factor was responsible for the 5–7°C changes characterising Ice Ages. However, Simpson, in the early part of the twentieth century, incorporated solar variation into a theory for Ice Ages that gained wide popularity. Simpson's theory involved two cycles of glaciation–deglaciation for each cycle in solar radiation. Increasing radiation first led to glaciation because of increased evaporation, more cloud, greater planetary albedo and increased snowfall, especially in mountainous regions. As temperatures continued to rise, snowfall turned to rain, leading to deglaciation and an interglacial. As insolation then decreased, the second glacial period began. Further falls in radiation decreased evaporation and precipitation. This starved glaciers, leading to a period of deglaciation that resulted in a cool interglacial. While the lag in time between lowest temperatures and the maximum extent of alpine glaciers on timescales of centuries can be explained by Simpson's theory, there is little other proof to support it. For instance, the Pleistocene record shows no evidence of any cool interglacial (Figure 5.1).

Changes in orbital parameters (Milankovitch hypothesis)
(Imbrie and Imbrie, 1979; Clemens and Prell, 1991; Imbrie et al., 1992, 1993)

While the sun's output can be ruled out as controlling glacial cycles, there is good evidence that variations in the degree of radiation reaching various latitudes on the Earth's surface due to orbital perturbations, does control these cycles. The Earth has three fundamental orbital attributes (Figure 2.4 and 2.5), each of which can vary over time. These orbital fluctuations are due to changes in the shape of its elliptical orbit about the sun, termed eccentricity; in the tilt of the Earth, termed obliquity; and in the timing of the Earth's closest approach to the sun, termed precession of the equinoxes. The index of eccentricity of the Earth's orbit can vary between 0.0 and 0.07. Whatever this value, the total annual amount of radiation received by the Earth remains constant; however, it can vary between hemispheres. When the Earth's orbit is circular, eccentricity equals zero, and both hemispheres receive the same amount of radiation over the year. At present eccentricity has a value of 0.0174, and the southern hemisphere receives 6.7% more solar radiation than the northern hemisphere. Eccentricity can obtain a maximum value of 0.07, with the difference in solar radiation between the two hemispheres reaching 28%. The cycle of eccentricity is approximately 96,000 years, although it can vary between 90,000 and 100,000 years because of slight distortion by the gravitational effect of the larger planets, mainly Jupiter and Saturn.

The tilt of the Earth, known as the angle of obliquity, presently has a value of 23.47°. However it can vary between 21.39° and 24.36° over a period of 40,000 to 41,000 years. When the Earth is highly tilted, there are pronounced seasons, resulting in significant annual variation in the receipt of solar radiation at high latitudes, in both hemispheres. Cooling is exacerbated in the northern hemisphere, where landmasses are concentrated, when the degree of tilt is smallest. At these times, snowfall which has accumulated during winter may not completely melt during a summer that is shorter, with less radiation input. The difference is enhanced, if the tilt of the Earth is maximum, when the northern hemisphere is furthest from the sun. Thus the interrelationship between eccentricity and obliquity can lead to one hemisphere being substantially cooler than the other.

Precession refers to the timing of the seasons. Each time the Earth revolves around the sun, it does not come back to its original location, but tends to move forward slightly in its orbit. If a complete cycle is 360°, then the rate of precession of the Earth is 50.2564" per year. The Earth presently has its closest approach to the sun twelve days after the December solstice. Exact coincidence last occurred in the year 1250. The precession cycle is 25,780 years; however the phenomenon is distorted by the fact that the other planets, mainly Jupiter, also affect the cycle, leading to a cycle of between 19,000 and 23,000 years. Precession does not affect the annual amount of radiation received by each hemisphere, just its timing. When one hemisphere has a cooler summer because of precession, the opposite hemisphere should have a warmer one. Precession is the only orbital parameter that should be noticeable at timescales of centuries. Over a century, the seasons are shifted forward by about 1.4 days. Over recent centuries, this has led to cooler springs, prolonged summers and a delay in the onset of winter in the northern hemisphere.

Solar radiation received at any latitude on the Earth's surface is constantly undergoing fluctuations over cycles of 90,000–100,000, 40,000–41,000 and 19,000–23,000 years. One of the more crucial regions where these fluctuations have an appreciable effect on air temperature is between 60° and 70° N, where continentality has the greatest effect. The fluctuations in solar radiation at the top of the atmosphere at 30° and 60° N and S latitude are presented in Figure 6.1, for summer. This figure shows that variations in radiation over thousands of years are more substantial towards the poles. These fluctuations were hypothesised in the 1860s as causing cycles of glaciation. In 1920, the Serbian mathematician, Milutin Milankovitch, accurately calculated the sum of their effects, and it has since been shown that global temperature

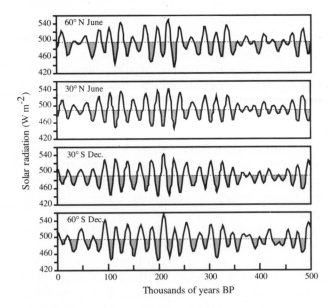

Figure 6.1 Insolation values in watts per square metre (W m⁻²) over the last 500,000 years, for mid-June at 60° N and 30° N, and mid-December at 60° S and 30° S. Calculations include eccentricity, obliquity and precession orbital parameters (data from Berger and Loutre, 1991).

over the past 700,000 years manifests these orbital periodicities (Figures 5.1 and 5.2). The 96,000-year eccentricity cycle accounts particularly well for the spacing of the last six interglacials; the 40,000 to 41,000-year obliquity cycle accounts for the timing of glacial cycles before 600,000 BP; and the 19,000 to 23,000-year precession cycle explains the occurrence of interstadials, and the temperature swings of the last 150,000 years. Spectral analysis of oxygen isotope ratio records (∂¹⁸O‰) shows the prominence of these cycles (Figure 6.2), and the shift towards the dominance of eccentricity and precession towards the Late Pleistocene.

The 23,000-year precession cycle has been shown to influence also the strength of the Indian monsoon during Ice Ages (Figure 5.6). Glaciation in the northern hemisphere reaches a minimum, not when the radiation effect of the precession cycle is strongest at high latitude, but about five thousand years later when sensible heating over the Tibetan Plateau is maximum. This point in time corresponds to the Holocene Climatic Optimum 6,000 years ago. Four thousand years later in the cycle, insolation due to precession is greatest at 30° S, and sea surface temperatures in the southern subtropical Indian Ocean reach their maximum value. This enhances latent heat transfer from the ocean to the atmosphere, and the Indian

monsoon reaches its maximum intensity. Up to 70% of the latent heat released over the Tibetan Plateau and Himalayas comes from this part of the southern Indian Ocean. Over the next twelve thousand years of the precessional cycle, insolation decreases away form 30° S, and the monsoon weakens as conditions for ice expansion in the northern hemisphere increase.

Two paradoxes exist in the linking of orbital perturbations to the timing of Ice Ages. First, while the 96,000-year cycle of eccentricity dominates glaciation, the variation in radiation throughout this cycle is very small, around 2 watts per square metre. This amplitude can only account, at most, for global temperature variations of 1–2°C, far less than the 7–10°C temperature swings actually observed. Second, glaciations were synchronous worldwide, however the radiation deficit triggering them occurred only in the northern hemisphere. Changes in other factors of the Earth–ocean–atmosphere system, in combination, are required to amplify these orbital variations to produce a global effect.

Finally, the apparent 96,000-year cycle in the ∂¹⁸O‰ records may be an illusion, or at least a very subtle effect. Non-linear harmonic generation, created by the interaction of shorter cycles of precession and obliquity, can also produce a cycle of around 100,000 years. This effect is described in Chapter 1. The first subharmonic and some of the more common periodicities generated by the interaction of pairs of these orbital parameters are listed in Table 6.1. The 29,000 and 35,000 year cycles appear in the Indian monsoon dust record (Figure 5.6). Most of the temperature fluctuations of the Pleistocene owe their origin to a combination of precession and obliquity. Eccentricity only appears because its effects are longer lasting, or enhanced by non-linear response of the Earth–ocean–atmosphere system. For instance, the peak of glaciation should lag any peak in the effect of minimum radiation by about 15,000 years. It takes that long for an icesheet to grow. Similarly, the response time of large continental icesheets to climatic change is about 10,000 years. When an icesheet centred on 65–70° N becomes large enough, it alters the nature of atmospheric and oceanic circulation, and becomes self-sustaining (Figure 5.5). Once this critical size is reached, the icesheet has thermal inertia that can withstand the melting induced by fluctuations in insolation generated by the shorter precession and obliquity cycles. Thermal inertia is controlled by the amount of ice that is locked into an icesheet, and the fact

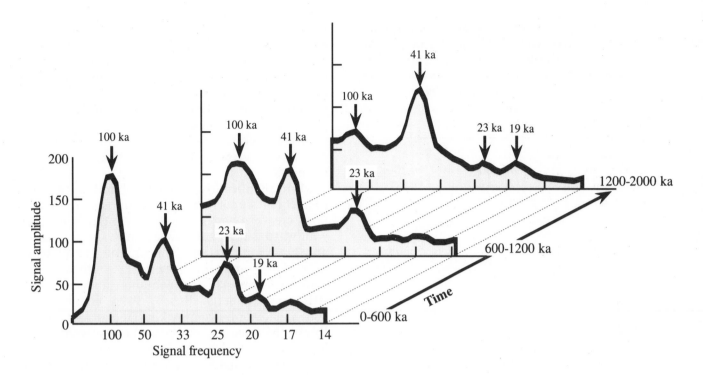

Figure 6.2 Spectral analysis of frequency of peaks in the oxygen isotope ratio ($\partial^{18}O$) temperature records over thousands of years (ka) (modified from Imbrie et al., 1993).

that the phase change of ice to water requires 334.72 joules of heat per gram of ice at 0°C. The shorter orbital cycles simply don't last long enough to produce enough heat to melt an icesheet, at high latitudes in the northern hemisphere. However, when the effects of these cycles are superimposed upon the eccentricity cycle, there is enough radiation to overcome thermal inertia and melt an icesheet. Thus, the precession and obliquity cycles only control temperature fluctuations during glacials and interglacials, while the eccentricity cycle controls the overall timing of Ice Ages.

GEOLOGICAL FACTORS

Continental drift
(Lamb, 1977; Fairbridge, 1987b)

Continental drift created the atmospheric and oceanic conditions favourable for global glaciation. An Antarctic landmass has existed near the South Pole for the last 120 million years, whereas a continual icecap has not. An icecap only developed after the separation of South America, Africa and Australia from that landmass. This separation opened up a southern ocean, permitting the establishment of the Antarctic Circumpolar Current about fifty million years ago. The current prevented the exchange of warmer water from the tropics with Antarctic water, dropping temperatures by several degrees. About forty million years ago, Drakes passage began to open up south of South America, leading to the development of ice in Antarctica. Deep water temperatures there dropped by 4–5°C, and land temperatures by 10°C, over a period of ten million years. At the same time, equatorial oceans were compartmentalised as continents drifted northwards. In addition, northward movement of continents restricted the exchange of ocean water with the Arctic Ocean, virtually creating a shallow inland sea that has been ice-covered since the Pliocene. Global glaciation reduced sea-levels, exposing more of this shallow sea's seabed to sub-aerial glaciation. While icesheets cannot originate over sea-ice, the freezing of such a large body of water near the North Pole established the strong cooling necessary for intensification of mobile polar highs during glacials. Finally, the collision of the Indian subcontinent into the Eurasian plate, in the last three million years, formed the Himalayas and the Tibetan Plateau. Such high relief disrupted jet

Table 6.1 Periodicities produced by the interaction of eccentricity, precession and obliquity (values rounded to whole numbers).

	Periodicity (in thousands of years)	First subharmonic
Main cycles:		
Eccentricity	90–100	180–200
Obliquity	40–41	80–82
Precession	19–23	38–46
Additive interactions:		
Eccentricity–obliquity	28–29*	56–58
Eccentricity–precession	16–19	32–38
Obliquity–precession	13–15	26–30
Subtractive interactions:		
Eccentricity–obliquity	67–75	134–150
Eccentricity–precession	24–31	48–62
Obliquity–precession	35–54*	70–108*

* Cycles referred to in text

stream circulation in the northern hemisphere. The formation of the Tibetan Plateau also played a role in establishing Walker circulation. Both atmospheric phenomena are important in circulating air from one hemisphere to another, and transmitting regional climate change to the rest of the globe.

Mountain building
(Lamb, 1977; Ruddiman and Kutzbach, 1991; Paterson, 1993)

Mountains can cause glaciation by providing cooler locations for icecaps to develop, by altering patterns of atmospheric circulation, especially in the upper atmosphere, and by changing the geochemical composition of the oceans. However, mountains cannot explain alternate glacial and interglacial changes in climate. Large scale mountain building can decrease temperature over a wide area, because of adiabatic cooling effects with increasing altitude. Alpine glaciers, forming locally as the result of slight global cooling, can coalesce into a piedmont icesheet, such as the Cordilleran icesheet that developed over the Canadian Rockies during the Last Glacial. Ice flow onto adjacent lowlands enhances reflectivity over a wide area. Such a process cooled a substantial section of North America during the Last Glacial, and had a similar effect over the European Alps and Himalayas, but to a much lesser degree. However, it is doubtful if piedmont glaciation can generate a global effect. Certainly, there is no evidence that formation of the Cordilleran icesheet ever preceded the growth of other North American icesheets.

More significantly, mountain building can alter the general circulation of the atmosphere. Mountains create zones of heavy precipitation on their windward sides that favour snow accumulation and glacier formation. The uplifting of plateaux over Tibet and the American southwest also enhanced seasonal continentality, and caused jet streams to undergo Rossby wave enhancement. When glaciation occupied the Tibetan Plateau and Rocky Mountains, cool conditions were transferred downwind by the jet stream. Climates over wide areas around the plateaux became spatially and seasonally more variable. Similarly continental icesheets, when they obtained thicknesses of 3–5 kilometres, began to act as plateaux, and alter mid-latitude atmospheric circulation (Figure 5.5).

Finally rapid uplift, on timescales of millions of years, alters the geochemical regime of the globe. Weathering of bedrock is exacerbated by the heavier rainfall on windward slopes of mountains. The weathering process removes carbon dioxide from the atmosphere, reducing the 'greenhouse' effect of this gas, and enhancing overall global cooling. The level of carbon dioxide in the atmosphere over the past six hundred million years parallels the degree of mountain building, except for the last fifty million years. The formation of the Tibetan Plateau was such a significant event, creating steep terrain and more surface exposure for weathering, that all of the atmosphere's carbon dioxide should have been removed within 100,000 years. However, the Earth's climate cooled over this period, enough to slow chemical weathering and the removal of carbon dioxide.

Volcanism
(Hammer et al, 1980; Bluth et al., 1993; Sutton and Elias, 1993; Jones et al., 1995; Gagan and Chivas, 1995)

Over the past 250 million years, major volcanic activity has occurred every thirty-three million years, with a minor peak every 16.5 million years. Not all volcanism has led to glaciation. For instance volcanism during the Ordovician and

Devonian did not produce glaciation even though it was much more substantial than that during the Pleistocene. Finally, volcanism has to be sustained at least for a thousand years to produce the type of temperature cooling required to initiate Ice Ages. Even during the Pleistocene, volcanism cannot be associated with every Ice Age. The exception appears to be the eruption of Mount Toba, Indonesia about 78,000 years ago. The global consequence of this eruption probably terminated the Last Interglacial.

Volcanism can alter climate in three ways: increased carbon dioxide, increased sulphate aerosols, and enhanced dust ejection into the stratosphere. Presently volcanoes contribute insignificant amounts of carbon dioxide to the atmosphere. For example, Kilauea in Hawaii, which has erupted continually since 1983, emits only one seventeen-thousandth of the carbon dioxide produced annually by human activity. Sulphur dioxides and attendant aerosols fare little better. The total volcanic contribution of sulphur dioxide to the Earth's atmosphere is only 5–10% the annual anthropogenic flux of 190×10^6 tonnes of sulphur dioxide. On average, explosive volcanoes currently eject 4×10^6 tonnes of sulphur dioxide into the atmosphere annually, of which 65% reaches the stratosphere. Passively, degassing volcanoes release another 9×10^6 tonnes of sulphur dioxide annually, of which little reaches the stratosphere. Even the Mount Pinatubo eruption of 15 June 1991 released only 20×10^6 tonnes of sulphur dioxide to the atmosphere. In comparison, the single smelting city of Norilsk, Russia, with a population of only 300,000, emits more than 2×10^6 tonnes of sulphur dioxide into the troposphere each year.

Dust emissions, in the form of ash injected into the stratosphere, appear to affect climate more significantly, by attenuating incoming solar radiation over a wide area of the globe for two or three years after a major eruption. For example, dust from the Krakatoa, Indonesia eruption of 27 August 1883 reduced solar radiation in France by 10–20% over a three-year period. An index of stratospheric volcanic dust can be constructed since 1550 (Figure 6.3). This index has a base value of 1,000, referenced to the Krakatoa eruption of 1883. Also plotted in Figure 6.3 is the average temperature of the northern hemisphere, referenced to 1880. There have been some truly cataclysmic eruptions over the past three hundred years. The Tambora eruption on the island of Sumbawa in Indonesia, during its main eruption in 1815, and the Cosequina eruption in Nicaragua in 1835, both

released four times as much dust as Krakatoa. The 1760s and 1770s, corresponding to the Little Ice Age, were notable for sustained volcanic activity worldwide. Volcanic activity was noticeably quiet during the first half of the twentieth century. After the volcanic eruption of Katmai, Alaska in 1912, there was little activity until the eruption of Mount Agung in Bali in 1961. Recent activity has forced the dust veil index to climb above 2,000 in the 1980s and 1990s.

The correspondence between the dust veil index and global temperature is strong ($r = -0.65$ which is significant at the 0.01 level). Periods of increased volcanism decreased surface air temperatures and had a marked impact on society. For instance, volcanism in the 1780s produced extremely cold winters and drought throughout western Europe and Japan. In France, the adverse climatic events exacerbated the social conditions leading to the French Revolution in 1789. In 1816, the dust and sulphate emissions from the Tambora eruption in the previous year, produced temperature drops of 1°C over the northern hemisphere, resulting in the 'year without a summer'. Frost occurred in every month of the year in eastern North America, crops failed in Wales and central Europe leading to famine, and the Southeast Asian monsoon was very intense. This eruption, and a series of others, were conducive to climatic conditions that favoured the first global epidemics of typhus and cholera between 1816 and 1819.

The dust veil index is supported to a limited degree by two other proxy measures of volcanic activity, namely tree-ring dendrochronology and ash accumulation in ice cores. Volcanoes affect tree-ring growth by reducing solar radiation, lowering temperatures, and increasing the likelihood of frost. Following major volcanic eruptions, tree-rings should have increased density and diminished width. A record of the twenty lowest tree-ring density measurements for the last four hundred years, from Europe and North America, is also plotted in Figure 6.3. Many, but not all, of these measurements coincide with documented volcanic eruptions. By far the most widespread effect followed the eruption of Huaynaputina, in Peru in 1601. Significantly, the eruptions of Tambora and Krakatoa, in 1815 and 1883 respectively, did not have a major impact upon tree-ring growth in the northern hemisphere.

Volcanic dust and acids also accumulate in icesheets. The record for the Greenland icecap is plotted in Figure 6.4 for the last 10,000 years. The record is biased towards local events in

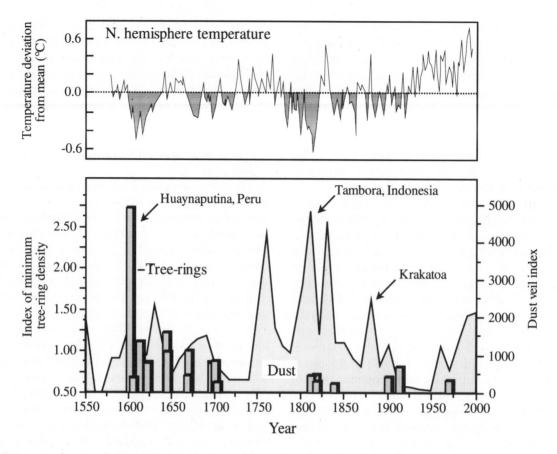

Figure 6.3 Recent volcanism (1600–1990) as determined by Lamb's dust veil index and tree-ring density measurements. The top panel plots northern hemisphere air temperature over the same timespan. (Based on Gribbin, 1978; Lamb, 1985 and Jones et al., 1995.)

Iceland and others in the northern hemisphere that penetrated the stratosphere. The very large eruption of Taupo in New Zealand, in the year 177 AD does not show up well in this record. The largest accumulation can be attributed to the eruption of Mount Mazama, Oregon around 6,400 BP. Except for the eruption of Mount Etna in Roman times, Hekla in Iceland accounts for many major peaks since then. There is also a negative correlation (r −0.52) between ice acidity and temperatures reconstructed back to 1,500 BP. There is also significant agreement between Figures 6.3 and 6.4. One of the largest concentrations of eruptions for the past 9,000 years occurred during the Little Ice Age, between 1600 and 1880. The last five hundred years may be witnessing a period of increased volcanism not seen since the beginning of the Late Holocene. This, and other volcanic activity, may have accounted for the 0.5–1.0°C temperature decrease

measured in the northern hemisphere over the last 2,000 years (Figure 5.9).

Not all explosive volcanoes inject dust into the stratosphere, nor do all stratospheric eruptions affect the globe. For example, ash from the Mount Saint Helens eruption of May 1980 in the United States affected only the troposphere, for a short period of time. Eruptions in the northern hemisphere also do not affect the southern hemisphere very much. Eruptions in the northern hemisphere occur mostly at mid-latitudes, leading to increased atmospheric stability and widespread cooling. The stability suppresses circulation needed to transmit cooling to the southern hemisphere. Volcanism in the southern hemisphere occurs mostly in the tropics. Here, eruptions increase the stability of the tropical atmosphere, reduce the intensity of Hadley circulation, and strengthen mid-latitude westerlies. This effect is transmitted to the

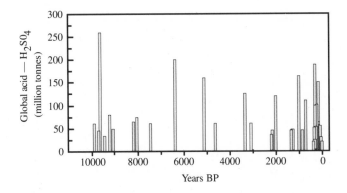

Figure 6.4 Late Holocene volcanism since 10,000 BP, as inferred from Greenland acidity records (based upon Hammer et al., 1980).

northern hemisphere where the amount of energy associated with the jet stream and Rossby waves is reduced.

Coincidentally, two of the last three major eruptions have occurred at the same time as major ENSO events. El Chichon, in Mexico in May 1982, erupted after the initiation of the 1982–1983 ENSO event which was the strongest in one hundred years. The eruption of Mount Pinatubo, in June 1991, together with the eruptions of Mount Hudson, in Chile in the same year, and Mount Spurr, in Alaska in 1992, produced the most significant volcanism of the twentieth century. Subsequent global cooling averaged 0.3–0.5°C coincided with the beginning of the prolonged 1990–1995 ENSO event. It has been hypothesised that, because Mount Pinatubo was located so close to the west Pacific warm pool, it cooled sea surface temperatures in this region, suppressing convection and preventing the reestablishment of 'normal' Walker circulation. Temperature determinations from corals, downstream of the Indonesian Throughflow, indicate a drop in sea surface temperature of 0.6–0.9°C between 1992 and 1993.

Geomagnetism
(Wollin et al., 1971; Fairbridge, 1987a,b; Hopgood and Barton, 1987)

The Earth's magnetic field is modulated by the core–mantle boundary. The field reverses about three times in any million year period. Reversals take 5,000 years to complete and affect climate. However, major changes in climate have occurred during the Pleistocene without any reversal of the Earth's geomagnetic polarity. The last major switch occurred 740,000 years ago, and there have been six glaciations since then. If the Earth's present dipole intensity is taken as 1.0, this intensity has fluctuated between 0.1 and 3.0 over the Pleistocene. Evidence has been found for a stronger geomagnetic field during glacial cycles, for increased storminess when the geomagnetic field undergoes rapid change, and for warmer temperatures at locations on the Earth where magnetic intensity has been decreasing. For example, over the Holocene period since 11,700 BP, stronger geomagnetic activity paralleled colder global temperatures as shown in proxy records derived from land-based pollen sequences, and $\partial^{18}O‰$ values from Greenland ice and deep sea sediments. Certainly during the Holocene Climatic Optimum, when the Earth was warmer by 1–2°C, the Earth's geomagnetic field was weaker than at any other time in the last 10,000 years.

The solar wind also modulates the Earth's geomagnetic field, and there is a strong correlation (r = 0.65) between sunspot number (stronger solar wind) and the Earth's geomagnetic field strength over the period from 1868 to 1986. The solar wind is electrically neutral and consists mainly of ionised hydrogen. It carries with it the effect of the sun's magnetic field, and distorts the dipolar pattern of the Earth's geomagnetic field. Vorticity in the atmosphere has been shown to increase in the northern hemisphere when geomagnetic activity surges upwards. Thus, as solar wind increases, so does the Earth's temperature. The solar wind can also be generated by solar flares independent of the sunspot cycle. Solar flare events lead to magnetic storms on the Earth, associated with a distortion of the magnetosphere, and disruption to the pattern of electric currents in the ionosphere. Following strong solar flare events, there is intensification of cyclonic activity to the west of the British Isles. The cyclonic enhancement occurs beneath divergence in the atmosphere associated with an upper atmospheric low pressure. Additionally, changes in air pressure in both the Gulf of Alaska and eastern North America lag variations in the intensity of the horizontal magnetic field by five days. Some of the greatest winter storms on the east coast of the United States have been associated with these changes. More broadly, much of the world's short-term climate change can be attributed to the sudden breakdown of strongly zonal circulation into cellular blocking patterns, dominated by large relatively immovable high pressure systems. The strongest and most frequent development of cellular blocking occurs when the solar wind, and its corresponding geomagnetic activity on the Earth, are strongest. At the same time, there is strong warming in the stratosphere leading to enhanced ozone production and absorption of solar radiation.

LAND-OCEAN-ATMOSPHERE SYSTEM

The 96,000-year cycle of eccentricity in the Earth's orbit, while matching the cyclicity of glaciations, is insufficient by itself to explain the magnitude of glacials. Internal factors of the Earth–ocean–atmosphere system, more than likely, act in combination to amplify this external forcing. Some factors are more important than others. The purpose of the following discussion is to highlight each possible mechanism, and to evaluate its role in this feedback process.

Carbon dioxide and methane changes
(Idso, 1989; Jouzel et al., 1993)

Both carbon dioxide (CO_2) and methane (CH_4) are 'greenhouse' gases that have a significant forcing effect upon longwave radiation emitted by the Earth's surface. The concentration of these two gases varies enough, between glacials and interglacials, that it has been proposed that they were responsible for transmitting, globally, the cooling triggered by insolation deficiencies over the northern hemisphere high-latitude landmasses during glacials. This hypothesis is supported by gas measurements in ice cores from Vostok, Antarctica spanning the last 175,000 years (Figure 6.5). The record shows a good correspondence between temperature, inferred from $\partial^{18}O‰$ measurements, and both CO_2 and CH_4 in the atmosphere (a correlation of 0.81 and 0.76 respectively). Carbon dioxide levels reached a low of 180–190 ppmv during the Last Glacial, a value that is very close to the lower threshold of 150 ppm required to sustain plant life via photosynthesis on Earth.

There are problems confirming cause and effect in these correlations. Theoretically, a reduction in partial pressure of CO_2 of approximately 180 ppmv below pre-industrial levels of 270 ppmv, would only lower the Earth's temperature by 0.2°C (Figure 1.1), and at most 0.5°C, while the reduction for methane would lower temperatures no more than 0.3°C. At most, the combined 'greenhouse' effect was no more than 1–2°C, a

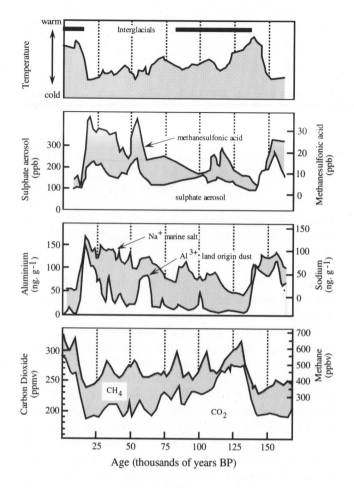

Figure 6.5 Time series of temperature; sulphate and methanesulfonic acid concentrations; sodium (Na^+) and aluminium (Al^{3+}) ions; and carbon dioxide (CO_2) and methane (CH_4); as measured from Vostok, Antarctica ice cores over the last 175,000 years (based on De Angelis et al., 1987; Fells and Liss, 1993 and Jouzel et al., 1993).

value far less than the 7–10°C temperature reduction recorded during the Last Glacial. Additionally, while the correlation between carbon dioxide and temperature is high, carbon dioxide often lags major temperature changes. Indeed, both carbon dioxide and methane differ in their behaviour over time, and show intrinsic differences between glacials and interglacials that cannot solely be linked to temperature. For instance, the time series for methane displays more defined precessional peaks than the carbon dioxide record, a fact that probably reflects the important role of tropical swamps in methane production. Decreases in carbon dioxide, on the other hand, probably reflect decreased overturning of ocean water linked to the behaviour of the North Atlantic thermohaline sink. Biological activity in the ocean, and the consequent fallout of dead organisms to the ocean bottom, also slowly removed carbon dioxide from surface waters and the atmosphere. Without significant upwelling of deep bottom water, this carbon dioxide was not returned efficiently to the ocean surface during glacials, and thus was removed slowly from the atmosphere.

Ocean heating and cooling

There are strong correlations between ocean and air temperatures on global, hemispheric and regional scales. For example, there is strong coherence in fluctuations between air and sea surface temperatures over intervals as little as two years, in both the northern and southern hemispheres since 1900. Many of these fluctuations are linked to the Southern Oscillation (Figure 5.11). On longer timescales, it is debatable whether or not the ocean temperature changes of the Southern Oscillation affected Holocene or Pleistocene temperature variations. For instance, the Southern Oscillation acted with virtually the same intensity and frequency throughout the Little Ice Age as it does at present. However, the type of sea surface temperature changes present during the Last Glacial Maximum (Figure 5.4), undoubtedly, had a major impact on air temperatures over adjacent landmasses. A cooler globe did not necessarily lead to a cooler ocean especially in the tropics. Land–ocean temperature relationships thus may have behaved very differently between glacials and interglacials.

Albedo changes in ice, snow and dust

Greater positive feedback effects upon planetary albedo, and resulting global cooling, were induced by changes in the amount of sea-ice, glacial ice and snow. Snow and ice have a reflectance of 90% that is particularly high at high latitudes, where the majority of the Earth's glacial ice and sea-ice formed. In combination with low insolation values at these latitudes, the Earth's surface received very little solar radiation, even in summer, to melt the accumulation of snow from the previous winter. This led to a positive feedback mechanism that maintained continental icesheets, and enhanced their thermal inertia against melting. Ultimately, if this process was so effective, the Earth should have entered into runaway glaciation, and become permanently locked into an Ice Age. Obviously, the

overall planetary albedo must be highly distorted by latitudinal effects, leading to glaciation, in order to avoid this scenario. The process also begs the question of how glaciation was eventually terminated. A threshold in albedo must exist that is crossed during periods of higher insolation and so allows melting to dominate ice accumulation.

Additionally, the glacial atmosphere was dramatically dustier and windier than the nonglacial one. The amount of dust has been inferred from Greenland ice cores using electric conductivity measurements (Figure 5.2B), and from Antarctic ice cores, using measurements of aluminium ion concentrations and total sediment amounts (Figure 6.5). Windier conditions are implied by increased sodium levels in the Antarctic cores during Ice Age peaks (Figure 6.5), a fact indicative of increased sea spray deposition on this icecap because of a more vigorous atmospheric circulation. These records imply that dust levels were at least ten to twenty times higher than present values during the Last Glacial. While the degree of reflectance by ice and snow depended upon latitude, the dust effects were global. This dust consisted of weathered material from regions in Australia, North America and Asia, that were undergoing increased aridity under colder and windier glacial climates. In addition, sea-levels were up to 130 metres lower during glacials, exposing the continental shelves to aeolian erosion. Despite large tracts of land being covered by glacial ice in the northern hemisphere, the area of arid land worldwide increased sevenfold during glacial periods, as climate became drier and vegetation cover less extensive.

Dust in the atmosphere both reflects and absorbs incoming solar radiation. Reflection increases the planetary albedo, while absorption heats the upper atmosphere. Both processes reduce the amount of solar radiation reaching the lower atmosphere, leading to cooler air closer to the ground. This cooler air tends to sink, enhancing atmospheric stability, and minimising convective processes that form clouds leading to rainfall. The relationship between dustiness and aridity is self-perpetuating. Increased volcanism may be the trigger that initiates this positive feedback mechanism, or it may compound the effect. Again a problem arises in terminating this cycle. However, when sea-levels begin to rise, accompanied by minor melting of icesheets, a point is reached whereby large tracts of relatively flat continental shelves are rapidly flooded, reducing the amount of land surface exposed to wind activity. Additionally, slight warming and increased rainfall near the front of icecaps increases the amount of vegetation covering previously wind-desiccated, periglacial landscapes.

Antarctic icesheet surges (Wilson's theory)
(Lamb, 1977)

The Antarctic icesheet has undergone substantial waxing and waning over time, as inferred by higher glacial limits in mountains. While temperature changes in the Antarctic were synchronous with those in the northern hemisphere, ice growth was not. When an icesheet the size of the Antarctic thickens, it traps geothermal heat emanating from the Earth. This can lead to icesheet instability, through a similar process to that which generated Heinrich ice raft events in the North Atlantic during the Last Glacial. The resulting icesheet thinning and surging in the Antarctic could have formed extensive ice shelves that spread as far north as 50° S. This floating ice increased albedo dramatically at these latitudes, leading to a 4% increase in the overall planetary albedo. As a consequence, icesheets in the northern hemisphere accreted. The process would have been terminated by the breakup of the ice shelf, because it covered the supply of moisture that fed the thickening icecap in the first place. Thus in the Antarctic, where an icecap has been a permanent feature of the Pleistocene, ice-cap thickening occurs during interglacials, while surging initiates worldwide glacials.

Unfortunately, evidence from $\partial^{18}O$ records suggests that the globe drifts slowly into a glacial epoch over thousands of years (Figure 5.1). In addition, surging should be matched by a final rapid rise of sea-level by 20–30 metres at the end of an interglacial. The surging should also carry continental sediment out into the ocean, where it eventually sinks to the seabed. There is little evidence of either effect. Finally, the mechanism occurs in the hemisphere where cooling and warming have the least effect in controlling the onset and demise of worldwide glaciation. Evidence exists for surging in the Antarctic, but it appears to occur in individual drainage basins, rather than contemporaneously across the continent. This surging hypothesis has only passing relevance for the rapid melting of the West Antarctic, and the associated rise in sea-level, if global

warming either causes accretion of the icecap, or melting of basal ice.

North Atlantic thermohaline sink
(Monastersky, 1994; Weaver and Hughes, 1994)

The rapid temperature swings of 6–7°C in the space of five to twenty years during the Last Glacial are characteristic of a chaotic system. Such 'flip-flops' require some very sensitive regional trigger mechanism that can reorganise hemispheric, if not global, atmospheric circulation quickly. The most likely mechanism appears to be the North Atlantic salinity sink.

The North Atlantic thermohaline sink is a crucial climatic mechanism that has global transmission. It forces the turnover of water throughout the ocean column, warms northern Europe, and provides heat to the waters of the Antarctic. It can be turned off by increased rainfall over the North Atlantic, melting of icebergs during Heinrich events, divergence of meltwaters directly to the North Atlantic, or by sudden catastrophic sub-glacial flood events. Any of these mechanisms make Western Europe cooler. The Younger Dryas between 11,500 and 12,900 years BP represents one of the times when the North Atlantic salinity sink was inoperative. The cooling took only one hundred years to become established, and ended in as little as twenty years. Because the conveyor belt spans the world's main oceans, what should be a regional effect can be transmitted to the Antarctic and the North Pacific very rapidly. However, warming in the northern hemisphere is transmitted less rapidly than cooling. Interstadials only appear in the Antarctic ice record if they lasted longer than 2,000 years.

There are four other areas where vertical overturning in the oceans also occurs: where water from the Mediterranean enters the Atlantic, in the Greenland and Labrador Seas, and in the Weddell Sea of Antarctica. In the Greenland Sea, a tongue of sea-ice, labelled the Odden feature, protrudes from the east coast of Greenland. Historically, ice here forms and melts rapidly, creating a pump that forces surface water to the ocean bottom and into the deep sea conveyor belt. Enhanced 'greenhouse' warming, by raising air temperatures, will decrease the frequency of ice formation in this region and shut down this pump. In addition, a warmer climate would increase rainfall over the ocean and make the surface waters of the North Atlantic less salty. Modelling of this situation indicates that the

North Atlantic conveyor belt could collapse repetitively, or alternate between a present-day and a stronger mode of surface-to-bottom water overturning. Either situation would increase the variability of the Earth's present climate, with changes occurring rapidly within a few decades, and lasting for intervals of up to a thousand years. Global warming could feasibly lead to periods of cooler air temperatures in the region. If the North Atlantic conveyor belt were to shut down today, winter temperatures in the North Atlantic and over Northern Europe would fall by 5°C or more in ten years or less. An 8°C warming in Greenland over the past century, and the recent occurrence of colder winters in northern Europe may be the harbingers of this change. The North Atlantic is one of the crucial areas of the world; it requires an increased research effort, in order to unravel the feedback mechanisms and linkages between the ocean and the atmosphere that affect climate change.

The freezing and thawing of the Arctic
(Lamb, 1977)

A popular theory for the onset of Ice Ages deals with the openness of the Arctic Ocean. An unfrozen ocean would permit profuse snow accumulation on surrounding landmasses. If this process operated long enough, then icesheets could become established before the Arctic Ocean froze over, cutting off the moisture supply to the growing icecaps. With the complete freezing of the ocean, glaciation would begin to wane because of a lack of snow. Eventually, melting would outstrip accumulation and rising sea-level would accelerate calving of the icesheets. This would lead to an interglacial and the cycle would begin anew. This theory ignores the fact that, today, much of the Arctic is frozen for significant parts of the year, and even when it is unfrozen, very little precipitation falls on adjacent landmasses because cool air holds little moisture. Most evidence also indicates that icesheets grow because of movement of moist air into the Arctic from southern latitudes (Figure 5.5). This theory had great popularity when the former USSR proposed to dam rivers draining into the Arctic Ocean. It was postulated that this would increase the salinity of the Arctic Ocean, lead to less freezing and more evaporation, and produce excessive snowfalls in the region. Such a scenario is impractical, when it is considered that most, major continental icesheets have developed well inland from the Arctic Ocean, at more southern latitudes.

Sea-level as a trigger for deglaciation
(Imbrie et al., 1993; Blanchon and Shaw, 1995)

The last deglaciation coincided with increased radiation over icesheets in the northern hemisphere. Despite the thermal inertia effects of a large icesheet, the last deglaciation was very rapid. Gradualist approaches to deglaciation invoke progressive melting of icecaps that raised sea-levels. Rising oceans impinged upon existing icesheets, causing them to float and calve. This supposedly led to rapid icesheet collapse, but still at timescales much slower than indicated in marine records. It is now known that melting of icesheets caused large quantities of meltwater to accumulate within the ice mass. Eventually this accumulation became so great that large sections of the icesheet floated, leading to catastrophic discharges of water. Three such events terminated the Last Glacial and raised sea-level by 6.5–13.5 metres in the space of a century each time. Two of the earliest events were associated with large iceberg discharges into the North Atlantic. During the last deglaciation, both catastrophic and slow rises in sea-level created massive icesheet calving over continental shelves; this led to significant decreases in planetary albedo, and finally, to a degree of icesheet wasting that overcame the positive feedback relationship, between icesheets and air circulation, that was maintaining glacial conditions.

The role of dimethylsulphides and cloudiness
(Idso, 1989; Fells and Liss, 1993; Restelli and Angeletti, 1993)

Ice cores in the Antarctic also imply that sulphates in the atmosphere were more abundant during glacials (Figure 6.5). Sulphates enhance cloud formation because they are hygroscopic, and provide the nuclei required for water condensation. Volcanic eruptions produce sulphates that can be thrown either into the stratosphere or, more likely, into the troposphere where they seed cloud formation. Recent satellite measurements show that middle level cloud has a forcing effect upon incoming solar radiation amounting to 12–30 watts per square metre. The reflection of shortwave radiation by clouds over North America in summer can reduce absorbed solar radiation by 50 watts per square metre. And, over the mid-latitude and polar oceans, cloud reflectance associated with storms and extensive layers of stratus, may reduce the radiative heating by as much as 100 watts per square metre. However, the process of cloud formation also scavenges sulphate aerosols from the sky. Unless volcanic sulphates are continually replenished, their cooling effect in the troposphere is short-lived.

Sulphates can also be produced by phytoplankton (microscopic algae) in the oceans, through the emission of dimethylsulphides (DMS). The oceans are mostly biological deserts, but phytoplankton grow wherever there are nutrients and relatively warm water. These conditions are usually met on continental shelves where water can be warmed in relatively shallow water, and nutrients can be supplied by runoff from the continents. Each spring in the northern hemisphere, a bloom of phytoplankton occurs utilising nutrients brought to the surface by mixing during the previous winter, and triggered by warm water and increased sunlight. Phytoplankton contain a salt, dimethylsulphoniopropionate (abbreviated as DMSP, with the chemical formula $(CH_3)_2S^+(CH_2)_2COO^-$) which is used by cells to maintain osmotic balance with seawater, thus preventing dehydration. When phytoplankton die or are eaten by zooplankton, DMSP breaks down to form dimethylsulphide $((CH_3)_2S)$, of which about 10% diffuses through the water column into the atmosphere. Various algae produce different amounts of DMS. For instance, *Coccolithophores*, which form huge blooms in the open ocean in nutrient deficient water, can produce a hundred times more dimethylsulphide than diatoms growing in abundance on continental shelves. Overall, algae transfer 27–56 million tonnes of sulphur from the oceans to the atmosphere each year. (The present industrial contribution is eighty million tonnes.) Dimethylsulphides convert hydroxyl (OH^+) and nitrate $(NO3^+)$ radicals, in the form of gases, to sulphate and methanesulfonic acid (MSA) particles which form cloud condensation nuclei. This produces marine stratus and altostratus clouds and a higher albedo over the oceans. As surface water temperature in the ocean increases, the productivity of marine phytoplankton should also rise, resulting in a concomitant increase in cloud nuclei. Ultimately, there should be a negative feedback from the cloud on any enhanced warming.

Methanesulfonic acid concentrations have also been measured from Antarctic ice cores (Figure 6.5), and found to be up to twenty-five times greater during the last glaciation than they are today. As MSA can only be produced by algae, the higher values suggest that the phytoplankton–DMS–cloud reflectivity relationship was very much turned on during the Last Glacial. Logic implies that phytoplankton growth accelerates in warm water, so this relationship appears dichotomous for glacials. It is possible that different, faster DMS-producing algae dominated the southern hemisphere oceans during the Last Glacial; or that algae produced more dimethylsulphoniopropionate to compensate for the increased salinity of the oceans because more of the Earth's water was locked in icecaps; or that atmospheric circulation delivered more DMS to the Antarctic continent. Alternatively, the warmer oceans of the tropics that existed during the Last Glacial (Figure 5.4), may have provided a fertile region for phytoplankton growth. Because DMS has a lifespan of less than one day, the cloud formed in the tropics needs to be dispersed to higher latitudes. Mobile polar highs generate the vigorous atmospheric circulation required to achieve this dispersal. As tropical cloud moves to polar regions, cooler atmospheric temperatures extend the lifetime of clouds, ensuring maximum augmentation of the planetary albedo.

Increased cloudiness does not necessary imply greater precipitation amounts. By increasing the available cloud condensation nuclei, there is increased competition for water in the atmosphere, such that cloud droplet size decreases. This restricts the amount of drizzle falling from clouds, a process that can be a major source of moisture. Drizzle also stabilises the lower atmosphere, because it tends to evaporate as it falls, thus cooling the atmosphere between the cloud base and the ground. Thus a cloudy globe during glacials could also have been an arid one.

Iron enrichment of the oceans
(Idso, 1989; Fells and Liss, 1993)

Much of the dust that was blown around during glacials originated from weathered material containing significant amounts of iron. Iron is a stimulant for photosynthesis. However, this iron was of little use to land-based vegetation during glacials, because flora was stressed by aridity and strong winds. Phytoplankton also use photosynthesis and were probably stimulated by iron fallout into the ocean. As a result, the dimethlysulphide–methanesulfonic acid–cloud feedback process should have been turned on more during the Last Glacial. Organic accumulation rates in the world's oceans at the peak of the Last Glacial support this hypothesis, being two to five times greater than during interglacials. Experiments also show that phytoplankton activity increases tenfold in nutrient rich water when iron is added. Enhanced photosynthesis would have drained more carbon dioxide from the atmosphere, a scenario supported by the lower carbon dioxide levels coinciding with increased iron concentrations which have been measured in the Antarctic ice cores. Thus, lower carbon dioxide would have been the result of global cooling and glaciation, rather than the cause of it. Reduced carbon dioxide levels would have exacerbated the limited growing conditions for vegetation on land.

The same hypothesis also accounts for the acceleration of deglaciation. As climate warmed and enough landmass was deglaciated, rapid recolonisation by vegetation would have dramatically reduced the sources for wind-blown dust and iron. The feedback mechanism was aided by the fact that rising sea-levels flooded continental shelves, thus lessening the area of land subject to wind erosion. The decrease in iron input into the oceans would have slowed down photosynthesis in phytoplankton, eventually leading to clearer skies and more solar radiation reaching the Earth's surface.

STOCHASTIC RESONANCE
(Monastersky, 1994)

The concept of stochastic resonance (Figure 1.6) may also explain how eccentricity, the weakest of the Milankovitch orbital forcing mechanisms, was able to control the timing of Ice Ages. Stochastic resonance implies that a long-period cycle can force warming, if that cycle is superimposed upon noise, and if there is a threshold above which positive feedback mechanisms lock climate into an Ice Age. Glaciation will not occur if this threshold is crossed only briefly. However, when long cycles, such as orbital eccentricity, are involved, this threshold can be crossed repeatedly, no matter how weak the magnitude of the underlying cycle may be. The effects of crossing the threshold are prolonged by positive feedback processes favourable to Ice Ages. These positive feedback processes may include volcanism, a change in the North Atlantic thermohaline sink, or increased snow coverage at high latitudes. Stochastic resonance, in combination with positive feedbacks, is particularly apt at explaining how climate always

returned to glacial conditions during the 'flickerings' and Dansgaard–Oeschger oscillations of the Last Ice Age. Each time temperatures increased, the slightest cooling in the presence of persistent, positive feedback mechanisms favourable to glaciation, plunged climate back into Ice Age conditions. The crucial parameters in this theory appear to be the length of the cycle, rather than its magnitude, and the length of time that positive feedback processes can operate once they are established.

CYCLES AND THE HISTORICAL RECORD

Sunspot cycles
(Gilliland, 1982; Friis-Christensen and Lassen, 1991; Magny, 1993)

Over the past century, sunspots have been associated with a range of climatic fluctuations that include pressure, temperature and rainfall. The majority of studies attempt to model the effect of sunspots upon recent temperature records. Some of the model results, singly or in combination with other variables, are shown in Figure 6.6 for the period from 1860 to 1984. Sunspot numbers are poorly correlated to temperature, for the simple reason that temperature changes precede sunspots! A model combining volcanic activity, solar luminosity, sunspot cycles and inferred carbon dioxide levels in the atmosphere fits much better, and accounts for 93% of the variance in the northern hemisphere land temperature record between 1881 and 1975. The magnitude of temperature responses by these variables, in order of importance, are as follows: for volcanic aerosols, 0.4°C; for solar luminosity, 0.27°C; for CO_2, 0.23°C; and for sunspot cycles, 0.21°C. The forcing due to carbon dioxide was only prominent after 1940, a timing whose significance is discussed in more detail in Chapter 7. The best model describing the temperature time series is based upon the length of the sunspot cycle. The length of a sunspot cycle varies inversely with the degree of solar activity. A shorter cycle is associated with more solar activity, and produces warmer global temperatures. Figure 6.6 shows that this model can even explain the cooling between 1940 and 1978. The model has subsequently been shown to apply back to 1740. However, over this longer timespan, sunspot number becomes more important. These inconsistencies have been one of the main reasons why

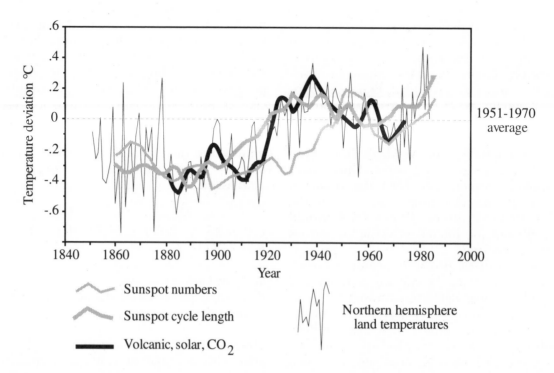

Figure 6.6 The temperature record of the northern hemisphere (from Figure 5.11). Superimposed on this time series are the results from various models, incorporating some component of solar activity, and purporting to explain temperature fluctuations over time (based on Gilliland, 1982; Friis-Christensen and Lassen, 1991).

solar behaviour has not been treated seriously, as showing a cause-and-effect relationship with climate change.

Sunspots and the Quasi-Biennial Oscillation
(Labitzke and van Loon, 1989; Varotsos, 1989)

If sunspot cycles are affecting climate, then they should have a stronger correlation than often found in analyses. It now appears that this signature may have been masked by interaction with a third variable, namely the Quasi-Biennial Oscillation (QBO). The QBO is a quasi-periodic switch in direction of winds in the tropical stratosphere that repeats every 2.2 years. The signature of the QBO has been found in a range of climate variables including the Southern Oscillation (Figure 4.9), the northern hemisphere temperature record (Figure 5.11) and rainfall records across eastern North America. The QBO also controls the response of temperature and ozone fluctuations in the stratosphere to sunspot cycles, between 1956 and 1985 (Figure 6.7). Crucially, the data need to be separated into two groups based upon the direction of tropical stratospheric winds. If the data are combined, there is virtually no correlation

between solar radiation fluctuations over a sunspot cycle, and temperature or ozone. However, when winds at the top of the atmosphere are westerly in the northern hemisphere winter, Arctic air temperatures in the stratosphere increase by 20°C, and total global ozone increases by 7% at sunspot peaks. If stratospheric winds are easterly, values in each parameter peak at sunspot minima. Both stratospheric air temperature at the poles and ozone amounts are interrelated, and affect the amount of UVB radiation reaching the Earth's surface, as well as near-surface climatology. While the relationships in Figure 6.7 appear strong, there is still much debate as to the mechanism transferring stratospheric behaviour in these variables to climate changes in the lower troposphere.

Astronomical cycles and worldwide drought and rainfall
(Tyson et al., 1975; Lamb 1977; Currie, 1981, 1984; Bryant, 1991)

Many studies also have examined the effect of sunspot cycles upon the timing of major floods and droughts. One of the better relationships between

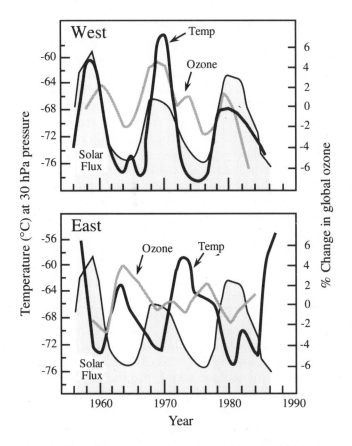

Figure 6.7 Stratospheric temperature and the percent change in global ozone during sunspot cycles, 1956–1985. The data have been separated into the west and east phases for stratospheric winds, at an elevation of 30 hectopascals over the tropics (modified from Labitzke and van Loon, 1989; Varotsos, 1989).

sunspot cycles and drought occurs on the United States Great Plains, approximately every twenty years (the 22 year Hale sunspot cycle). Statistically, significant 22- and 11-year cycles in the frequency of droughts and floods have also been found for the whole of China since 1440. A 20 to 25-year cycle in drought is also present in the Dnieper basin from 3,650 years BP onwards, while a 10 to 15-year cycle appears from 750 onwards. The Indian flood record evidences a 22-year cycle with peaks at sunspot maxima, while the drought record displays an 11-year sunspot-related periodicity. In the southern hemisphere, ten and twenty year periodicities appear in the occurrence of drought in South Africa. In all cases, these cycles account for about 15% of the variance in rainfall records.

The 18.6-year M_N lunar cycle may also affect rainfall. This cycle represents a fluctuation of about 5° in the orbit of the moon relative to the sun's equator.

The moon does not return to the same location relative to the sun after each orbit of the Earth. Instead precession displaces the moon relative to the solar equator over an 18.6-year cycle. The magnitude of the cycle appears trivial until one considers the moon's orbit relative to the Earth. The solar equator is displaced seasonally 23.47° north and south of the Earth's equator. If the moon's orbit relative to the sun is superimposed upon this oscillation, then there will be a time when the moon is displaced 28.5° (23.5° + 5°) north and south of the equator. In the following 9.3 years, the moon drifts to the opposite end of the sun's equator, and takes up a position relative to the Earth's equator of 18.5° (23.5° − 5°). This overall change causes a variation in the gravitational attraction of the moon that is 3.7% that of its daily component. The last maximum in this cycle occurred in 1991.

The relative importance of these astronomical cycles has been difficult to unravel over the past century, because the Hale sunspot cycle has peaked at times of the 18.6-year lunar maximum. In fact only the 18.6-year M_N lunar tide and 11-year sunspot cycles dominate rainfall. Cycles of rainfall at the 18.6-year frequency are highly coherent amongst the western United States and Canada, northern China, India, the Nile region of Africa, and mid-latitudes of South America. Not only is the 18.6-year cycle dominant, but the data also show temporal and spatial bistable phasing. In this process, drought may coincide with peaks in the M_N lunar cycle over a two hundred or three hundred year timespan, but then suddenly switch to times of the mid-cycle trough. The last occurrence of this bistable 'flip-flop' occurred at the turn of the century for South America, China, Africa and India. Historical trends since 1657 indicate that the United States Great Plains was locked into drought shortly after peaks in the M_N lunar cycle. The latest peak in 1991 did not lead to subsequent drought, but to some of the severest flooding of the Mississippi River ever recorded. The 1993–1994 wet period on the United States Great Plains may be evidence that this region's rainfall pattern has undergone a bistable 'flip-flop', and is in phase with other major regions of the world manifesting an 18.6-year cycle in rainfall.

Droughts in the regions of the northern hemisphere affected by the M_N lunar tide, are controlled by resonance of the planetary Rossby wave that is locked topographically into position by the Tibetan Plateau and Rocky Mountains. The wider coincidence in the timing of floods and drought implies that the amplitudes of other planetary waves, such as that produced by the Southern Oscillation, may be enhanced by astronomical cycles. There is strong coherence in the timing of rainfall amongst Australia, Argentina and South Africa in the southern hemisphere, linked to this latter phenomenon. However, correlations between the Southern Oscillation and both sunspot and the M_N lunar cycles are very weak ($r < 0.35$). It must be concluded that, at present, associations between climate and astronomical cycles for much of the southern hemisphere are of minimal practical significance.

CONCLUDING COMMENTS

The following scenario summarises the positive feedback mechanisms producing Ice Ages, as well as accounting for each subsequent deglaciation. The 23,000- and 41,000-year precession and obliquity cycles in the Earth's orbit control solar radiation changes in the northern hemisphere, and drive Ice Ages. Whenever winter snowfall around 65° N cannot consistently be melted in the following summer, icesheets form. At present, interglacial circulation is powered by the North Atlantic thermohaline sink, reinforced by outflow of water from the Arctic Ocean between Greenland and Scandinavia, and from the Labrador Sea. These sinks act as heat pumps, causing warm surface waters to be brought into the North Atlantic, while at the same time causing heat to be exported, via the ocean bottom, to the Southern Ocean around Antarctica. With the onset of glaciation, evaporation decreases in the Arctic Ocean, and the Arctic outflow to the North Atlantic thermohaline sink shuts down. The immediate consequence is a decrease in the export of heat to the Antarctic, resulting in expansion of sea-ice and increased sinking of brine. Sinking bottom waters around the Antarctic no longer become part of the conveyor belt leading to the Pacific, but begin to force their way into the Atlantic, further blocking the ocean-bottom conveyor belt. As a result, the Antarctic cools, and the latitudinal temperature gradient in the southern hemisphere increases, shifting cold fronts and phytoplankton communities northwards. Phytoplankton begin to thrive on slightly warmer subtropical waters, 'fertilised' by iron blowing off subarid continents. These phytoplankton increase the amount of dimethylsulphides in the atmosphere, resulting in increased cloudiness, and an enhanced planetary albedo. At the same time, increased competition for water

vapour in clouds produces smaller sized water droplets, that mitigate against more rainfall and aggravate atmospheric stability and aridity. Mobile polar highs direct this tropical cloud to polar regions, where cooler temperatures extend the lifetime of the clouds. Phytoplankton also withdraw carbon dioxide from the atmosphere, adding a 'greenhouse' factor to global cooling. As northern icesheets continue to grow, they divert wind systems, extend cold temperatures, and intensify aridity and atmospheric dust production. Sea-levels now fall, allowing continental icesheets to ground themselves on continental shelves, and exposing more land to aridity and wind desiccation. The thermal inertia of icesheets overwhelms the radiation effects of the 23,000- and 41,000-orbital perturbations, and brings out the dominance of the weaker 96,000-year eccentricity cycle.

Deglaciation is initially driven by the insolation effects of the short-term Milankovitch cycles, superimposed upon the longer eccentricity cycle. Meltwaters flowing into the Gulf of Mexico force warmer, briny surface water into the North Atlantic, turning on the North Atlantic thermohaline pump. This pump can also be turned on during shorter warming events or interstadials. However, if the timespan is less than two thousand years, the warming can never be transmitted to the Antarctic to become a global phenomenon. If catastrophic flooding from the North American icesheets causes rapid rises in sea-level, especially when seas can flood the relatively flat portions of continental shelves, rapid ice calving and substantial glacial retreat can take place. Finally, as sea-levels continue to rise, the Arctic Ocean outflow into the Atlantic is reestablished, and the North Atlantic heat pump reaches maximum efficiency. This leads to rapid retreat of sea-ice in the Antarctic, and sets deglaciation into an interglacial mode.

The iron 'fertilisation' aspect of this scenario raises two questions: can experiments adding iron to surface waters trigger algal growth, and does iron in polluted runoff also activate algal production? A joint British–US experiment in 1994 'fertilised' a patch of ocean between Tahiti and the Galápagos Islands with iron, and found a huge increase in algal growth with an attendant drawdown in carbon dioxide. While this result may provide a technical fix for anthropogenically enhanced 'greenhouse' warming due to increased carbon dioxide in the atmosphere, it strongly supports the hypothesis for increased cloudiness during glacial cycles, due to increased dimethylsulphide production by phytoplankton, exacerbated by iron enrichment of ocean waters. Furthermore, there is growing evidence based upon the proliferation of toxic coastal algal blooms, known as red tides, that the increase of nutrients from sewage into coastal waters, coupled with enhanced iron levels in this waste and in runoff, can also trigger proliferate algal growth. The algae–cloudiness–cooling hypothesis certainly bears further research.

The above scenarios ignore the possibility of fluctuating solar activity or increased volcanism as major factors in the cooling, leading to Ice Ages. Both parameters certainly have affected Holocene climate, and are more than likely to play an important role in future climate change. The Little Ice Age, between 1645 and 1715, was a time when sunspots were rare, and global volcanic activity was twice as high as during the following seventy-year period. Solar activity appears to be more important than volcanic activity, because the coldest temperatures of the Little Ice Age preceded the volcanic peak. The global effect of volcanic eruptions upon temperature also depends upon the location, nature and type of explosion. All solar–climate associations suffer from several flaws. The first and most serious is the lack of a physical mechanism explaining the relationships between the solar cycle and surface temperature. Not only does this include the sunspot cycle, but also other solar cycles involving rotation, flares and prominences. Secondly, many of the correlations lack statistical significance or, in some cases, are only defined subjectively. Solar cycle oscillations must be detectable above the spatial and temporal background of regular climatic variability, before they can be judged as a significant forcing mechanism for climate change. These two criticisms are responsible for much of the vitriol associated with the question of solar influences on climate change.

There is almost universal agreement that increased solar activity during the Holocene led to increased global air temperatures. Solar activity is predicted to increase in the early part of the twenty-first century, concomitantly with any enhanced, anthropogenic 'greenhouse' warming effect. This does not necessarily diminish the importance of any increased 'greenhouse' effect due to human activity. It simply implies that solar forcing effects may be masking any present 'greenhouse' warming signature. This blurring amongst solar, volcanic and human influences adds considerable 'noise' to temperature records, and makes the task of unravelling their relative impacts all but impossible. The impact of humans upon climate is assessed against this background of natural climate variability and forcing, in Chapter 7.

REFERENCES AND FURTHER READING

Berger A. and Loutre M.F. 1991. 'Insolation values for the climate of the last 10 million years'. *Quaternary Sciences Review* v. 10 pp. 297–317.

Blanchon, P. and Shaw, J. 1995. 'Reef drowning during the last deglaciation: evidence for catastrophic sea-level rise and ice-sheet collapse'. *Geology* v. 23 pp. 4–8.

Bluth, G.J.S., Schnetzier, C.C., Krueger, A.J. and Walter, L.S. 1993. 'The contribution of explosive volcanism to global atmospheric sulphur dioxide concentrations'. *Nature* v. 366 pp. 327–329.

Bryant, E.A. 1991. *Natural Hazards*. Cambridge University Press, Cambridge, 294p.

Clemens, S.C. and Prell, W.L. 1991. 'Late Quaternary forcing of Indian Ocean summer-monsoon winds: a comparison of Fourier models and general circulation model results'. *Journal Geophysical Research* v. 96 pp. 22683–22700.

Currie, R.G. 1981. 'Evidence of 18.6 year M_N signal in temperature and drought conditions in N. America since 1800 A.D'. *Journal Geophysical Research* v. 86 pp. 11055–11064.

Currie, R.G. 1984. 'Periodic (18.6) and cyclic (11-year) induced drought and flood in western North America'. *Journal Geophysical Research* v. 89 pp. 7215–7230.

De Angelis, M, Barkov, N.I. and Petrov, V.N. 1987. 'Aerosol concentrations over the climatic cycle (160 k yr) from an Antarctic ice core'. *Nature* v. 325 pp. 318–321.

Fells, N. and Liss, P. 1993. 'Can algae cool the planet?'. *New Scientist* 21 August pp. 34–38.

Fairbridge, R.W. 1987a. 'Climatic variation, historic record'. *In* Oliver, J.E. and Fairbridge, R.W. (eds). *Encyclopedia of Climatology*. Van Nostrand Reinhold, New York, pp. 305–323.

Fairbridge, R.W. 1987b. 'Ice Age theory'. *In* Oliver, J.E. and Fairbridge, R.W. (eds). *Encyclopedia of Climatology*. Van Nostrand Reinhold, New York, pp. 503–514.

Friis-Christensen, E. and Lassen, K. 1991. 'Length of the solar cycle: an indicator of solar activity closely associated with climate'. *Science* v. 254 pp. 698–700.

Gagan, M.K. and Chivas, A.R. 1995. 'Oxygen isotopes in western Australian coral reveal Pinatubo aerosol-induced cooling in the Western Pacific Warm Pool'. *Geophysical Research Letters* v. 22 pp. 1069–1072.

Gilliland, R.L. 1982. 'Solar, volcanic, and CO_2 forcing of recent climatic changes'. *Climatic Change* v. 4 pp. 111–131.

Gribbin, J. 1978. *The Climatic Threat*. Fontana, Glasgow, 206p.

Gribbin, J. 1994. 'Fire from the stars could spell global disaster … '. *New Scientist*, No. 1918, p. 16.

Hammer, C.U., Clausen, H.B. and Dansgaard, W. 1980. 'Greenland ice sheet evidence of postglacial volcanism and its climatic impact'. *Nature* v. 288 pp. 230–235.

Hopgood, P.A. and Barton, C.E. 1987. 'The great magnetic storm of February 1986'. *Search* v. 18 pp. 26–30.

Idso, S.B. 1989. *Carbon Dioxide and Global Change: Earth in Transition*. Institute for Biospheric Research, Tempe, 292p.

Imbrie, J. and Imbrie, K.P. 1979. *Ice Ages: Solving the Mystery*. Macmillan Press, London, 224p.

Imbrie, J., Boyle, E.A., Clemens, S.C., Duffy, A., Howard, W.R., Kukla, G., Kutzbach, J., Martinson, D.G., McIntyre, A., Mix, A.C., Molfino, B., Morley, J.J., Peterson, L.C., Pisias, N.G., Prell, W.L., Raymo, M.E., Shackleton, N.J. and Toggweiler, J.R. 1992. 'On the structure and origin of major glaciation cycles 1: Linear responses to Milankovitch forcing'. *Paleoceanography* v. 7 pp. 701–738.

Imbrie, J., Berger, A., Boyle, E.A., Clemens, S.C., Duffy, A., Howard, W.R., Kukla, G., Kutzbach, J., Martinson, D.G., McIntyre, A., Mix, A.C., Molfino, B., Morley, J.J., Peterson, L.C., Pisias, N.G., Prell, W.L., Raymo, M.E., Shackleton, N.J. and Toggweiler, J.R. 1993. 'On the structure and origin of major glaciation cycles 2: The 100,000-year cycle'. *Paleoceanography* v. 8 pp. 699–735.

Jones, P.D., Briffa, K.R. and Schweingruber, F.H. 1995. 'Tree-ring evidence of the widespread effects of explosive volcanic eruptions'. *Geophysical Research Letters* v. 22 pp. 1333–1336.

Jouzel, J., Barkov, N.I., Barnola, J.M., Bender, M., Chappellaz, J., Genthon, C., Kotlyakov, V.M., Lipenkov, V., Lorius, C., Petit, J.R., Raynaud, D., Raisbeck, G., Ritz, C., Sowers, T., Stievenard, M., Yiou, F. and Yiou, P. 1993. 'Extending the Vostok ice-core record of palaeoclimate to the penultimate glacial period'. *Nature* v. 364 pp. 407–412.

Labitzke, K. and van Loon, H. 1989. 'Associations between the 11-year solar cycle, the QBO and the atmosphere. Part 1: the troposphere and stratosphere in the northern hemisphere in winter'. *Journal of Atmospheric and Terrestrial Physics* v. 50 pp. 197–206.

Lamb, H.H. 1977. *Climate: Present, Past and Future*: v. 1&2. London, Methuen, 613p.

Lamb, H.H. 1985. Volcanic loading: The dust veil index. *Carbon Dioxide Information Analysis Center Numeric Data Package Collection*, Dataset No. NDP013.DAT, Oak Ridge National Laboratory, Oak Ridge, Tennessee.

Magny, M. 1993. 'Solar influences on Holocene climatic changes illustrated by correlations between past lake-level fluctuations and the atmospheric 14C record'. *Quaternary Research* v. 40 pp. 1–9.

Monastersky, R. 1994. 'Staggering through the Ice Ages'. *Science News* v. 146 pp. 74–76.

Paterson, D. 1993. 'Did Tibet cool the world?' *New Scientist*, 3 July, pp. 29–33.

Restelli, G. and Angeletti, G. (eds) 1993. *Dimethylsulphide: Oceans, Atmosphere and Climate*. Kluwer Academic Publishers, Dordrecht, 399p.

Ruddiman, W.F. and Kutzbach, J.E. 1991. 'Plateau uplift and climatic change'. *Scientific American*, March, pp. 42–48.

Schönwiese, C-D., Ullrich, R., Beck, F. and Rapp, J. 1994. 'Solar signals in global climatic change'. *Climatic Change* v. 27 pp. 259–281.

Sutton, J. and Elias, T. 1993. 'Volcanic gases create air pollution on the island of Hawaii'. *Earthquakes and Volcanoes* v. 24 pp. 178–196.

Tyson, P.D., Dyer, T.G.S. and Mametse, M.N. 1975. 'Secular changes in South African rainfall 1880–1972'. *Quarterly Journal Royal Meteorological Society* v. 101 pp. 817–833.

Varotsos, C. 1989. 'Comment on connections between the 11-year solar cycle, the Q.B.O. and total ozone'. *Journal of Atmospheric and Terrestrial Physics* v. 51 pp. 367–370.

Weaver, A.J. and Hughes, T.M.C. 1994. 'Rapid interglacial climate fluctuations driven by North Atlantic ocean circulation'. *Nature* v. 367 pp. 447–450.

Wollin, G., Ericson, D.B., Ryan, W.B.F. and Foster, J.H. 1971. 'Magnetism of the Earth and climatic changes'. *Earth and Planetary Science Letters* v. 12 pp. 175–183.

7

Human Effects on Climate

INTRODUCTION

The purpose of this chapter is to outline the mechanisms of enhanced 'greenhouse' warming, and to describe the computer-based predictions for that warming. However, many popularised views of future climate change do not acknowledge the uncertainties that lie behind these scenarios. Nor do they acknowledge the fact that other mechanisms may be responsible for the climate change that is currently being measured. These aspects also are examined, specifically issues related to: 'greenhouse' gas sinks; the nature of temperature changes; the importance of negative feedback mechanisms such as clouds, sulphate aerosols and dust in controlling temperature; and the efficiency of computer simulation in modelling the climatic impact of enhanced 'greenhouse' gases. The 'greenhouse' debate has been obscured by speculation about the possible deterioration of stratospheric ozone through the production of manufactured halocarbons. In many respects, both scenarios overlap, with some of the effects of 'greenhouse' warming exacerbating the potential for ozone destruction in the upper atmosphere. These aspects are described. Finally, in the 1980s, nuclear war was hypothesised as having the potential profoundly to cool the globe. Few scientists would support that hypothesis now because it is seriously flawed. The scenario for 'nuclear winter' is described as a cautionary lesson that not all scientific proposals for climate change are valid.

ENHANCED 'GREENHOUSE' WARMING

Discovery
(National Research Council, 1983; Bryant, 1993)

The potential effect of atmospheric changes in carbon dioxide induced by human activity has been the subject of scientific enquiry for the past 130 years. In 1861, John Tyndall suggested that carbon dioxide changes in the atmosphere might be responsible for changes in climate. In 1896, Svante Arrhenius proposed that a doubling of carbon dioxide might warm the atmosphere by 5°C. Thomas Chamberlain, in 1899, proposed that carbon dioxide fluctuations could cause large variations in the Earth's climate, including ice ages. For the next forty years, the question of how much carbon dioxide humans were adding to the atmosphere, and its effect on climate, were debated. In 1938, Callender showed how carbon dioxide and water vapour absorb longwave radiation at different wavelengths. He suggested that anthropogenic production of carbon dioxide would alter the natural balance between incoming solar radiation and outgoing longwave radiation, and cause global warming that would exceed natural variations in historical times. Over the next twenty-three years, Callender also collated the records of carbon dioxide in the atmosphere, and showed that it had been rising steadily since the Industrial Revolution. In 1957, Revelle and Suess showed that carbon dioxide was not rapidly removed from the atmosphere by the oceans, and thus could accumulate in the atmosphere to significant levels.

At the same time, international efforts recognised that human activities could affect climate. In 1957, regular monitoring of carbon dioxide at Mauna Loa and the South Pole was established as part of the International Geophysical Year (IGY). At the end of the 1960s, the World Meteorological Association (WMO) established the World Weather Watch and the Global Atmospheric Research Programme to monitor trends in the trace gases, including both carbon dioxide and ozone, that could have potential long-term impacts on climate. International concerns about enhanced 'greenhouse' warming first received widespread publicity at the First World Climate Conference in 1979. However, the seminal conference, the one that catalysed world attention on the significant ramifications of enhanced 'greenhouse' gas concentrations in the atmosphere, was the WMO-sponsored Villach, Austria, conference of 9–15 October 1985. This conference brought together a limited number of experts in various fields associated with climatic change and its impact. Contributing to the success of this conference was the release of a policy statement that was readily disseminated to the world's media. This statement pointed out that most current planning and policy decisions assumed a constant climate, when, in reality, increases in 'greenhouse' gases by the 2030s would warm the globe between 1.5 and 4.5°C, leading to a sea-level rise of 20–140 centimetres. This conference was followed in 1992 by the Rio Conference, where over one hundred nations pledged to fix carbon dioxide production at 1990 levels by the year 2000. At the same time, the WMO and the United Nations Environment Programme established the Intergovernmental Panel on Climate Change (IPCC) which published state-of-the-art scientific reports about 'greenhouse' gases, the impacts of enhanced 'greenhouse' warming upon society, and the responses required to mitigate such impacts.

Mechanisms
(Bridgman, 1990; Cotton and Pielke, 1995)

The term 'greenhouse' is a misnomer, originally based on the mistaken notion that a glass-encased building or greenhouse warms up by trapping longwave radiation. Greenhouses actually warm by preventing the loss of heat via convection to the outside atmosphere. Nor do 'greenhouse' gases trap radiation. Such gases intercept more longwave radiation radiated from the Earth's surface than they do shortwave radiation transmitted downward through the atmosphere (Figure 2.7) The longwave radiation that is intercepted is re-radiated, both downwards and upwards in the atmosphere. About 20% of outgoing longwave radiation is absorbed within 80 centimetres of the Earth's surface, and virtually all within the lower four kilometres of the troposphere. The downward radiation warms the lower layers of the atmosphere, until eventually a higher temperature is reached, whereby energy can be radiated upwards through the atmosphere to balance the incoming shortwave radiation flux (Figure 2.8). As no energy is created by enhanced 'greenhouse' warming in the Earth's atmosphere, the stratosphere must cool as the surface warms.

The pre-industrial, natural degree of this forcing maintains the Earth's surface temperature at a comfortable 15°C, rather than a more life-threatening −18°C. Most of the effect is due to water vapour. Other gases such as carbon dioxide (CO_2), methane (CH_4), ozone (O_3) and nitrous oxide (N_2O), occurring naturally in the atmosphere, are also 'greenhouse' gases. Humans have the ability to increase the concentration of all of these gases, plus manufacture new gases which have a significant 'greenhouse' radiative forcing. This human contribution is commonly known as the enhanced 'greenhouse' effect.

Anthropogenic 'greenhouse' gases
(Ramanathan et al., 1989; Leggett, 1990; Elkins et al., 1993; Houghton, 1994; Houghton et al., 1990, 1992, 1996)

There are more than thirty-five trace gases and non-methane hydrocarbons produced by humans that can cause enhanced 'greenhouse' warming. The concentrations, rate of increase and relative warming effects of the major gases, current to 1995, are listed in Table 7.1. Historical and recent rates of increase of the main anthropogenic 'greenhouse' gases are presented, graphically, in Figures 7.1 and 7.2 respectively. The global warming potential of each gas has a degree of uncertainty approximating about 35%.

While carbon dioxide tends to be used synonymously for all 'greenhouse' gases, it is responsible for about 64% of the enhanced 'greenhouse' effect. The pre-industrial level of carbon dioxide in the atmosphere was 280 ± 5 parts per million (ppm). Interestingly, carbon dioxide levels did not fall much below 270 ppm during the Little Ice Age, when global mean temperatures were about 1°C lower, a fact suggesting that the natural level of

Table 7.1 'Greenhouse' gas concentrations, trends and degree of forcing relative to carbon dioxide (mainly from Houghton et al., 1996)

Gas	Formula	Concentrations 1994	Pre-industrial levels	Observed trend per year (%)[1]	Projected concentration 2050	Magnitude of forcing to the year 2014 relative to a molecule of CO_2	% of total warming effect	Lifetime in atmosphere (years)[2]
Naturally occurring:								
carbon dioxide	CO_2	356 ppmv	280 ppmv	0.46	400–600 ppmv	1	63.8	30–100
methane	CH_4	1,725 ppbv	0.8 ppmv	0.7	2.1–4.0 ppmv	56	19.2	12–17
nitrous oxide	N_2O	311 ppbv	280 ppbv	0.25	350–450 ppbv	280	5.7	120
carbon monoxide	CO	50–120 ppbv	?[3]	1.0	90–200 ppbv	?[3]	?[3]	0.25
ozone	O_3	10–100 ppbv[4]	7–80 ppbv	−1.8 to 2.0	11–150 ppbv	1969	?[3]	0.1
sulphur dioxide[5]	SO_2	24–90 pptv	3–6 pptv	–	–	–	−22	0.02
Halocarbons:								
CFC–11	CCl_3F	280 pptv	0	0	410 pptv	5,000	2	50
CFC–12	CCl_2F_2	503 pptv	0	1.4	850 pptv	7,900	7	102
CFC–113	CCl_2FCClF_2	82 pptv	0	0	140 pptv	5,000	1.5	85
CFC–114	$CClF_2CClF_2$	20 pptv	0	4.0	20 pptv	6,900	0.2	300
CFC–115	CF_3CClF_2	4 pptv	0	2.0	10 pptv	6,200	0.1	1,700
carbon tetrachloride	CCl_4	132 pptv	0	−0.4	–	2,000	0.3	42
methyl chloroform	CH_3CCl_3	135 pptv	0	−7.0	–	360	0.2	4.9
HCFC–22	$CHClF_2$	100 pptv	0	5	940 pptv	4,300	0.4	12
Perfluorinated compounds (PFCs):								
perfluoromethane	CF_4	70 pptv	0	1.7	–	4,100	–	50,000
perfluoroethane	C_2F_6	4 pptv	0	–	–	8,200	–	10,000
sulphur hexafluoride	SF_6	3.2 pptv	0	7.0	–	3,200	0.3	16,500

[1] quoted values for CFCs and their substitutes can vary by a factor of 2 or more
[2] values refer to disequilibrated residence time and not to recycling time
[3] these components are significant but still uncertain
[4] troposphere only
[5] SO_2 is not a 'greenhouse' gas. It is added to this table for comparison
ppmv = parts per million by volume
ppbv = parts per billion by volume
pptv = parts per trillion by volume

carbon dioxide in the atmosphere may be insensitive to temperature changes. As of 1995, carbon dioxide had reached a concentration of 362 ppm in the atmosphere, a 29% increase over pre-industrial values. The amount of carbon dioxide is now rising by 0.46% or 1.8 ppm per year, a rate that appears to be accelerating slightly over the last twenty years (Figure 7.2). The rate of fossil fuel burning is much higher, approximately 4.0% per year. The sources of increased carbon dioxide have been subjected to considerable speculation, but appear to be partitioned as follows: deforestation (mainly tropical), 3.5%; other land use changes, 19.1%; coal burning, 31.0%; petroleum burning (mainly combustion engines), 31.4%; gas burning, 12.9%; and cement manufacture, 2.0%.

Carbon dioxide concentrations can also fluctuate with time due to biological activity. In the northern hemisphere, there is a very noticeable change from summer to winter as deciduous leaves stop photosynthesis. This seasonal variation amounts to 6–7 parts per million. The effect is not noticeable in the southern hemisphere or tropics, because plants there show little seasonal variation in leaf activity. If the seasonal signature is removed, then carbon dioxide fluctuations reflect the intensity of the Southern Oscillation. Landmasses affected by moisture deficits during ENSO events have reduced plant growth, and fail to absorb their normal quota of carbon dioxide. As well, the natural absorption of carbon dioxide from the atmosphere into the ocean is suppressed in the east Pacific. During a severe ENSO event, the excess carbon dioxide left in the atmosphere can overwhelm the anthropogenic signature.

The second most important 'greenhouse' gas is methane. A single molecule of methane gas has fifty-six times the radiative forcing of a molecule of carbon dioxide over a twenty-year period. Pre-industrial levels of methane were 800 parts per billion by volume (ppbv), and currently stand at 1,725 ppbv, a 115% increase. Methane concentrations in the atmosphere are rising at a much faster rate, 0.7% per year or 8 ppbv per year, than those for carbon dioxide. However the rate of increase of methane has slowed in the past decade, decreasing from 20 ppbv per year to its present value. The cause of the decrease is uncertain, but has been attributed to the effects of the Mount Pinatubo eruption in June 1991, or even to the fixing of leaking natural gas pipelines in the former Soviet Union. Methane is responsible for 19% of the enhanced 'greenhouse' effect. The current atmospheric sources of methane are variable. Methane is a by-product of biomass burning (currently 8% of annual methane gas emissions to the atmosphere), rice cultivation (22%), and enteric fermentation and subsequent flatulence in ruminant animals (15%). The latter source is attributed to cattle, which have increased in number by forty times since the Second World War. Methane is also released in coal mining (7%), natural gas drilling and transmission (9%) and as a fermentation product in landfills (8%). Natural, and potentially unstable, reservoirs of methane exist with the anaerobic decomposition of peat bogs, swamps and marshes (21%); and in clathrates consisting of ice-like hydrates of methane in permafrost (1%). Termites, and even some bacteria in bedrock, are also copious producers of methane (8%). The remaining methane entering the atmosphere is dissolved in water. Clathrates also exist on the seabed where they form an enormous reservoir. Fortunately, methane is a chemically reactive molecule in the atmosphere. With no additional anthropogenic input, methane would return to pre-industrial levels within ten to fifteen years. In contrast, the disequilibrated residence time for carbon dioxide is thirty to one hundred years, although the actual turnover time in the atmosphere is only 3.4 years. Unfortunately, if methane reaches the stratosphere, it can be chemically broken down into water vapour which can react with ozone.

Although halocarbons (any carbon gas containing fluorine, chlorine, bromine or iodine) are better

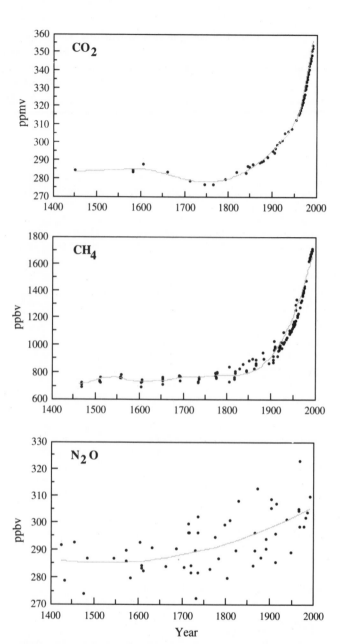

Figure 7.1 The historical record, 1400–1993, for atmospheric concentrations of carbon dioxide (CO_2), methane (CH_4) and nitrous oxide (N_2O). (Based on ice core data compiled from various sources in Boden et al., 1994, supplemented with atmospheric measurements after 1950.) Lines represent average trend lines.

known for their effect on ozone destruction in the Antarctic stratosphere, they form the third most significant category of 'greenhouse' gases. One group, chlorofluorocarbons, or CFCs, occupy a very narrow absorption band, between 8 and 13 micrometres, that is presently open to longwave transmission through the atmosphere. For this reason,

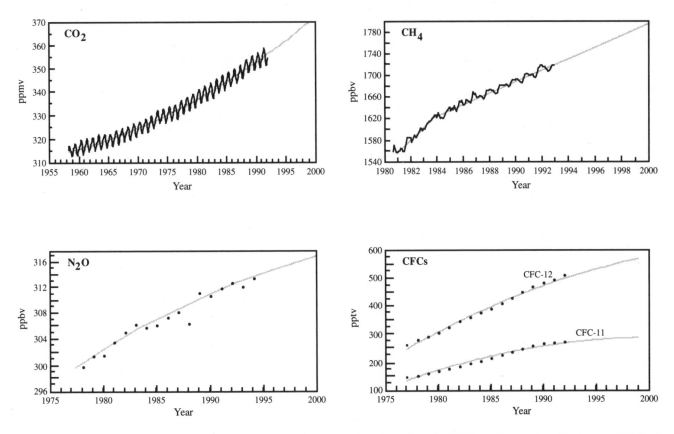

Figure 7.2 Recent measurements of atmospheric concentrations of carbon dioxide (CO_2) at Mauna Loa (Keeling and Whorf, 1994); global methane (CH_4) (Steele and Lang, 1991 and Dlugokencky et al., 1994); nitrous oxide (N_2O) at Mauna Loa (based on Watson et al., 1990); and global CFC-11 and CFC-12 (various sources). Lines represent average trend lines.

many of the CFCs have radiative forcing on a molecule-for-molecule basis that is up to eight thousand times greater than that of carbon dioxide. CFCs, because they are stable gases, also have exceptional residence times in the atmosphere, in some cases up to four hundred years. Until the Montreal Protocol of 1987, CFC-11 and CFC-12 were increasing by 4% per year. Peak rates were reached between 1985 and 1988. By 1995, CFC-11 had stopped accumulating in the atmosphere, while CFC-12 was increasing by only 10.5 parts per trillion by volume per year. To date, CFCs have been responsible for 10% of the enhanced 'greenhouse' effect; however, over the period from 1980 to 1990, the contribution to enhanced warming by CFC-11 and -12 together was 17%. The lifetimes for CFCs reported in Table 7.1 may be exaggerated. Research now indicates that CFCs are absorbed in soils, rice paddies, termite mounds and the oceans; are destroyed by soil bacteria and ocean biota; and are captured efficiently by plant lipoproteins.

CFCs are being replaced quickly by halogens which apparently do not destroy stratospheric ozone. While these compounds have similar radiative forcing potential as CFCs, they have much lower lifetimes in the atmosphere, and should not become significant contributors to enhanced 'greenhouse' warming. Some substitutes for CFCs are particularly worrisome as 'greenhouse' gases. One group, perfluorinated compounds (PFCs), presently accounts for less than 1% of enhanced 'greenhouse' warming, but has exceedingly long disequilibrated lifetimes of up to 50,000 years. Sulphur hexafluoride (SF_6) is one of most potent 'greenhouse' gas molecules, and is increasing at the rate of 7% per annum.

The fourth most important 'greenhouse' gas is nitrous oxide. Nitrous oxide has two hundred and eighty times the radiative forcing potential of a single molecule of carbon dioxide. The gas had a pre-industrial concentration of 288 ± 5 parts per billion by volume (ppbv), and currently has a concentration of 311 parts per billion by volume (ppbv) in the atmosphere, an increase of 8% over pre-industrial levels. It is currently increasing at the rate of 0.25% per year or 0.8 ppbv per year. The greatest anthropogenic source of this gas is nitrate and ammonium fertilisers, followed by biomass burning and fossil fuel combustion. Unfortunately, relative estimates are uncertain, and may underestimate by 50 to 75% the total amount of nitrous oxide added to the atmosphere over the past century. Stimulation of biological activity, because of agricultural expansion, may account for some of the missing emissions; however, natural fluctuations cannot be ignored. Nitrous oxide concentrations have a long residence time in the atmosphere, one hundred and fifty years. By itself, nitrous oxide accounts for 6% of the enhanced 'greenhouse' effect to date. When combined with other nitrogen oxides, the overall contribution to enhanced 'greenhouse' warming by this family of gases rises to 10%.

The last 'greenhouse' gas of major significance, and the one with the greatest degree of uncertainty, is ozone. Ozone not only absorbs outgoing longwave radiation, but also intercepts incoming shortwave radiation. The latter effect is described in more detail subsequently. Ozone is a very reactive gas, with a lifetime of weeks. It also varies with latitude, season and altitude. In the lower troposphere, ozone is produced through photochemical reactions over urban and industrial areas, during the formation of smog. It is often a by-product of petrochemical refining. It is also a by-product of biomass burning; ozone values over tropical rainforests reach values usually associated with industrialised areas at mid-latitudes, in summer. The actual radiative forcing of tropospheric ozone is uncertain. Few long-term measurements of ozone exist from the lower troposphere; however, there is some indication that ozone has increased by two to three times since the end of the nineteenth century. Ozone in the lower troposphere is currently increasing by 1–2% per year, but overall in the atmosphere it has been decreasing. Over the Antarctic, ozone has been decreasing on average by 1.8% per year.

Uncertainties
(Daly, 1989; Idso, 1989; Michaels, 1992; Santer et al., 1996)

The degree by which enhanced 'greenhouse' gas concentrations can raise the surface air temperature of the globe is often expressed as tropospheric radiation forcing, and measured at the tropopause or top of the atmosphere. These values for specific,

Table 7.2 Radiative forcing by increases in different 'greenhouse' gases, 1850–1994, measured as watts per square metre (based on Houghton et al., 1990, 1992).

Gas	Radiative forcing (W m^{-2})
Carbon dioxide	1.56
Methane	0.50
Nitrous oxide	0.10
CFCs, methyl chloroform, CCl$_4$	0.30
Halocarbon substitutes	0.05
Total	2.51

anthropogenically enhanced 'greenhouse' gases are listed in Table 7.2, over the period from 1850 to 1994. All totalled, the radiative forcing to date has been estimated at 2.51 watts per square metre (W m^{-2}). Note that ozone has not been included because of uncertainties regarding its concentration in the atmosphere. There are some data suggesting that the radiative forcing of ozone in the stratosphere and troposphere is -0.15 and 0.40 W m^{-2} respectively. If these values are correct than the total forcing of human 'greenhouse' gases, so far, may be as high as 2.76 W m^{-2}. If there were a doubling in equivalent carbon dioxide, this value would increase to 4 W m^{-2}. Before the 1990s, three-dimensional computer simulation models of climate predicted that the Earth's average surface air temperature should have risen by 0.6–1.0°C, in response to the increases in 'greenhouse' gases that had already occurred. The models also predicted that a doubling of carbon dioxide would lead to a global average warming of 3.0–4.5°C. This warming would cause melting of sea-ice and snow cover, and lead to a decrease in planetary albedo equivalent to 0.9–1.5 W m^{-2} for each 1°C of temperature increase.

At present, it is not possible to detect with certainty any of these changes in radiation forcing, because the resolution of satellite measurements is only ± 5 W m^{-2}. More importantly, the magnitude of the values of radiation forcing listed in Table 7.2 is questionable. The 'greenhouse' warming potential of any 'greenhouse' gas is dependent upon its partial pressure. As shown in Figure 1.1, a doubling of the partial pressure of carbon dioxide would only raise surface air temperature by 0.4°C, and not the values predicted by computer models. The historical temperature record also does not support computer simulations for the warming that should have occurred already. For instance, there should be no difference between night and day rates of temperature increase, because the 'greenhouse' effect is dependent upon the longwave radiation emission flux, and virtually independent of incoming solar radiation. However nighttime temperatures have been increasing faster than daytime temperatures. These discrepancies are discussed in more detail subsequently.

Missing carbon dioxide
(Calder, 1991; Schimel, 1995)

There are four main reservoirs in the carbon cycle: fossil carbon, the atmosphere, the oceans and the terrestrial biosphere. Two of these reservoirs are being depleted by human activity: the fossil reservoir through fossil fuel burning, and the terrestrial reservoir through deforestation. One of the most puzzling aspects about anthropogenic inputs of carbon dioxide to the atmosphere is the fact that about 50% of the carbon depleted from these two reservoirs cannot be accounted for. In the 1980s, fossil fuel burning was adding 5.5 gigatonnes (Gt) of carbon in the form of carbon dioxide to the atmosphere each year. Inputs from forest clearing ranged between 0.3 and 2.6 Gt. However, measurements in the atmosphere can account for only 3.2 Gt of this carbon. Between 2.6 and 4.9 Gt of carbon dioxide go missing each year.

Earlier suggestions that the oceans were absorbing the excess were proven to be incorrect. Detailed measurements indicate that the equatorial ocean, especially the eastern Pacific, contributes, rather than absorbs, carbon dioxide from the atmosphere. The North Atlantic Ocean is a strong absorber of carbon dioxide in winter and spring, while the North Pacific Ocean tends to have a balanced carbon dioxide budget. The rest of the world's oceans, and importantly the mid-latitude oceans of the southern hemisphere, absorb carbon dioxide all year round. This poses an interesting budget problem. The chief oceanic sinks tend to be in the southern hemisphere, while the majority of carbon dioxide is emitted by industrial activity in the northern hemisphere. Thus, there should be a noticeable export of carbon dioxide from the northern to the southern hemisphere. No such transfer has yet been detected. Overall, absorption of carbon dioxide by the oceans amounts only to 0.7–2.0 Gt per year.

Part of the missing carbon can be accounted for by refinements in the estimates of tropical deforestation. By 1988, the amount of land cleared of forest in the Brazilian Amazon basin had been put as high as 44%. However satellite measurements have revealed this to be a gross over-estimation. In 1975, 0.6% of the Amazon region had been cleared. In 1989, detailed study by the Brazilian government showed that this value had risen to only 5%. Measurements of carbon dioxide confirm this calculation. Tropical deforestation at the higher rates would produce a strong signature of carbon dioxide input near the equator. This cannot be substantiated by the observed pattern, that shows a declining north-to-south global gradient in carbon dioxide production–absorption. It now seems likely that only 1.0–1.6 Gt of carbon are being released to the atmosphere each year due to tropical deforestation. The best guess estimate of missing carbon is 1.4 Gt averaged yearly during the 1980s.

Because the global transfer of carbon dioxide is small, and no large equatorial repository can be measured, the only obvious sink for carbon dioxide must occur locally near sources of emissions in the northern hemisphere. This unknown sink has been termed 'The Great Northern Absorber'. Recent studies show that this sink is caused by the rapid absorption of carbon dioxide by growing forests, soils and river estuaries. Detailed measurements of forest volume, based upon tree-ring data, indicate that, before 1890, boreal forests were a source of carbon dioxide, as fires and clearing outstripped growth. After 1920, the forests grew faster and became carbon sinks. The suppression of wildfires, especially in North American forests, may have aided carbon storage in the northern hemisphere forests. Forest growth slowed dramatically after 1970, probably because of the effects of acid rain upon tree growth. However, the extent of this deterioration is questionable, and it would appear that boreal forests are absorbing up to one gigatonne of carbon each year. A high degree of uncertainty still surrounds this estimate.

UNCERTAINTIES ABOUT THE GLOBAL WARMING OF THE PAST CENTURY

Description of temperature changes
(Houghton et al., 1990, 1992)

The Earth's climate underwent a significant change in the latter half of the nineteenth century.

Air temperature has been emphasised as the main component of this change, however, substantial shifts worldwide in rainfall regimes, pressure patterns and sea surface temperature have also occurred. The land- and ocean-based temperature records for the past 100 to 140 years show that temperature has increased 0.40–0.45°C per century (Figure 5.11). This increase varied across this timespan. Temperatures cooled slightly between 1945 and 1977, before accelerating again in the last two decades of the twentieth century. While the rates of warming are similar between hemispheres, the change has not been contemporaneous. The southern hemisphere temperature record shows a steady increase over time that substantiates a warming hypothesis better than the northern hemisphere data. The 'greenhouse' warming effect on the globe has not been consistent with major differences between the equator and poles, and between the land and the oceans.

If the Earth's surface was warming, there should be a corresponding decrease in stratospheric temperatures to maintain the Earth–atmosphere's radiation budget. Figure 7.3 plots global stratosphere temperatures since 1958, at a level where atmospheric pressure ranges between 50 and 100 hectopascals. The record shows a tendency for cooling by −0.5 to −1.0°C over this timespan. Recently, when surface temperatures have attained their highest values on record, stratospheric temperatures have undergone their greatest cooling, by −1.0 to −1.5°C. As a first approximation, the stratospheric temperature record supports enhanced 'greenhouse' warming better than does the surface record.

The bias of surface temperature records
(Bryant, 1987; Houghton et al., 1990, 1992)

The surface temperature records are biased. Early temperature records are subject to instrumental error and improper siting. The spatial representativeness of both land and sea stations making up the record is also uneven. There are too many northern hemisphere records, and not enough from the southern hemisphere and Antarctic. Many of the trends shown in Figure 5.11 are not spatially coherent. Warming was most pronounced between 1967 and 1986 over Australia; southern parts of South Africa and South America, Alaska and northwest Canada; and most of the Soviet Union. However, cooling occurred over a large part of Europe, northeast and eastern Canada, and parts of the Antarctic coastline. The temperature changes ranged from a 2°C warming over western Canada to a 0.6°C cooling over Scandinavia. There is also evidence of substantial regional variability in both land and marine temperature records, annually and seasonally, that has not been adequately explained.

Temperature time series contain five important features that are inherent in most geophysical time series. First, although a trend may be statistically significant, if the signal-to-noise ratio is low, the trend may be simply an aberration of the record. Short-term fluctuations in temperature records,

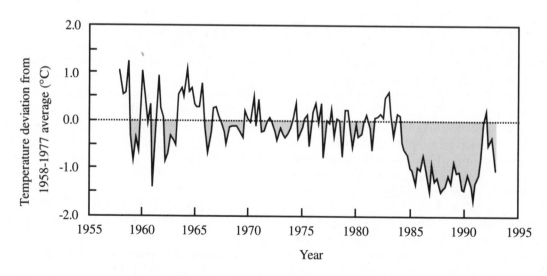

Figure 7.3 Stratospheric temperature record, between 50 and 100 hectopascals, for the period 1958–1992 (based upon Angell, 1994). Values are referenced to the mean for the period 1958–1977.

such as those in Figure 5.11, range in amplitude from 0.1 to 1.0°C. A temperature record with a standard deviation of 0.5°C requires forty-five years of data to determine a rise in temperature of 0.5°C at the 95% confidence level. If the standard deviation doubles to 1.0°C, then the length of record must be increased more than fourfold. Second, the amplitude of these fluctuations may vary over time (non-stationarity). Fluctuations are appreciable in the northern hemisphere time series between 1850 and 1890, and again between 1975 and 1985. In reality, these long-term fluctuations may be more representative of temperature behaviour than any trend. This vacillation characterises most aspects of climate, especially since 1948. Cold spells have become colder, storms more intense, rainfall events heavier, and droughts more severe. This variability has increased since 1970. In statistical terms, warming trends on the order of 0.5°C per century, determined from temperature time series containing non-stationarity, must be viewed as approximate. Predicting trends beyond the timespan of the data thus becomes virtually meaningless. Third, trends may not be continuous, but show sudden and random changes or discontinuities over relatively short periods of time. These step-functional jumps represent shifts in the mean of the record, and may occur without a change in variance. Fourth, time series may evidence periodicity. Northern hemisphere temperature records display a long-period cyclicity of approximately 65 to 70 years, whereas the southern hemisphere manifests shorter, 9 to 20-year cycles. This characteristic imparts to the record, substantial long-term serial- or autocorrelation that presents statistical problems. That is, in order to determine precisely the significance level of any trend line through the data, one must begin and end the time series at the same point in these oscillations. Otherwise, a significant rate of change can be calculated, when in fact none exists. Autocorrelation also increases the number of data points required to obtain a statistically significant trend. Additionally, sampling lower frequency fluctuations over inappropriate timespans can generate trends that are not representative of the true behaviour of the time series, even if they are long-term. Finally, geophysical time series may not be spatially contemporaneous. The peak of the Little Ice Age, which reached Europe in the 1700s and 1800s, appears to have shifted slowly across Asia from China, beginning in the 1400s (Figure 5.10).

The combined marine and land record presented in Figure 5.11 masks a major discrepancy in the marine records that has been subjected to considerable debate. Ocean air temperatures were significantly warmer than land temperatures before 1900. As a result, ocean air temperatures do not show a well defined, long-term tendency for warming or cooling. Before the Second World War, sea surface temperatures were measured using various types of buckets, while after the war they were measured by sampling the water being sucked through the cooling intake tubes to a ship's engine. Correction of the early records for this discrepancy apparently resolves the early inconsistency. However, the method of measuring ocean temperature through intake cooling tubes is also subject to bias. Engine rooms on ships plying cold waters are heated, otherwise they would be unbearable places to work. Even though water is pumped through heated intake tubes quickly, there is still the possibility that the water could be heated slightly before measurements are taken.

Solar activity versus 'greenhouse' warming
(Thomson, 1995)

In Chapter 6, it is shown that solar parameters, namely the length of the sunspot cycle, can explain a high proportion of the noise in temperature records. Also, variables such as volcanic dust are shown to be more important than carbon dioxide increases. A cause-and-effect relationship between solar parameters and temperature has not been substantiated, while the insignificance of carbon dioxide in explaining temperature changes before 1940 has been challenged. Logically, if solar activity controlled temperature, then summers should be warming more than winters, because a greater proportion of the annual solar input is received during the former season. However, present records reveal that winters are warming more than summers. Additionally, the precession cycle in the Earth's orbit is short enough that it should affect temperature timescales as short as a century. The precession cycle affects the day of the year that the Earth is tilted at its greatest angle towards the sun. The rate of precession in the Earth's orbit is 50.256" per year. At present, this should be leading to cooling in the northern hemisphere. When temperature data are replotted annually from perihelion to perihelion, rather than according to the equinoxes, long-term records in Central England, between 1650 and 1950, show a distinct cooling trend which, when converted to a phase angle, agrees within 1.0" with the rate of precession.

The northern hemisphere record of land and marine temperature, shown in Figure 5.11, has been replotted in a similar fashion in Figure 7.4. The effect of precession has been subtracted from the record. From 1854 to 1915, temperatures followed the precession cycle in agreement with the central England record. However, in the fifty years since 1940, temperatures have suddenly increased, and reached values that are up to seven standard deviations greater than those predicted by precession effects. The effect of ENSO events, volcanic activity, stratospheric aerosols and solar variability explain only a small amount of the discrepancy. However, if the exponential increase in carbon dioxide over the past century is superimposed on this adjusted temperature record, then the carbon dioxide curve fits the rate of temperature changes over the last forty years remarkably well. Such a result implies that anthropogenic 'greenhouse' enhancement of temperature may be exponential, rather than linear, as assumed in most models. If

these results are correct, then the next decade should witness temperature increases that not only are without precedent in the instrumental record, but also will continue to intensify.

But are temperatures warming?
(Bryant, 1987; Michaels, 1992; Christy and McNider, 1994)

The accelerated temperature change shown in Figure 7.4 may be simply an artifact of temperature records collected within the 'heat islands' of major urban areas, especially ones that are growing. Rather than reflecting an accelerated increase in temperature, the record may only be reflecting the quickening growth of cities. This 'heat island' effect applies to urban centres in both hemispheres with populations as low as 2,500 to 10,000 people (Figure 3.17). For instance, relative to small adjacent provincial centres, urban temperatures have increased 0.12°C per decade this century in the United States, and 0.45°C per decade since the Second World War in Australia. These rates are three to ten times the global rate, and account for all of the accelerated warming shown in Figure 7.4. It is noteworthy that cities in China do not show a consistent warming. This may be due to the shading effect of industrial pollutants, a factor that is discussed later.

There is also a major discrepancy between satellite measurements and the combined land and marine temperature record traditionally used to construct global temperature records. The lowest seven kilometres of the troposphere have been monitored, since January 1979, by microwave sensors on polar orbiting satellites. This temperature record is presented in Figure 7.5 together with the surface-based record. Also included is the satellite record, after being adjusted for the effects of the Southern Oscillation and volcanic activity. The former includes three ENSO and two La Niña events, while the latter includes two major volcanic eruptions, El Chichon in March 1982, and Mount Pinatubo in June 1991. The surface-based record displays a warming trend of 0.05°C per decade over this period. The unadjusted fifteen-year satellite record shows global cooling since 1979 at a rate of −0.04°C per decade. This result is not significantly different from zero. In fact since 1991, one of the NOAA satellites used to compile this record has been affected by upward sensor drift. After correcting for this satellite, the fifteen-year global temperature trend is about −0.06°C per decade. Considerable latitudinal variability exists in the satellite record. For

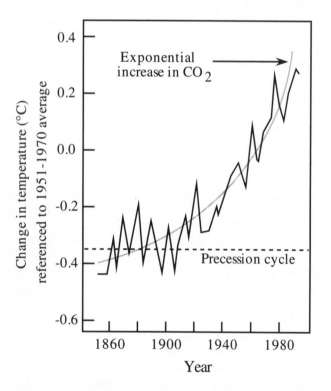

Figure 7.4 Northern hemisphere land and sea temperature record, 1854–1990, adjusted for the effect of precession of the Earth's orbit. Temperatures are plotted as deviations from the 1951–1970 average. Superimposed on the data is a best-fit projection of temperatures caused by an exponential increase in CO₂ (based upon Thomson, 1995).

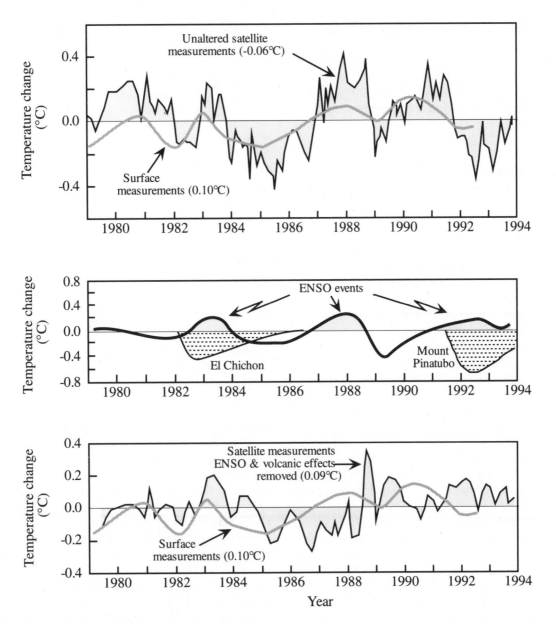

Figure 7.5 Global satellite temperature data, 1979–1993 (based on Christy and McNider, 1994). Corrections for the Southern Oscillation effect and volcanic activity are also shown. Global surface measurements are superimposed for comparison (from Figure 5.11). The changes in temperature per decade are shown in brackets for each record.

instance, while the northern hemisphere shows warming of 0.11°C per decade, the southern hemisphere shows decadal cooling of −0.15°C. If the Southern Oscillation signature is removed from the satellite record, the globe has cooled by −0.09°C per decade. If the volcanic signature is then removed, the resulting satellite record indicates a warming trend of +0.09°C per decade. However, this latter result is still about one fourth the rate predicted by climate models for the anthropogenic increases in 'greenhouse' gases

measured to date. The satellite results show clearly that ENSO events cause global warming of about 0.2–0.3°C, while La Niña events cause global cooling of about 0.2–0.4°C. Major volcanic eruptions produce global cooling of about 0.4–0.7°C. None of these perturbations to the Earth's longwave radiation budget appears to last more than two years, although Southern Oscillation changes are certainly more frequent than volcanic eruptions. More significantly, the unadjusted global surface record (Figure 5.11), for the same timespan, shows a temperature change of +0.10°C per decade, compared to a satellite record that is decreasing by −0.04 to −0.06°C per decade. This is a statistically significant difference (0.05 level of significance), and clearly indicates that surface-based measurements are either biased, or underrepresenting globe temperatures.

The role of clouds
(Henderson-Sellers, 1992; Michaels, 1992; Karl et al., 1993)

Over the past century, the diurnal temperature range (the difference between maximum and minimum daily temperatures) has decreased at the same rate as the mean temperature has increased. Analysis of 2,000 stations distributed globally, over a shorter timespan of four decades, indicates that, on average, nights are 0.84°C warmer in contrast to days which are only 0.28°C warmer. This implies that almost all of the warming detected by ground-based stations has been due to a rise in nighttime temperatures. If the warming were due to the enhanced 'greenhouse' effect, then both night and day temperatures should have risen in parallel reflecting the impact of these gases on the longwave radiation flux. The nighttime warming effect cannot easily be put down to urban 'heat island' effects, because the trend occurs even for rural stations free of development and having populations of less than 10,000 people.

Statistical relationships indicate that cloud cover, and probably low cloud cover, is responsible for much of this pattern. The correlation between cloud cover and diurnal heat range in the United States is seasonally dependent, ranging from −0.72 in spring to −0.96 in autumn. Cloud cover is the key regulator of nighttime temperatures, because of its control on longwave emission. Cloud cover also decreases the amount of shortwave radiation entering the atmosphere during daylight hours, and hence can minimise any daytime temperature increases.

Cloud cover changes, representing the world's oceans, industrialised areas, and semi-arid regions, are plotted since 1900 in Figure 7.6. Overall in this century, cloudiness has increased 3.4–9.4% wherever records have been analysed. While the margin for error is large, observations are consistently detecting more cloud. This feature could occur without global warming; however, it is difficult to hypothesise global warming without an increase in cloudiness. All records, except for Europe, show a sudden increase in cloudiness during the rapid

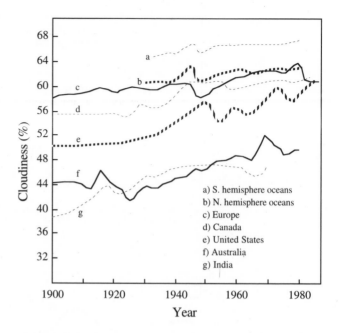

Figure 7.6 Change in cloudiness this century over the oceans, and at various locations on land (based on Henderson-Sellers, 1992 and Michaels, 1992).

warming of the late 1930s and early 1940s. The oceans are generally cloudier than continents. Additionally, there is more cloud as one moves into cooler climates (Canada and Europe being cloudier than United States or Australia). This is logical given the lower dew point of cooler air. In the United States, cloudiness increased by 3.5% between 1950 and 1988, while for North America the changes have averaged 8% this century. Except for Europe, the increase in cloudiness appears to have occurred at lower rates over the oceans than over the continents. Ship records indicate that there has only been a 2–3% increase in cloudiness since 1940 over the oceans. This difference alludes to an additional factor enhancing cloud formation over land, namely increased nuclei due to air pollution.

The role of sulphate aerosols
(Bridgman, 1990; Karl et al., 1993; Benkovitz et al., 1994; Pearce, 1994; Hadley Centre, 1995)

Until the 1990s, anthropogenic sulphate aerosols were considered to be a minor factor in forcing long-term climate change, mainly because their residence time in the atmosphere was limited to seven to ten days. However, the presence of acid sulphate rain outside industrial areas of the northern hemisphere indicated a more widespread dispersal of anthropogenic sulphates than previously realised. The atmosphere of the North Atlantic Ocean has become measurably more polluted over the past fifty years, in contrast to the air over the oceans of the southern hemisphere, which are still relatively pristine. Sulphur compounds in the atmosphere take four different chemical forms (Table 7.3). Dimethylsulphides are solely a product of phytoplankton emissions. Of the 138 × 10^6 tonnes of natural sulphate found in the atmosphere, only 2% can be related to anthropogenic sources. The rest is emplaced by sea spray. Sulphur dioxide and hydrogen sulphide are injected into the atmosphere by volcanoes and human industrial activity. The anthropogenic flux of sulphur dioxide into the Earth's atmosphere is now ten to twenty times greater than that of volcanoes.

Industrial sulphur compounds take two forms in the atmosphere: dry and wet states. These sulphates are produced mainly by chemical reactions of sulphur dioxide in the lower troposphere. Dry sulphates form when sulphur dioxide undergoes chemical reactions over several days, in a cloud-free atmosphere. Solid or liquid particles are formed that are highly reflective. Wet sulphates form when sulphur dioxide, which is hygroscopic, interacts with water droplets, mainly in clouds. When the water evaporates, then ammonium sulphate is left behind, and is deposited on other particles in the atmosphere.

This makes cloud denser and brighter. The effect is controlled by the nature of emplacement in the atmosphere. Before the Second World War, most industrial sulphates were released into the atmosphere from low smokestacks. Particulates fell out of the atmosphere locally, making urban areas heavily polluted. After the Second World War, taller smokestacks ejected sulphates higher into the atmosphere, and spread pollutants over ever larger areas downwind. If sulphates form clouds within their seven- to ten-day residence time, then the cooling effect can persist longer, and be spread over a much wider area. Some of the largest increases in cloudiness (Figure 7.6) are downwind of major industrial sulphate source regions in North America and Asia. Satellite pictures show plumes emanating from point sources in North America, and spreading over the Atlantic and into the Arctic. Darker, low-level hazes often characterise skies over industrial regions, as well as over rural areas in Europe, eastern Asia, South America and Africa.

The high reflectivity of most industrial sulphates provides a continuous mechanism, under both clear and cloudy skies, for the cooling of the troposphere below two thousand metres. Table 7.4 estimates the magnitude of radiative forcing for both natural and anthropogenic sulphates in each hemisphere. Globally, the magnitude of the human component of sulphate radiative forcing now equals the natural component. In the northern hemisphere, the anthropogenic contribution is now 50% greater than the natural effect. Over

Table 7.3 Sulphur (S) compounds affecting climate (based on Bridgman, 1990).

Compound	Chemical formula	S produced (10^6 tonnes y^{-1})	Concentration	Lifetime in days	Climatic effects
Dimethylsulphide	$(CH_3)_2S$	27–56	<10 pptv	0.6	cloud formation cooling
Sulphur dioxide	SO_2	103	24–90 pptv	2–4	shortwave reflection cooling
Natural sulphate	SO_4	138	0.1 μg m^{-3}	7	shortwave reflection cooling
Hydrogen sulphide	H_2S	–	30–100 pptv	4.4	converts to SO_2

pptv   parts per trillion by volume
μg m^{-3} micrograms per cubic metre

Table 7.4 Radiative cooling due to sulphate aerosols, expressed in watts per square metre (Kiehl and Briegleb, 1993).

Location	Natural	Anthropogenic	Total
Northern hemisphere	−0.29	−0.43	−0.72
Southern hemisphere	−0.25	−0.13	−0.38
Globe	−0.26	−0.28	−0.54

very extensive industrial areas, forcing may exceed −4.0 W m⁻², equal, but opposite in sign, to the value postulated for an effective doubling of carbon dioxide. In all, human and natural sulphates in the northern hemisphere account for an average forcing by −0.72 W m⁻². Globally, the effect of all sulphates is −0.54 W m⁻² or about 20% of the calculated forcing of all anthropogenically enhanced 'greenhouse' gases at present (2.5 W m⁻²).

The regional effect of emissions of anthropogenic sulphates upon temperature is shown in Figure 7.7. Superimposed on this figure is the calculated shading effect, in degrees Celsius, caused by these sulphates. Figure 7.7 still underestimates the extent of sulphate aerosols, because there is clear evidence that sulphate production from the burning of coal and oil, over eastern North America and eastern Asia, is dispersed by easterly moving weather systems over adjacent oceans. In addition, sulphate cooling at the surface can alter the path of the wintertime jet stream over Canada, causing more frequent and stronger storms in eastern North America. The sulphate plume from North America often loops over the North Atlantic, and mixes with sulphate-rich air over western Europe. The western European plume tends to drift southeasterly covering the Black and Caspian Sea regions. Even in Africa and South America, anthropogenic sulphates due to mining and smelting activities cool climate. Overall, significant areas of the northern hemisphere landmasses have a negative radiation forcing in excess of 2 W m⁻², because of anthropogenic sulphate production. Over eastern China, the figure can rise to −7.2 W m⁻² in some locations, while in central Europe it can reach a value of −11 W m⁻². Presently anthropogenic sulphate aerosols are depressing temperatures in eastern North America, southwestern

Europe and eastern China by 1.0–1.5°C.

The increase in sulphate aerosols parallels the increase in carbon dioxide over the past century. The industrial sulphur record, analogous to sulphur dioxide air pollution, is plotted in Figure 7.8. The graph clearly shows that sulphur dioxide emissions have risen exponentially over time. Production briefly dropped during the Second World War, and slowed after 1960, because of either the introduction of air pollution control measures in the 1970s in westernised countries, or the collapse of polluting economies in eastern Europe at the end of the 1980s. Increases in temperature at these times, in North America and western Europe, lend support to the role that sulphate aerosols have in depressing temperatures. Finally, sulphates by themselves do not explain all the cooling over this timespan. Because sulphates increase the reflection of solar radiation, their effect should be greatest during the daylight hours, in summer, and in the northern hemisphere. The slower increase of daytime and summer temperatures in records supports this sulphate cooling scenario. However, nights are also warming much faster than days in Australia, where there is no major sulphate aerosol problem. This fact suggests that sulphate aerosols are only exacerbating a more widespread cooling process, most likely due to increased cloudiness.

The role of dust
(Bryson and Murray, 1977; Bryant, 1991; Pearce, 1994; Tegen et al., 1996)

Human activity is also a major contributor of dust to the atmosphere. This effect was espoused in the 1970s during a debate about global cooling. At this time, it was postulated that humans were enhancing the dust content of the atmosphere at an accelerating rate, through industrial and agricultural activity. The evidence for the scenario was impressive. Anthropogenically enhanced 'greenhouse' warming could be observed in temperature records, similar to those plotted in Figure 5.11, before 1945. However, after 1945, this warming was overwhelmed by the cooling effect of dust, and, as a result, surface temperatures dropped by 0.5°C.

Both agricultural and industrial dust is easily suspended and mixed through the atmosphere. At present, substantial amounts of dust are blown from subarid and desert regions, and deposited over the oceans five thousand to ten thousand kilometres from these source regions. Dust rises into the upper troposphere where it absorbs heat, but

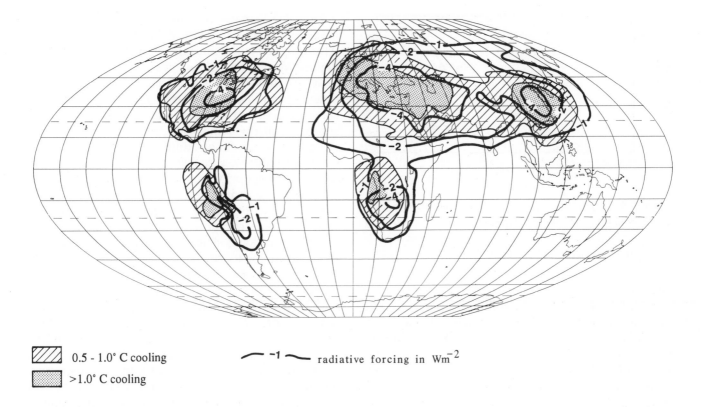

0.5 - 1.0° C cooling

\>1.0° C cooling

— ⁻¹ — radiative forcing in Wm⁻²

Figure 7.7 Global distribution of radiative forcing by anthropogenic sulphates, averaged over the period 1970–1990, and the magnitude of resulting cooling effect in °C (based on Hadley Centre, 1995 and Houghton et al., 1995).

blocks incoming solar radiation. At the top of the atmosphere, the effect of dust on solar radiation is -0.25 W m⁻², while its effect on longwave radiation is $+0.34$ W m⁻². The global mean forcing of dust, at the ground, is -0.96 W m⁻², rising to more than 8 W m⁻² over such arid locations as the Arabian peninsula and adjacent sea (Figure 7.9). As a result of dust, the ground surface is slightly cooler than normal during daytime, resulting in less convection. At night, high altitude dust radiates longwave radiation in the upper atmosphere, cools and sinks. At the ground however, the dust traps in longwave radiation that would normally escape from the Earth's surface, causing the air above the surface to remain warmer than normal, thus preventing dew formation. The sinking air aloft leads to conditions of stability, while the lack of dew keeps the ground surface dry and friable, conditions favouring aridity. The recent increases in nighttime temperatures relative to daytime ones support this hypothesis.

In the 1970s, archaeological evidence was used to show that some of the earliest civilisations and societies, reliant upon agriculture in semi-arid regions, collapsed because their poor land management practices led to increased atmospheric dust and aridity. Insidiously, arable topsoil was blown away, and the slow process of marginalisation or desertification of semi-arid land took place. The long-term social consequences involved the dislocation of communities, famine, and ultimately the destruction of civilisations dependent upon such areas for their existence. This was the fate of early, agriculturally dependent civilisations in Mesopotamia, reliant upon the Euphrates and Tigris Rivers for their water. It was also argued that, in the present era, there were deserts in humid regions maintained by this dust effect. Nowhere was this exemplified more than in the Rajputana Desert on the Indian subcontinent. Historically, the Rajputana was one of the cradles of civilisation, with a well developed agrarian society based on irrigation around the Indus River. Successive cultures occupied the desert, but always they collapsed. The Rajputana area is affected by monsoons, and much of the air is humid; however, rainfall generally amounts to less than 400 millimetres per year. At the time of the development of

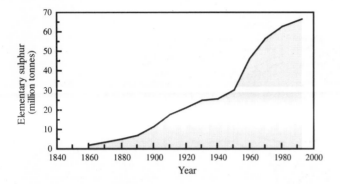

Figure 7.8 Proxy measure of sulphur dioxide emissions 1860–1990 (data from Dignon and Hameed, 1989). Note the exponential increase in industrial sulphur compared to that for carbon dioxide plotted in Figure 7.4.

civilisation by the Harappan culture around 4,500 BP, rainfall exceeded 600 millimetres per year. The dust put into the atmosphere by agricultural activities inhibited convection and rainfall, leading to the collapse of this, and successive, civilisations. Extrapolated to the modern world, these case studies implied that anthropogenic dust from industrialisation, intensifying agriculture, and accelerated burning of tropical rainforests in developing countries, was creating a global cooling effect.

Historically, human activity at the margins of agriculture has resulted in dust storms of enormous proportions. Between 1951 and 1955, 3,882 dust storms were recorded in central Asia. In Turkmenistan, 9,270 dust storms have been recorded over a twenty-five-year period. Dust storms have had a notable and extensive impact. For instance in April 1928, a dust storm affected the whole of the Ukrainian steppe, an area exceeding one million square kilometres. Up to fifteen million tonnes of black chernozem soil was removed and deposited over an area of six million square kilometres in Rumania and Poland. In the affected area, soil was eroded to a depth of 12–25 centimetres in some places. Globally, between 130 million and 800 million tonnes of dust, with amounts as high as 5,000 million tonnes, are presently entrained by winds each year. It is debatable whether or not the incidence of dust storms is increasing as a result of growing populations. Chinese data show no such increase; but intense cultivation of the United States Great Plains in the 1920s, the opening up of the Virgin Lands in the USSR in the 1950s, and the Sahel drought of the 1960s–1970s, have led to an increase in the frequency of dust storms in these locations.

Figure 7.9 maps the present frequency, source areas and tracks of main dust storms worldwide. Also mapped is the amount of dust currently being deposited in the world's oceans. A broad band of dust, amounting to over ten tonnes per square kilometre per year, occurs up to 1,000 kilometres off the east coast of Asia. A similar swathe extends southeast of the Indian subcontinent and through Southeast Asia. High dust transport rates also occur off the west coast of North Africa. These tracks correspond to areas with the highest modelled negative radiative forcing due to atmospheric dust. Significant fluxes exceeding one tonne per square kilometre per year surround most of the coastline of Asia, Australia, tropical Africa, and the eastern coasts of the United States and South America. Fluxes of this magnitude cover the central Atlantic Ocean between South America and Africa. These areas are not necessarily downwind of deserts or cultivated semi-arid regions. They represent the global signature of dust output due to both industrial and agricultural activities. Carried with this dust are substantially increased amounts of iron, both from industrial activity and soil weathering. In addition to atmospheric cooling caused directly by dust, windblown material indirectly may be cooling the atmosphere by 'fertilising' the growth of phytoplankton, and with it, increasing dimethylsulphides and cloudiness in the marine atmosphere.

MODELS OF 'GREENHOUSE'-WARMED CLIMATE

General circulation using computer models

Model results
(Mitchell et al., 1990; Washington, 1992)

Most of the forecasts of the climatic impacts of 'greenhouse' warming have been based upon computer simulation models, or GCMs. For an equivalent doubling in carbon dioxide, all GCMs predict five outcomes: an increase of mean surface temperature which is more pronounced at night (although not as great as that which has been measured recently), in winter or at the poles; an increase in precipitation, especially during the cooler season; more severe and longer lasting droughts during the warmer part of the year; a greater portion of warm-season precipitation originating from heavy convective showers or thunderstorms; and a decrease in the spatial and

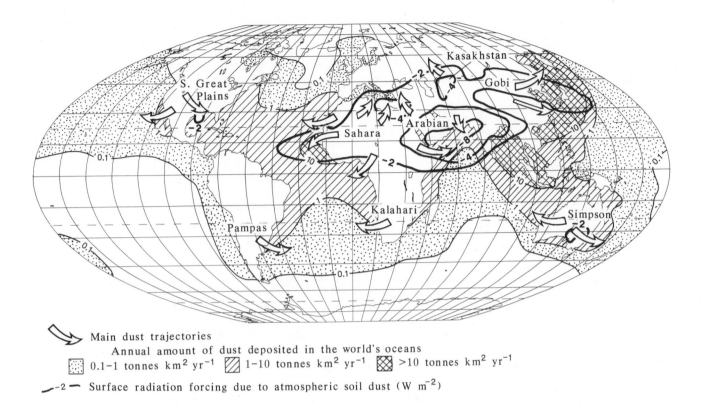

Figure 7.9 Trajectories of dust storms and annual amount of atmospheric dust transported to the world's oceans (based on Bryant, 1991 and Pearce, 1994). Also mapped is the radiative cooling effect at the Earth's surface due to soil dust (from Tegen et al., 1996).

temporal variability of temperature, for example, a lessening in the temperature gradient from the pole to the equator, or day-to-day temperature fluctuations. Specifically, the IPCC predictions for a doubling of carbon dioxide indicate an average surface warming of 1.5–4.5°C over the next forty to sixty years, with a best guess estimate of 2.5°C (Figure 7.10).

Temperatures predicted by a typical model, the United Kingdom Meteorological Office high resolution GCM (UKHI), are shown for winter and summer in Figure 7.11. This model forecasts an average global warming of 3.5°C for an equivalent doubling of carbon dioxide, a value that is about average for the seventeen different models used in the IPCC 1990 predictions (Figure 7.10). In the boreal winter/austral summer (December, January, February) temperatures will warm by as much as 8°C over the North American Great Lakes, Siberia, the Baltic Sea and the European Arctic. Almost all

landmasses in the northern hemisphere will undergo at least a 4°C warming. Very little warming above 2°C will occur at this time of year over the oceans, or in the southern hemisphere. In the austral winter/boreal summer, (June, July, August) warming by 8°C will occur in the middle of Asia, and in the southern Atlantic offshore from Antarctica. Parts of the Antarctic icecap will warm by 4°C. Elsewhere in the southern hemisphere, warming will be less than 2°C. Large sections of northern Africa, North America and Asia will undergo at least a 4°C increase in temperature in their summertime. These projections exhibit considerable inconsistency, regionally and amongst different models. They also do not include the negative feedback effect of anthropogenic dust and sulphate aerosols.

Because warmer surface temperatures enhance evaporation, global average precipitation will increase. The relationship between the magnitude

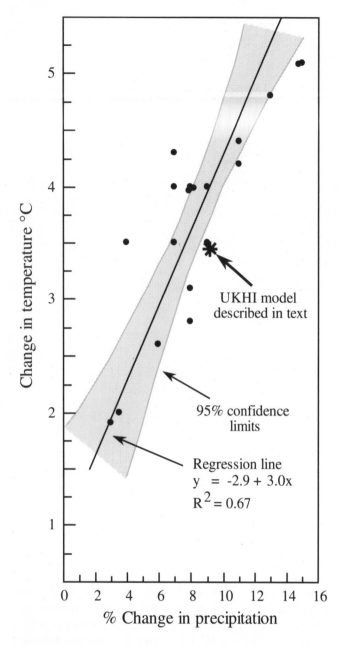

Change in temperature °C

UKHI model
described in text

95% confidence
limits

Regression line
$y = -2.9 + 3.0x$
$R^2 = 0.67$

% Change in precipitation

Figure 7.10 Linear relationship, with 95% confidence limits, between temperature and precipitation changes using seventeen different GCM models (based on Mitchell et al., 1990).

of warming and the increase in precipitation is linear (Figure 7.10). Increases in precipitation will range from 3 to 15%. Historical records indicate a global increase of 3.2% for each 1°C increase in temperature. The computer models support this relationship. While the variability in predictions is high amongst different models, generally, precipitation will increase throughout the year at high latitudes and in winter at mid-latitudes. There is only a small increase in precipitation in the tropics,

and little if any change in the subtropics beneath the Hadley pressure belt. The simulated changes in global precipitation for the United Kingdom Meteorological Office high resolution GCM (UKHI) are shown for winter and summer in Figure 7.12. This model forecasts a 9% increase in precipitation for a doubling of carbon dioxide. In the austral summer/boreal winter (December, January, February), there is a tendency for rainfall to decrease slightly over tropical and subtropical oceans. Decreases will be as great as 5 millimetres per day over the mouth of the Amazon basin, and the west Pacific warm pool. Generally, the landmasses of the northern hemisphere will undergo minimal changes in winter precipitation. In the boreal summer/austral winter, minor decreases of up to one millimetre per day will take place over the subtropical regions of northern hemisphere landmasses. Decreases greater than two millimetres per day will occur in the Caribbean Sea, and the equatorial west Pacific Ocean. There is substantial inconsistency regionally amongst different models, as well as considerable spatial variability in the predictions.

Because the mean temperature and precipitation generally will increase everywhere, heavier rainfall events and heat waves will increase under enhanced 'greenhouse' warming. Many 'greenhouse' forecasts also predict an increase in the frequency of tropical cyclones and mid-latitude storms. While increases in convective cells will occur because of enhanced surface heating, the occurrence of extreme regional storm events is not necessarily a certainty. For one thing, the temperature gradient that presently exists between the tropics and the poles (Figure 7.12) will be reduced under a warmer globe. Tropical storms and mid-latitudes storms transfer excess energy from the tropics to the poles. If this necessity to balance the heat deficit at the poles diminishes, than logically so must the number of these storms. However, their frequency may increase in the transition from a cooler to a warmer globe. The ocean lags considerably behind the atmosphere in its response to surface warming. Thus, it is possible for cooler water to persist at mid- to high latitudes, while the air above has warmed. This situation will lead to an increased probability of the formation of mid-latitude depressions and east coast lows. The record of severe storms in northern Europe, leading to the Little Climatic Optimum of the Middle Ages supports this possibility (Figure 5.10). Storms appear more likely to occur during

Part A

Part B

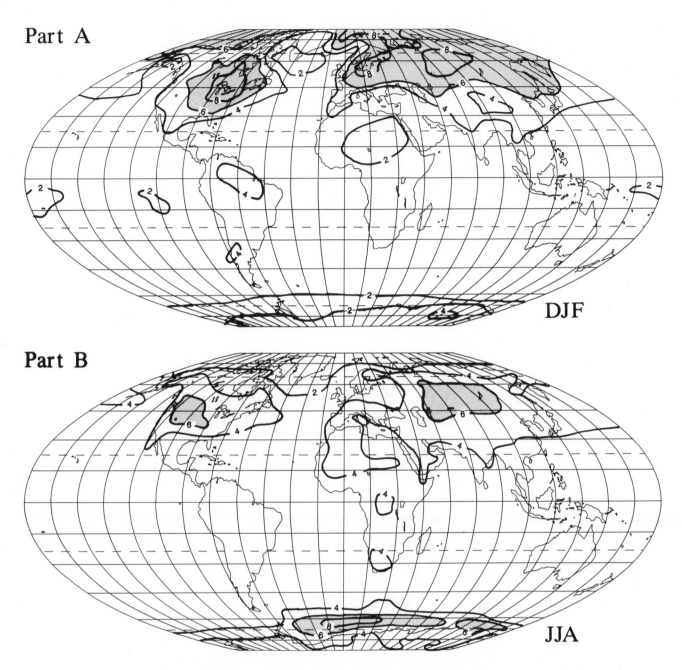

Figure 7.11 Simulated global temperature increases (°C) for winter and summer seasons for an equivalent doubling of carbon-dioxide, using the United Kingdom Meteorological Office high resolution model (based upon Mitchell et al., 1990): (A) December, January, February; (B) June, July, August. Note that these predictions include some of the warming that has occurred over the last hundred years. Increases in temperature greater than 6°C are shaded.

changeovers in climate. Thus the initial signature of the onset of enhanced 'greenhouse' warming may be the current global increase in storminess.

For another thing, while the area of ocean with warmer temperatures conducive to tropical cyclone formation will increase, this does not necessarily imply more and bigger tropical cyclones.

Chapter 3 outlines seven requirements for the formation of tropical cyclones. Within an enhanced 'greenhouse'-warmed world, winds in the upper troposphere will increase, a condition that tends to dissipate the vertical air circulation necessary for tropical cyclone eye development. Moreover, the frequency of tropical cyclone formation is not

Part A

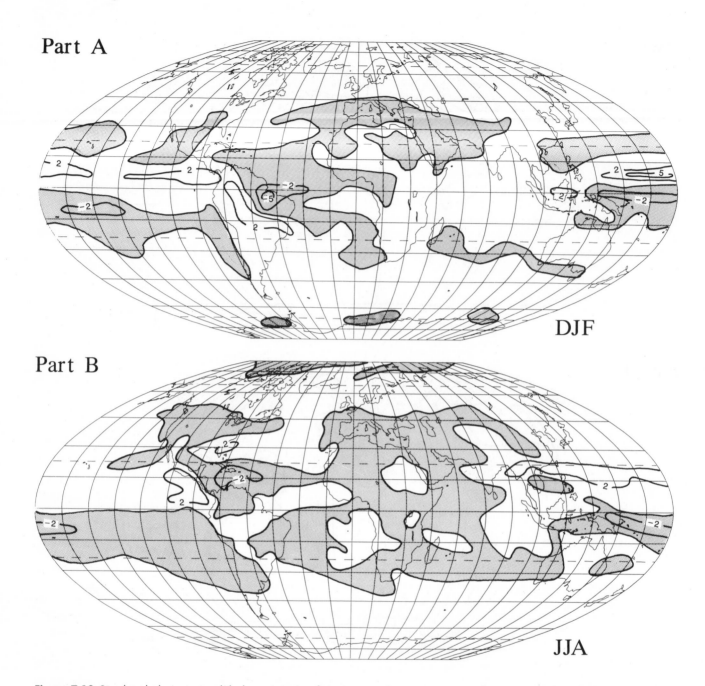

DJF

Part B

JJA

Figure 7.12 Simulated changes in global precipitation for winter and summer seasons for an equivalent doubling of carbon dioxide, using the United Kingdom Meteorological Office high resolution model (based upon Mitchell et al., 1990): (A) December, January, February; (B) June, July, August. Values represent average changes in millimetres per day. Reductions in precipitation are shaded.

a direct linear response to increased ocean temperature. Above 30°C, convection becomes too intense, and widespread, to allow for the formation of a coherent organised cell with the dimensions required to produce a tropical cyclone of 400–1,000 kilometres in diameter.

Because of increased evaporation, soil moisture deficiencies should be more severe. However, the trend is offset in many locations by increased precipitation. The extent of snow cover should diminish, despite slightly increased precipitation at higher latitudes. In addition, as the oceans warm up, sea-ice cover will diminish in extent. Both effects cannot be modelled adequately at present, and there could be significant feedback mechanisms on regional climates as a result of these changes. The Antarctic icecap is one area that may not witness less snowfall. The average temperature of this icecap is −22°C. Temperature increases of 4°C will not melt much more ice than is presently melted there in summer. However, computer simulation indicates that the oceans around the icecap could warm by up to 8°C. This will increase evaporation from open ocean waters, increase the amount of moisture in the atmosphere falling as snow, and enhance icesheet accretion. For this reason, early projections of a 1.5-metre rise in global sea-level by the year 2030, caused by a partial melting of the Antarctic icecap, later were lowered significantly to rises of only 8–29 centimetres, with a best-guess estimate of 18 centimetres. Even these rises are questionable, given the inherent variability of sea-level and the fact that sea-level responds so erratically to a myriad of climatic parameters, all of which will be changing on a forecast warmer globe.

Regional predictions
(Bryant, 1990; Karl et al., 1995)

Regional predictions of climate under enhanced 'greenhouse' warming often rely upon GCM output. However, because the spatial resolution of GCMs is low, these are often supplemented by scenarios based on past analogues. For instance, in the United States, possible climatic signatures of enhanced 'greenhouse' warming include an increase in mean temperature, especially in winter, and a warming of nighttime temperatures relative to daytime ones, mainly in summer. In addition, there will be an increase in the number of stations having long-term records reporting minimum temperatures within the upper tenth percentile. While extremes at the warmer end of the temperature scale may increase, the day-to-day variability will decrease. These projected changes can be considered benign. Precipitation will increase mainly in the US winter months, October through April, because of an increase in atmospheric instability under warmer temperatures. Extremes in precipitation will also increase, with more and heavier convective rainfall (more than five centimetres per day), especially in the warm season. Since 1976, several of these indicators, when integrated over time, have increased by 2.8% above the average of previous years. This trend is consistent with enhanced 'greenhouse' warming, but not yet statistically above what could be expected under a stable temperature regime.

In Australia, climate by the year 2030 will display increases in mean temperatures of 1–4°C, with rises being greater in the south, in the inland and in the dry season. Rainfall should increase by 10–20% in the summer rainfall region, which includes most of the northeast part of the continent, but decrease by 10% in the winter rainfall region, mainly in the extreme southwest. Generally these changes bode well for the agricultural productivity of the continent. In the tropics, rainfall events and the northern monsoon would be more intense. Because of the tropical and subtropical position of the continent, evaporation would increase by 5–15%. Tropical cyclones could become more variable and possibly intensify; however, this point has been intensely debated. Trade winds would become weaker, but westerlies at the southern extremity of the continent, and winds associated with convective storms, would become stronger.

In Europe, one possibility is that precipitation will increase in the North Atlantic. As a consequence surface water will become less salty and the North Atlantic salinity sink will slow down. A doubling of equivalent carbon dioxide in models led to a 30% reduction in the rate of formation of deep water in this region. Any change of this magnitude would have an impact on the temperature of northern Europe.

Limitations of GCM
(Idso, 1989; Ramanathan et al., 1989; Schlesinger, 1991 Santer et al., 1996)

While GCMs appear credible, they are far from perfect. Many of their limitations are discussed in Chapter 3. Other imperfections relevant to enhanced 'greenhouse' warming are presented here. Presently, GCMs have poor spatial resolution, and fail to account adequately for the dampening effect of the ocean and sea-ice upon air temperature fluctuations. Models in the 1980s often considered the ocean as a swamp, giving it a depth of 50 metres instead of the 4,000 metres it really has. The first attempts to couple the ocean and the atmosphere in a GCM led to a cool southern hemisphere, because of strong upwelling of cold bottom water close to Antarctica. In the North Atlantic, rain tended to dampen any warming caused by an influx of warmer subtropical surface waters to that region.

Computer simulation models have also underestimated the degree of evaporation from the ocean waters induced by 'greenhouse' warming. Significantly, the models simulate inadequately the negative feedback mechanisms produced by the additional cloud, and moisture, present in the atmosphere under a warmer climate. There has been an erroneous perception that increased cloudiness will enhance the 'greenhouse' effect. However, when realistic cloud parameters are incorporated into GCMs, the results always lead to suppression of extreme temperatures. As pointed out already, cloud forcing is fifteen to twenty times greater than the 2.5 watts per square metre presently attributed to enhanced 'greenhouse' warming. Even in polar regions, where computer models indicate that 'greenhouse' warming will be greatest because of diminished reflectivity from ice, the exposure of open ocean to the atmosphere should lead to more cloud, which will lessen the tendency towards increased temperatures. It now appears that a 4% increase in global average cloudiness, or a 10% increase in the amount of low-level cloud, is all that is needed to offset the warming associated with an equivalent doubling of carbon dioxide. The trends presented in Figure 7.6 indicate that such changes are quite feasible over timescales of fifty to eighty years. These type of observations, at their lowest level, led to a reduction, from 5.2°C, to 3.5°C, of the increase in simulated global mean temperature for an equivalent doubling of carbon dioxide using an earlier version of UKHI computer simulation model.

More fundamentally, the carbon dioxide-induced warming effects simulated by present GCMs are driven by positive water vapour feedback (Figure 7.12). Between 1973 and 1990, tropospheric water vapour increased globally by as much as 13% per decade. Thus, much of the observed warming of recent years could be attributed to increased water vapour in the atmosphere, rather than to increased anthropogenic 'greenhouse' gases. Finally, there is little attempt in GCMs to model realistic atmospheric dust patterns; similarly, diurnal variations in solar radiation, ocean heat capacity and ocean circulation are also poorly modelled.

Combined 'greenhouse' gas and sulphate aerosol models
(Hadley Centre, 1995)

The effect of sulphate aerosols on temperatures has recently been investigated using a newer GCM model at the Hadley Centre United Kingdom Meteorological Office, and the sulphate aerosol distribution mapped in Figure 7.8. The model was calibrated by comparing it to observed surface air temperature changes since 1900. Global temperature changes produced in the model showed an erratic rise that averaged 0.5°C over the past century (Figure 7.13). This is slightly greater than the 0.45°C per century rate actually measured, but less than the 0.6–1.3°C estimates produced using only anthropogenic 'greenhouse' gases. The inclusion of sulphates reduces the effects of enhanced 'greenhouse' warming by about 30% globally. Specifically, the revised model shows temperatures remaining constant between 1860 and 1920, rising quickly by 0.3°C during the 1930s and 1940s, stabilising with the post-war industrial boom until 1970, and then taking off again with a 0.2°C rise since then. Industrialisation triggered a carbon dioxide warming effect early on, followed by sulphate aerosol temperature suppression after the Second World War. Continuing 'greenhouse' gas emissions have recently overwhelmed the negative influence of sulphate aerosols. Note that the agreement between simulated and actual temperatures over the past century, shown in Figure 7.13, was achieved without considering any of the volcanic, solar or Earth orbital effects, presented in Figures 6.6 and 7.4, that were incorporated into the earlier models.

Figure 7.14 plots the globally averaged increases in temperature due to an equivalent doubling of carbon dioxide and predicted sulphate aerosol production for the year 2050, simulated by the

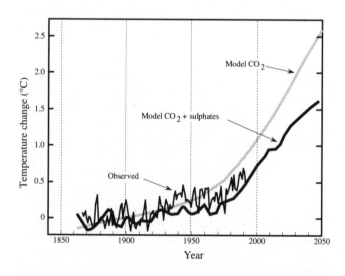

Figure 7.13 Predicted temperature changes, 1860–2050, from the United Kingdom Meteorological Office Hadley Centre GCM for an equivalent doubling of carbon dioxide, with and without sulphate aerosols (based on Hadley Centre, 1995). The measured global record, 1854–1993, is presented for comparison (data from Figure 5.11).

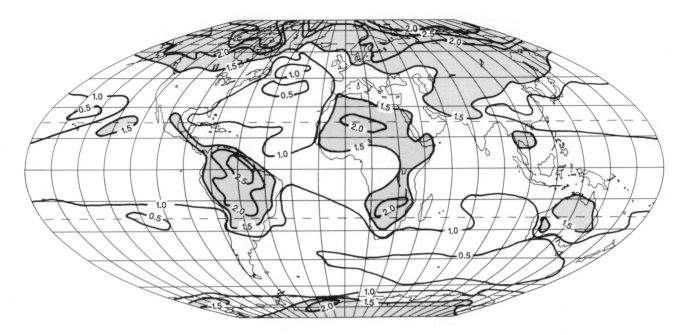

Figure 7.14 Simulated global annual temperature increases (°C), 2030–2050 versus 1970–1990, using the United Kingdom Meteorological Office Hadley Centre GCM (based on Hadley Centre, 1995). Model incorporates an equivalent doubling of CO_2, plus projected sulphate emissions. Increases in temperature greater than 1.5°C are shaded.

Hadley GCM. Sulphate aerosols will increase less rapidly over present-day industrial areas, and undergo notable increases over developing countries, with aerosol forcing increasing by a factor of three times over southern Asia. Without including sulphate aerosols the model predicts an average 2.9°C temperature increase for the globe by the year 2050. This represents a 2.4–2.5°C temperature increase above present 1990 values. Inclusion of sulphates reduces this projected value to 1.2–1.3°C, equivalent to an increase of about 0.2°C per decade. These temperatures surpass those characterising the Mediaeval Warm Period, and begin to match those of the Holocene Climatic Optimum. The results also indicate that it will take another decade before any global warming signal can be detected above the ±0.2°C 'noise' level, inherent in global temperature records.

Specifically, the dominant forcing in the GCM is produced by the increase in enhanced 'greenhouse' gases, which have much longer lifetimes than sulphate aerosols. The effect of sulphate aerosols are regional, and lower the enhanced 'greenhouse' effect most noticeably over industrial areas. The whole globe undergoes some degree of warming; however, warming is less over mid-latitude oceans, and greatest over the northern hemisphere pole. The highest temperature

increases occur in the latter region. Much of Asia will warm by 1.0–1.5°C, with significant warming also occurring over the Amazon basin, the Sahara desert and over north-central Australia. Thus the impact of enhanced 'greenhouse' warming will be most apparent over rural and underdeveloped agricultural regions, rather than over more populated areas. Local reductions in aerosols, because of the implementation of stricter pollution control measures, could lead to considerable local warming. This effect may already be occurring in northern Europe, mainly as a result of the recent closure of inefficient, and heavily polluting, industry in Eastern Europe.

General circulation using analogues
(Leroux, 1993)

Three different circulation models are used in this text, to explain a significant portion of air movement: the Palmén–Newton model (Figure 3.4), the model of mobile polar highs (Figure 3.11), and the Southern Oscillation model (Figure 4.7). Most GCMs, in presenting output in terms of averages, link climate change to the Palmén-Newton model. Verification of enhanced 'greenhouse' warming conditions has been assessed by comparing expected changes to past analogues of warmer climate such as the Holocene Climatic Optimum, and the peak of the Last Interglacial, around 6,000 and 125,000 years BP, respectively. However, these comparisons suffer from a scarcity of information about atmospheric circulation at these times.

From first principles, both the temperature gradient and energy imbalance in the atmosphere, between the poles and the equator, will diminish under enhanced 'greenhouse' warming. Thus there will be less need for dramatic transfers of heat by mid-latitude storms or cyclones. This is in direct contrast to many forecasts of increased storminess within an enhanced 'greenhouse'-warmed world. The concept of mobile polar highs, introduced in Chapter 3, incorporates the spectrum of climates that have dominated the Pleistocene (Figure 7.15). Rapid modes are characteristic of severe winters or Ice Age climates, while slow modes are characteristic of warm summers or the Holocene Climatic Optimum. Under 'greenhouse' warming, circulation will operate in the slow mode (Figure 7.15C), with the most dramatic change in atmospheric circulation occurring in the northern hemisphere, because of its larger landmass. In addition, warming should be greater at the poles than at the equator. Mobile polar highs will be greatly weakened and displaced poleward. As a consequence, cyclogenesis on the eastward sides of polar highs will be weak. This implies that the Aleutian and Icelandic Lows of the Palmén-Newton model would weaken considerably. The average position of westerly winds in mid-latitudes will also shift poleward. Weakened mobile polar highs should lead to a reduction in the intensity of trade winds, and a dramatic increase in the area of the globe affected by monsoons. However the development of stronger continental thermal lows, which follow the sun's zenith position, will amplify the Indian, African and Indo-Australian monsoons. Enhanced 'greenhouse' warming should thus lead to wetter climates in the tropics, and hence more catastrophic flooding, reduction of the arid zone of Africa, and expansion of the Amazon rainforest. Similar changes occurred in the tropics during the Holocene Climatic Optimum. Evidence from the United States and Australia indicates that floods, at this time, were over four times greater than anything experienced in the last hundred years. The concept of mobile polar highs foreshadows an enhanced 'greenhouse'-warmed climate that is benign, with less severe polar outbreaks, greatly diminished extratropical depressions, weakened tropical cyclone activity, retreating deserts, and an expanding tropical monsoon influence. The only negative climatic aspect would be the reappearance of catastrophic flooding in tropical and subtropical regions.

CAVEATS ON ENHANCED 'GREENHOUSE' WARMING

Regional climate change scenarios under enhanced 'greenhouse' warming lack the refinement required for governments to make long-term commitments in policy. For instance, the behaviour of extreme events can be judged important in terms of its impact on human health. Yet, in the United States, the incidence and pathways of tropical cyclones are considered to be too variable at present for any prediction to be made about future changes. Short-lived extreme events, such as floods, droughts and heatwaves, normally occur as part of random fluctuations about some statistical mean. A shift towards a warmer climate will lead to a disproportionate increase in the frequency of occurrence of such events. However in Australia, the present occurrence of these events is subject to the vagaries of the Southern Oscillation. The frequency and magnitude of the Southern Oscillation has not varied much over the past four hundred years, a period representing the peak of the Little Ice Age and the global warming of the past century.

Six points can be made about the climatic scenarios used in predicting a warmer globe. First, climate could change as predicted. If the scenarios for enhanced 'greenhouse' warming are correct, they represent greater average shifts in climate than witnessed over the past six thousand years. Adaptation to these shifts will require considerable readjustment by society in response to the natural environment. Second, for decision makers to set policy, without room for modification and adaptation, would constrain future responses should predicted climatic changes not occur as forecasted. Third, the scenarios do not take into account cycles, changes in the variance of records and any sudden shift in the mean of climatic parameters. To ignore these characteristics, especially the step-functional nature of climatic change, would be to underestimate severely the magnitude and frequency of extreme events that occur under the present climate regime. Fourth, and following on from the third point, somewhere in our past, an event has happened that was worse than anything documented in our historical written

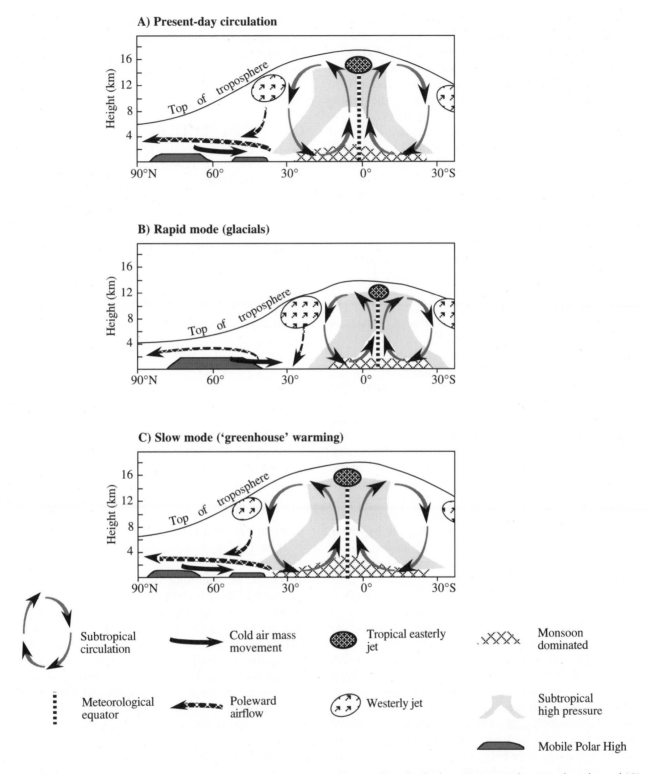

Figure 7.15 Analogues of general atmospheric circulation induced by mobile polar highs: (A) present-day, (B) glacials, and (C) enhanced 'greenhouse' warming (based on Leroux, 1993).

records. This event was not forced by enhanced or diminished concentrations of 'greenhouse' gases. For instance, corals from the Great Barrier Reef in Australia contain a signature of a flood event on the Burdekin River in Queensland within the last thousand years that exceeds anything witnessed by Europeans since colonisation of the continent. Fifth, events worse than the extremes documented in the written historical record will probably occur in the future. These events, when they do happen, may not necessarily be the signature of an enhanced 'greenhouse' warming. Finally, a close examination of climate records over the past forty years will show that many of the predicted outcomes of climate warming have already occurred. The extreme flood events in eastern Australia in 1950, the heatwave in Melbourne in 1959, Cyclone Tracy in Darwin in 1974, HurricaneAndrew in Florida in 1993, the Ash Wednesday bushfires of southern Australia in 1983, and the global droughts of 1982–1983 and 1990–1991, are all events that could represent the signature of a warmer globe. If they do, then society has already begun to adapt to enhanced 'greenhouse' warming.

It would be foolish for policy-makers to assume that enhanced 'greenhouse' warming will be the only climate outcome over the next fifty years. One of the certain outcomes derived from climatic monitoring over the past century is that urban climates have become warmer. The nature of the world's population distribution indicates that more and more people will be affected by these latter changes as cities grow in size, and as rural-to-city migration continues, especially in developing countries.

Finally, global climate may not warm, but cool. In the 1970s this was the preferred prediction of climate change. Many of the signatures proposed as evidence of a warming globe could just as easily be interpreted as indicators of a cooling globe. Snow and ice accumulation has recently been found to be accelerating on the Greenland icecap. This has been interpreted as a response to warmer air temperatures, and greater moisture capacity of air masses, moving over an icecap that is still below freezing. However, any view favouring the onset of new Ice Age would claim support from similar evidence. In northern Europe, the storms of the late 1980s, and flooding of the early 1990s have similarities with the events signalling the onset of either warm and cold climate in the Middle Ages. Most climatologists would agree that the Earth's climate is presently changing, and

almost all insurance companies recognise that the frequency and magnitude of climatic hazards is increasing. Not everyone would agree that these changes are the definitive signature of global warming.

OZONE DEPLETION
(World Meteorological Organization, 1989; Toon and Turco, 1991; Maduro and Schauerhammer, 1992; Graedel and Crutzen, 1993)

While ozone is pervasive throughout the atmosphere, the present concern is that human activity is destroying this gas in the stratosphere, and increasing ultraviolet radiation reaching the Earth's surface to harmful levels. Stratospheric ozone is located at altitudes between ten and fifty kilometres where it is formed by photolysis of molecular oxygen in the presence of sunlight as follows:

$$3O_2 + h\upsilon \rightarrow 2O_3 \qquad \textbf{7.1}$$

where $h\upsilon$ = solar radiation at a wavelength of $\leqslant 242$ nm

It is destroyed through reactions with atomic oxygen as follows:

$$2O_3 + h\upsilon \rightarrow 3O_2 \qquad \textbf{7.2}$$

where $h\upsilon$ = solar radiation at a wavelength of $\leqslant 1.14$ μm

In 1970, it was discovered that ozone can be destroyed in catalytic reaction cycles. In this process, some ozone is destroyed by a range of disassociated molecules without the aid of solar radiation. After the chemical reactions have finished, the original destructive radical is reformed. Unless these radicals are stabilised, it is feasible that they can react with ozone ad infinitum, or until ozone becomes totally depleted. These reactions are expressed chemically as follows:

radical reacts with ozone

$$X\bullet + O_3 \rightarrow XO\bullet + O_2 \qquad \textbf{7.3a}$$

ozone also disassociated by sunlight producing a free oxygen ion

$$O_3 + h\upsilon \rightarrow O + O_2 \qquad \textbf{7.3b}$$

free oxygen ion releases radical

$$O + XO• \rightarrow X• + O_2 \qquad \textbf{7.3c}$$

summary equation

$$X• + 2O_3 + hv \rightarrow X• + 3O_2 \qquad \textbf{7.3d}$$

where X• = radical

Catalysing radicals consist of hydrogen, nitrogen, chlorine and bromine found respectively in water vapour, nitrogen oxides, CFCs and halogens. In all, over seventy chemical reactions have been identified with ozone in the stratosphere, based on equation 7.3. In the case of CFCs, which are chemically stable in the troposphere, the molecules themselves must be chemically broken down by other radicals or by photodissociation. Some CFCs are more potent then others, because of the number of radicals they produce per molecule when broken down as follows:

$$CFC\text{-}11: CFCl_3 + hv \rightarrow 3ClO• \qquad \textbf{7.4a}$$

$$CFC\text{-}12: CF_2Cl_2 + hv \rightarrow 2ClO• \qquad \textbf{7.4b}$$

$$CFC\text{-}113: C_2F_3Cl_3 + hv \rightarrow 3ClO• \qquad \textbf{7.4c}$$

$$\text{Methyl chloride: } CH_3Cl + HO• \rightarrow 1ClO• \qquad \textbf{7.4d}$$

$$\text{Methyl chloroform: } CH_3CCl_3 + HO• \rightarrow 1ClO• \qquad \textbf{7.4e}$$

$$\text{Carbon tetrachloride: } CCl_4 + hv \rightarrow 4ClO• \qquad \textbf{7.4f}$$

Complex reactions can also occur amongst different catalysts thus neutralising some. Sunlight can subsequently photodissociate these molecules to form the original catalysts again. Finally, in the lower stratosphere the role of oxygen in equation 7.3 can be replaced by chlorine peroxide as follows:

chlorine radical reacts with ozone

$$Cl + O_3 \rightarrow ClO + O_2 \qquad \textbf{7.5a}$$

chlorine peroxide formed

$$ClO + ClO \rightarrow Cl_2O_2 \qquad \textbf{7.5b}$$

chlorine peroxide photodisassociated by ultraviolet

$$Cl_2O_2 + hv \rightarrow 2Cl + O_2 \qquad \textbf{7.5c}$$

summary equation

$$2Cl + 2O_3 + hv \rightarrow 2Cl + 3O_2 \qquad \textbf{7.5d}$$

The reactions based upon equation 7.3 are continual in the upper polar stratosphere, even in the dim ultraviolet light of winter. However, the reactions involving chlorine peroxide, in equation 7.5, require more ultraviolet light, and only commence in the Antarctic spring, September–November. As a result, chlorine and chlorine peroxide radicals can form throughout winter in the upper stratosphere, whereupon they drift to, and accumulate at, lower levels of the stratosphere, triggering massive ozone destruction in spring. During summer, ozone moves from the tropics to replenish this deficit.

Theoretically, the process should be the same over both poles; however the Antarctic undergoes a continual, but natural, depletion of ozone in winter that has not been apparent in the northern hemisphere. This difference between the Antarctic (Dobson measurements 1957–1958) and Arctic (Spitzbergen) is shown in Figure 7.16. In both regions, total ozone can vary seasonally by 30–40%. In winter, intense longwave emission cools the atmosphere at both locations, causing air to subside and flow away from the poles as active mobile polar highs. Coriolis force sets up a strong

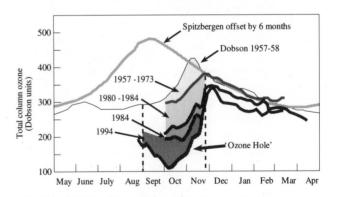

Figure 7.16 Seasonal ozone levels in the Arctic and Antarctic. Enhanced ozone depletion in spring is shown for the Antarctic (based on Dobson, 1968 and Farman et al., 1985).

westerly circulation in the upper atmosphere at the edge of this air, forming a polar vortex. This isolates the polar stratosphere from the rest of the globe, and inhibits ozone replenishment from the tropics in winter. The continuous reactions, characterised by equation 7.3, should slowly deplete the stratosphere of ozone; however, these reactions occur on ice particles making up stratospheric clouds of nitric acid and water. Such particles form at temperatures below −80°C, temperatures that only develop in the colder Antarctic stratosphere poleward of 55° S. Thus, while a polar vortex can be established around both poles in winter, only the Antarctic stratosphere becomes naturally depleted of ozone during winter.

Anthropogenic chemicals capable of forming catalysing radicals reach the stratosphere in airplane exhaust, or drift upwards from the troposphere. The 'stratospheric fountains' formed by intense thunderstorms at the western end of Walker circulation, and by tropical cyclones, are two mechanisms aiding this infusion. Anthropogenic activities have significantly increased the halogen content of the stratosphere since 1950, through the release of chlorofluorocarbons used as refrigerants, degreasers and foaming agents; and halons used in fire fighting and crop fungicides. Monitoring of stratospheric ozone began in 1957 (Figure 7.17) showing global fluctuations of ± 4% over time. Over the last twenty years, ozone globally has decreased by 1–2% in an

erratic fashion. Between 40° and 60° latitudes, ozone has decreased 4–5% per decade since the mid 1970s. During the winter of 1994–1995, ozone decreased by as much as 10% over Europe, and 35% over Siberia.

More importantly, ozone has decreased dramatically over the Antarctic in spring, since 1978. In 1984, the British Antarctic Expedition noted, for the first time, a dramatic decrease of 35% in ozone levels between the spring of 1968 and 1984. This latter phenomenon was termed the 'Ozone Hole'. About 70% and 25% of Antarctic ozone depletion has subsequently been linked to excessive levels of anthropogenically produced chlorine and bromine, respectively. The extent of ozone depletion, and the intensification of the 'Ozone Hole' over time, is shown in Figure 7.16. As sunlight illuminates stratospheric clouds between September and November, the photochemical reactions outlined in equation 7.5 commence, utilising the abundant accumulations of chlorine and chlorine peroxide that have been created on crystals of nitric acid trihydrate higher in the stratosphere, over winter. With the depletion of ozone, less ultraviolet light is absorbed in the stratosphere, and it does not warm up. This maintains cold temperatures throughout the spring, perpetuating ozone depletion. Since 1973, ozone amounts have dropped from a high of 400 Dobson units in 1958, to a low of 100 Dobson units in 1994 (Figure 7.17). About 95% of the ozone in the lower stratosphere, between fifteen and twenty kilometres, and about 50% above twenty-five kilometres is now destroyed each spring. Unfortunately, as the Antarctic polar vortex breaks down in summer, it can waft ozone-depleted air over inhabited areas in Chile and Australia. Ozone levels rapidly return to near-normal levels during the southern hemisphere summer.

Temperatures cold enough to form stratospheric clouds are favoured by enhanced 'greenhouse' warming. Because the Earth's total radiation budget at the top of the atmosphere must remain constant, any increase in surface temperature must be balanced by a decrease in the upper atmosphere. Evidence that such a process is operating is presented in Figure 7.3. Thus, it is possible that the 'Ozone Hole' is as much a signature of enhanced 'greenhouse' warming, as it is of increased halocarbon production. Since the 1980s, ozone depletion has also been found over the Arctic in spring, and over major urban areas of the northern hemisphere. Compared to the Antarctic 'Ozone Hole', many of these are insignificant, and

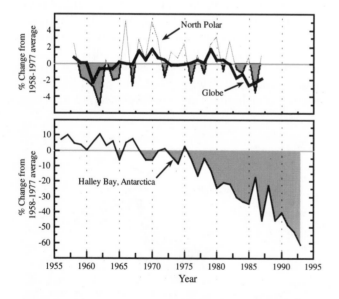

Figure 7.17 Trends in ozone for the globe and the pole regions since 1957 (data from Dr Robert Parson, Colorado University, and Angell et al., 1991). Halley Bay, Antarctica data are only for the month of October.

represent small localised perturbations in ozone. However, in the spring of 1993, ozone depletion amounted to 10–20% in the northern hemisphere in winter, leading to the suspicion that the type of reactions characterised by equation 7.5 were becoming more common at high northern latitudes.

While the theory for the formation of the Antarctic 'Ozone Hole' appears elegantly valid, it has uncertainties. For instance, halocarbons are not the only mechanism that can decrease stratospheric ozone. Many of the radicals that can scavenge ozone in the stratosphere occur naturally. OH and NO radicals can be dramatically enhanced in number during solar flare events, which tend to be more common at times of solar sunspot activity. In addition, volcanoes, through either slow degassing or eruptions, contribute five times more chlorine to the atmosphere than that bound up in CFCs. Chlorine is also a byproduct of coal burning, brush fires, seaweed decomposition and salt spray. Salt spray accounts for over one hundred times more chlorine in the atmosphere annually than CFCs. While much of this chlorine is scavenged out in the lower atmosphere, major volcanic eruptions can throw substantial quantities of chlorine, in the form of hydrochloric acid (HCl), into the stratosphere. Tambora and Krakatoa, erupting in 1815 and 1883 respectively, discharged 211 and 3.6 million tonnes of chlorine into the atmosphere. The eruption of El Chichon, in Mexico in 1982, led to a 2% loss of total ozone, while the recent major eruption of Mount Pinatubo, in June 1991, decreased tropical ozone by 20% within three to six months. Volcanoes can also contribute significant sulphate aerosols to the upper atmosphere. Sulphates assist the release of chlorine radicals conducive to ozone depletion, while at the same time enhancing 'denoxification' of the lower stratosphere. This removal of nitrogen radicals permits the existence or more chlorine peroxide ions, involved in ozone depletion via equations 7.3 and 7.5. Proliferation in the number of ice crystals in the upper stratosphere is also related to enhanced volcanic activity, and water vapour, both of which have increased in recent decades. Water vapour is a byproduct of the oxidation of methane, whose concentration in the atmosphere has been enhanced anthropogenically (Table 7.1).

Given these facts, definite conclusions concerning ozone depletion should be qualified. It can be concluded that the enhancement of the 'Ozone Hole' in the southern hemisphere spring has accelerated in the 1980s, and continued into the 1990s. While the most probable cause of this depletion is the photodissociation of halocarbons in the stratosphere, there are a range of other factors that can rapidly, and significantly, alter ozone levels. Many of these factors are a consequence of enhanced 'greenhouse' warming.

'NUCLEAR WINTER'
(Turco et al., 1983; Seitz, 1986; Bridgman, 1990; Sagan and Turco, 1990; Cotton and Pielke, 1995)

One of the more controversial debates of the 1980s was the hypothesised impact of a nuclear war upon climate. With the demise of the Soviet Union in the early 1990s, this 'nuclear winter' scenario all but disappeared from public consciousness. However, the credibility of a 'nuclear winter' had waned before then, because of serious doubts about its accuracy. An atmospheric nuclear explosion perturbs atmospheric circulation up to 30–45 kilometres above the ground. Each megatonne of nuclear material is estimated to produce 3,000 tonnes of nitrogen oxides, which can persist for over four years in the stratosphere, before being scavenged by chemical reactions. Nitrogen oxides in the stratosphere absorb shortwave radiation, destroy ozone and cool surface air temperatures. Nuclear bomb testing in the early 1960s produced 980 megatonnes of nitrogen oxides in the stratosphere. This should have produced a 2.5% reduction in the flux of solar radiation to the Earth's surface, a value subsequently supported by balloon measurements. Before the test ban treaty of 1963, accelerated testing put 1,500 megatonnes of nitrogen oxides into the atmosphere. As a result, the winters of 1962 to 1966 were some of the coldest recorded in the twentieth century. Concomitantly, the upper stratosphere warmed by up to 6°C. However not all of these changes can be ascribed to atmospheric nuclear testing, because Mount Agung erupted in Bali, Indonesia in 1963. It is estimated that at least half of the surface cooling was due to this eruption.

In 1982, it was discovered that nuclear explosions also could inject significant amounts of dust into the atmosphere. For each megatonne of nuclear explosive, 2,000 tonnes of surface dust are raised into the upper atmosphere. In addition, in a nuclear war, smoke from burning forests and cities would also be lifted into the upper atmosphere by convection. It was theorised that this would have a dramatic cooling effect upon the Earth's surface temperature.

Subsequent studies suggested that the smoke and soot effect would be catastrophic even in a limited nuclear war. Firestorms, lasting up to one week in forests and burning cities, would inject smoke and soot into the stratosphere where they could not be washed out easily by rain. Soot and smoke are much finer than volcanic dust, and would have a longer residence time in the atmosphere. In a limited war with 5,000 megatonnes of detonation, 225 million tonnes of carbon would be released, mainly into the upper stratosphere, in a few days. This is equal to the total smoke emissions by human activity for a whole year. While this smoke would block out incoming shortwave radiation, it would allow long-wave radiation to escape, much more than volcanic dust would. Soot also absorbs solar radiation and would warm the stratosphere. The net effect would create inversion conditions in the troposphere. Inversions in the troposphere would hamper the mixing, coagulation and eventual fallout of the soot. As a result, incoming solar radiation at the Earth's surface could be reduced to 1% of normal. The overall effect would lower temperatures by 30°C over continental landmasses. Computer simulations indicated that temperatures in summer would drop below freezing within two days over parts of North America and Asia. Within ten days, much of the interior of North America and Asia would be below freezing, and remain so for up to six months or more. The southern hemisphere would not be exempt from these effects, even though it contains few targets for nuclear attack. One-dimensional computer simulations indicated that the warm air in the stratosphere, mainly over mid-latitudes in the northern hemisphere where the targets were located, would circulate to the southern hemisphere, sinking back to the surface around 15–30° S latitude. The normal atmosphere circulation, consisting of two belts of Hadley circulation around the equator, would be supplanted by a single convection cell originating in the northern hemisphere, and spanning both hemispheres. This circulation would spread the clouds of soot and smoke globally. While temperature would decrease on average by 10–20°C in the interior of continents, in coastal areas, the falls would be only 5–10°C because of the thermal warming effect of the oceans. The interior of continents would be dominated by atmospheric stability and drought; but the presence of relatively warm water along coastlines would generate intense coastal storms, in a similar fashion to the present formation of east coast lows. Precipitation would increase in coastal areas, leading to more radioactivity in these areas through washout. This would be the prime mechanism for removing radioactive particles with a long residence time in the atmosphere.

This scenario was very reliant upon computer simulation models which, as discussed previously, have limitations. In addition, nuclear explosions near the ground often lead to the generation of rain, which tends to wash soot and dust from the lower troposphere. This localised scavenging may remove significant amounts of nuclear dust and soot from the atmosphere. Subsequent GCMs showed that the temperature outcomes were highly dependent upon season, with a nuclear war in summer producing temperature decreases of 5°C, compared to a 25°C drop in winter. Further refinement showed less dramatic temperature effects, until, by the end of the 1980s, the term 'nuclear autumn' was being used instead of 'nuclear winter' to describe the climatic impacts of nuclear war.

More importantly, aspects of the 'nuclear winter' scenario have occurred historically on at least two occasions, and yet the hypothesised climatic consequences did not eventuate. The first significant event occurred in July–August 1915, with a forest fire in Siberia, that was very intense, and generated the same amount of smoke (20–180 × 10^6 tonnes) as theorised for a large nuclear war. The fire burnt an area the size of West Germany. Visibility fell to as low as 4–20 metres in the fire zone, and dropped to 100 metres up to 1,500 kilometres away. The fire lasted fifty-one days. The smoke did not spread outside the continent of Asia. Cooling on the order of 10°C occurred in the immediate area because of the attenuation of solar radiation by the smoke. This cooling persisted for up to one month; however, harvests in the region were delayed by 10–15 days. It would appear that the smoke and cooling caused extreme atmospheric stability that did not disperse the smoke from the immediate area. Even with the smoke from this forest fire being confined to the region, and with some drop in temperature being recorded, the smoke-induced cooling had minimal impact upon agricultural production.

The second event involved the deliberate lighting of the Kuwaiti oil wells by the Iraqi army, at the end of the Gulf War in February 1991. Just under six hundred wells were lit within a two-day period, and by the beginning of June, only one hundred had been extinguished. In some places, temperatures under the smoke clouds were 11°C cooler than normal. The smoke had minimal impact outside the Gulf region, although sooty snow fell over Tibet and the French Alps, and smoke was detected in the atmosphere at a height of 7,000 metres above Japan. High levels of soot were eventually detected downwind over Mauna Loa, Hawaii, and in the troposphere over Wyoming, in late March. No smoke was detected in the atmosphere above 7,000 metres. The sulphur dioxide and ozone levels in the Gulf region remained below those measured in Los Angeles on a smoggy day. The smoke did not enter the stratosphere because the drop in surface temperature created local conditions of stability. In addition, the soot was hygroscopic, and easily washed out of the atmosphere. These two events indicate that a smoky nuclear war would not lead to the 'nuclear winter' as theorised. While any nuclear explosion would have a devastating impact upon a city, and while its radiation effects would be widespread, the climatic impact of a nuclear war would be relatively minor.

CONCLUDING COMMENTS
(Pearce, 1994)

It is opportune here to summarise what is certain and uncertain about recent climate change, and about human impact upon climate. There is little doubt that our climate has undergone change in the latter half of the twentieth century. Temperatures have probably warmed above background levels, although this warming is not uniform and is biased towards nighttime. Concomitant with these increases is a universal increase in cloudiness. Logically, the decrease in temperature, between night and day, should also witness a decrease in climatic extremes, including storms. This reduction is not happening. Climate hazards are increasing in frequency. Swings in climate have become more pronounced over shorter intervals, but this has not been validated statistically. As part of this latter change, ENSO events are becoming more frequent, larger and more persistent.

The human effects that are certain include: increased air pollution, mainly sulphate aerosols; the growth of cities, with an attendant increase in the temperature of urban nighttime 'heat islands'; accelerated production of 'greenhouse' gases; and minimal impacts on climate due to catastrophic possible events such as a nuclear war, forest burning or oil field sabotage. Undoubtedly industrial, urban and rural dust levels are increasing, as is dust originating from land degradation. However, the magnitude of these latter effects is poorly studied, and their climatic impacts have not

been modelled. In general, computer modelling, while supportive of some of the recent climate change, still needs refinement before it can become a predictive tool.

It is uncertain whether temperature records are measuring a general global warming, the effects of a changing ENSO regime, the impact of more hazards and extremes, or the growth of urban areas. The evidence does not conclusively support enhanced 'greenhouse' warming. Even if enhanced 'greenhouse' warming were occurring, the impacts appear benign. The largest threat will be the increased potential of catastrophic flooding in the tropics and subtropics. At present, the impact of climate change is being felt by city dwellers, by those experiencing increased hazards, by the rural sector, and by countries and people marginalised at the edge of society and international trade. Finally, the most likely reason for the 'Ozone Hole' over the Antarctic is the increase in manufactured halocarbons, although other factors, some of which are related to enhanced 'greenhouse' warming, cannot be ruled out.

Of all the parameters not modelled adequately in general circulation models, clouds pose the most serious problem. While increased cloudiness is logically a consequence of a warming globe, it is also a feature of an Ice Age globe. As well, clouds may be increasing because of industrial sulphate aerosols and anthropogenic dust production. Changes in cloud cover not only incorporate a myriad of positive and negative feedback mechanisms, but also involve many interrelated causative factors. Our knowledge of climate processes is not complete, and begs further research to resolve these dichotomies.

Reliance upon enhanced 'greenhouse' warming to explain recent climate change, or more importantly, reliance upon its predictions, ignores history. Climate has a high degree of unpredictability. In 1986–1987, proponents of enhanced 'greenhouse' warming suggested that the next decade would witness the signal of global warming rising above the inherent background noise of climate systems. No one could predict the eruption of Mount Pinatubo on the 15 June 1991. On that one day, 20×10^6 tonnes of sulphur dioxide were injected into the atmosphere. Global warming was reversed by 0.5–0.6°C over the next three years. While anthropogenic discharges of gas to the atmosphere may be predictable, the number, location and magnitude of volcanic eruptions are not. Forecasts of global warming, or cooling, must always recognise that nothing about climate change is certain.

Finally, climate change is also complicated by the spatial nature of many forcing mechanisms. 'Greenhouse' gas forcing is global, while aerosol forcing is regionally constrained to continents and adjacent areas. Ironically, fossil fuel burning may be responsible for both regional cooling and global warming. Were policy-makers to reduce carbon dioxide emissions by limiting industrial production, they would also be removing one of the main restraints presently suppressing global warming. However, the fact that industrial sulphates cool climate should not be taken as an open licence for uncontrolled air pollution. The production of dimethylsulphides by phytoplankton is still a major component in the global sulphur cycle, even though it represents only 20–40% of the current industrial sulphur production. The iron enrichment of oceans poses an alluring solution for the abatement of enhanced 'greenhouse' warming. At present, the southern hemisphere oceans contain more nutrients than phytoplankton can consume. It would be possible to 'fertilise' the oceans with iron to accelerate phytoplankton production, increasing ocean cloud, and removing carbon dioxide from the atmosphere, thus slowing global warming. But, as with all human tampering with climate, there are bound to be negative ramifications as well. There is a need for continual research into the effects of any human activity upon climate.

REFERENCES AND FURTHER READING

Angell, J.K. 1994. 'Global, hemispheric and zonal temperature anomalies derived from radiosonde records'. In Boden, T.A., Kaiser, D.P., Sepanski, R.J. and Stoss, F.W. (eds) *Trends'93: A Compendium of Data on Global Change*. ORNL/CDIAC-65. Carbon Dioxide Information Analysis Center, Oak Ridge National Laboratory, Oak Ridge, Tennessee, pp. 636–672.

Angell, J.K., Korshover, J. and Planet, W.G. 1991. Annual and Seasonal Global Variation in Total Ozone and Layer-Mean Ozone, 1958–1986. *Carbon Dioxide Information Analysis Center Numeric Data Package Collection*, Dataset No. NDP023.TOT, Oak Ridge National Laboratory, Oak Ridge, Tennessee.

Benkovitz, C.M., Berkowitz, C.M., Easter, R.C., Nemesure, S., Wagener, R. and Schwartz, S.E. 1994. 'Sulfate over the North Atlantic and adjacent continental regions: Evaluation for October and November 1986 using a three-dimensional model driven by observation-derived meteorology'. *Journal Geophysical Research* v. 99 pp. 20725–20756.

Boden, T.A., Kaiser, D.P., Sepanski, R.J. and Stoss, F.W. (eds) 1994. *Trends'93: A Compendium of Data on Global Change*. ORNL/CDIAC-65. Carbon Dioxide Information Analysis Center, Oak Ridge National Laboratory, Oak Ridge, Tennessee, 984p.

Bridgman, H. 1990. *Global Air Pollution: Problems for the 1990s*. Belhaven, London, 261p.

Bryant, E.A. 1987. 'CO$_2$ – warming, rising sea-level and retreating coasts: review and critique'. *Australian Geographer* v. 18 pp. 101–113.

Bryant, E. 1990. 'Climatic change predictions'. In Ewan, C., Bryant, E., Calvert, D. (eds) *Health implications of long term climate change. v II: Survey report and commissioned papers*. Australian Department of Community Services and Health, Canberra, pp. 6–35.

Bryant, E.A. 1991. *Natural Hazards*. Cambridge University Press, Cambridge, 294p.

Bryant, E.A. 1993. 'Global climate change and international response'. In Ewan, C.E., Bryant, E.A., Calvert, G.D. and Garrick, J.A. (eds) *Health in the Greenhouse: the Medical and Environmental Health Effects of Global Climate Change*. Australian Government Publishing Service, Canberra, pp. 131–153.

Bryson, R. and Murray, T. 1977. *Climates of Hunger*. Australian National University Press, Canberra, 171p.

Calder, N. 1991. *Spaceship Earth*. Penguin, London, 208p.

Christy, J.R. and McNider, R.T. 1994. 'Satellite greenhouse signal'. *Nature* v. 367 pp. 325.

Cotton, W.R. and Pielke, R.A. 1995. *Human Impacts on Weather and Climate*. Cambridge University Press, Cambridge, 288p.

Daly, J.L. 1989. *The Greenhouse Trap*. Bantam, Sydney, 192p.

Dignon, J. and Hameed, S. 1989. 'Global emissions of nitrogen and sulfur oxides from 1860 to 1980'. *Journal Air Pollution Control Association* v. 39 pp. 180–186.

Dlugokencky, E.J., Lang, P.M., Masarie, K.A. and Steele, L.P. 1994. 'Global CH$_4$ record from the NOAA/CMDL air sampling network'. In Boden, T.A., Kaiser, D.P., Sepanski, R.J. and Stoss, F.W. (eds) *Trends'93: A Compendium of Data on Global Change*. ORNL/CDIAC-65. Carbon Dioxide Information Analysis

Center, Oak Ridge National Laboratory, Oak Ridge, Tennessee, pp. 262–266.

Dobson, G.M.B. 1968. 'Forty years' research on atmospheric ozone at Oxford: a history'. *Applied Optics* v. 7 pp. 387–405.

Elkins, J.W., Thompson, T.M., Butler, J.H., Myers, R.C., Clarke, A.D., Swanson, T.H., Endres, D.J., Yoshinaga, A.M., Schnell, R.C., Winey, M., Mendonca, B.G., Losleben, M.V., Trivett, N.B.A., Worthy, D.E.J., Hudec, V., Chorney, V., Fraser, P.J. and Porter, L.W. 1994. 'Global and hemispheric means of CFC-11 and CFC-12 from the NOAA/CMDL flask sampling program'. In Boden, T.A., Kaiser, D.P., Sepanski, R.J. and Stoss, F.W. (eds) *Trends'93: A Compendium of Data on Global Change*. ORNL/CDIAC-65. Carbon Dioxide Information Analysis Center, Oak Ridge National Laboratory, Oak Ridge, Tennessee, pp. 422–430.

Elkins, J.W., Thompson, T.M., Swanson, T.H., Butler, J.H., Hall, B.D., Cummings, S.O., Gisher, D.A. and Raffo, A.G. 1993. 'Decrease in the growth rates of atmospheric chlorofluorocarbons 11 and 12'. *Nature* v. 364 pp. 780–783.

Farman, J.C., Gardiner, B.G. and Shanklin, J.D. 1985. 'Large losses of total ozone in Antarctica reveal seasonal ClO$_x$/NO$_x$ interaction'. *Nature* v. 315 pp. 207–210.

Graedel, T.E. and Crutzen, P.J. 1993. *Atmospheric Change: An Earth System Perspective*. Freeman, New York, 446p.

Hadley Centre 1995. *Modelling Climate Change*. United Kingdom Meteorological Office. 12p.

Henderson-Sellers, A. 1992. 'Continental cloudiness changes this century'. *GeoJournal* v. 27 pp. 255–262.

Houghton, J. 1994. *Global Warming: The Complete Briefing*. Lion Publishing, Oxford, 192p.

Houghton, J.T., Jenkins, G.J. and Ephraums, J.J. 1990. *Climate Change: The IPCC Scientific Assessment*. Cambridge University Press, Cambridge, 365p.

Houghton, J.T., Callander, B.A. and Varney, S.K. (eds) 1992. *Climate Change 1992: The Supplementary Report to the IPCC Scientific Assessment*. Cambridge University Press, Cambridge, 200p.

Houghton, J.T., Meira Filho, L.G., Bruce, J., Hoesung Lee, Callander, B.A., Haites, E., Harris, N. and Maskell, K. (eds) 1995. *Climate Change 1994: Radiative Forcing of Climate Change and an Evaluation of the IPCC IS92 Emission Scenarios*. Cambridge University Press, Cambridge, 339p.

Houghton, J., Meira Filho, L.G., Callander, B.A. Harris, N., Kattenberg, A. and Maskell, K. (eds) 1996. *Climate Change 1995: The Science of Climate Change*. Cambridge University Press, Cambridge, 572p.

Idso, S.B. 1989. *Carbon Dioxide and Global Change: Earth in Transition*. Institute for Biospheric Research, Tempe, 292p.

Karl, T.R., Jones, P.D., Knight, R.W., Kukla, G., Plummer, N., Razuvayev, V., Gallo, K.P., Lindseay, J., Charlson, R.J. and Peterson, T.C. 1993. 'Asymmetric trends of daily maximum and minimum temperature'. *Bulletin American Meteorological Society* v. 74 pp. 1007–1023.

Karl, T.R., Knight, R.W., Easterling, D.R. and Qualye, R.G. 1995. 'Trends in U.S. climate during the twentieth century. *Consequences* (**http://www.gcrio.org/CONSEQUENCES/ introCON.html**) v. 1 no. 1.

Keeling, C.D. and Whorf, T.P. 1994. 'Atmospheric CO$_2$ records from sites in the SIO air sampling network'. In Boden, T.A., Kaiser, D.P., Sepanski, R.J. and Stoss, F.W. (eds) *Trends'93: A Compendium of Data on Global Change*. ORNL/CDIAC-65. Carbon Dioxide Information Analysis Center, Oak Ridge National Laboratory, Oak Ridge, Tennessee, pp. 16–26.

Kiehl, J.T. and Briegleb, B.P. 1993. 'The relative roles of sulphate aerosols and greenhouse gases in climate forcing'. *Science* v. 260 pp. 311–314.

Leggett, J. (ed.) 1990. *Global Warming: The Greenpeace Report*. Oxford University Press, Oxford, 554p.

Leroux, M. 1993. 'The Mobile Polar High: a new concept explaining present mechanisms of meridional air-mass and energy exchanges and global propagation of palaeoclimatic changes'. *Global and Planetary Change* v. 7 pp. 69–93.

Maduro, R.A. and Schauerhammer, R. 1992. *The Holes in the Ozone Scare*. 21st Century Science Associates, Washington, 356p.

Michaels, P.J. 1992. *Sound and Fury: The Science and Politics of Global Warming*. CATO Institute, Washington, 196p.

Mitchell, J.F.B., Manabe, S., Meleshko, V. and Tokioka, T. 1990. 'Equilibrium climate change – and its implications for the future'.

In Houghton, J.T., Jenkins, G.J. and Ephraums, J.J. (eds) *Climate Change: The IPCC Scientific Assessment.* Cambridge University Press, Cambridge, pp. 93–130.

National Research Council, 1983. *Changing Climate: Report of the Carbon Dioxide Assessment Committee.* National Academy Press, Washington, 496p.

Pearce, F. 1994. 'Not warming, but cooling'. *New Scientist* 9 July, pp. 37–41.

Ramanathan, V, Barkstrom, B.R. and Harrison, E.F. 1989. 'Climate and the Earth's radiation budget'. *Physics Today* May 1989 pp. 22–32.

Sagan, C. and Turco. R. 1990. *A Path where No Man Thought: Nuclear Winter and the End of the Arms Race.* Century, London, 499p.

Santer, B.D., Wigley, T.M.L., Barnett, T.P. and Anyamba, E. 1996. 'Detection of climate change and attribution of causes'. In Houghton, J.T., Meira Filho, L.G., Callander, B.A., Harris, N., Kattenberg, A. and Maskell, K. (eds) *Climate Change 1995: The Science of Climate Change.* Cambridge University Press, Cambridge, pp. 406–443.

Schimel, D.S. 1995. 'Terrestrial ecosystems and the carbon cycle'. *Global Change Biology* v. 1 pp. 77–91.

Schlesinger, M.E. (ed.) 1991. *Greenhouse-gas-induced Climatic Change: A Critical Appraisal of Simulations and Observations.* Elsevier, New York, 615p.

Seitz, R. 1986. 'Siberian fire as "nuclear winter" guide'. *Nature* v. 323 pp. 116–17.

Steele, L.P. and Lang, P.M. 1991. 'Atmospheric methane concentrations–the NOAA/CMDL global cooperative flask sampling network, 1983–1988'. *Carbon Dioxide Information Analysis Center Numeric Data Package Collection*, Dataset No. NDP038. Oak Ridge National Laboratory, Oak Ridge, Tennessee.

Tegen, I., Lacis, A.A. and Fung, I. 1996. 'The influence on climate forcing of mineral aerosols from disturbed soils'. *Nature* v. 380 pp. 419–422.

Thomson, D.J. 1995. 'The seasons, global temperature, and precession'. *Science* v. 268 pp. 59–68.

Toon, O.B. and Turco, R.P. 1991. 'Polar stratospheric clouds and ozone depletion'. *Scientific American* v. 264 pp. 40–47.

Turco, R.P., Toon, O.B., Ackerman, T.P., Pollack, J.B and Sagan, C. 1983. 'Nuclear winter: global consequences of multiple nuclear explosions'. *Science* v. 222 pp. 1283–1292.

Washington, W.M. 1992. 'Climate-model responses to increased CO_2 and other greenhouse gases'. In Trenberth, K.E. (ed.) *Climate System Modeling.* Cambridge University Press, Cambridge, pp. 643–688.

Watson, R.T., Rodhe, H., Oeschger, H. and Siegenthaler, U. 1990. 'Greenhouse gases and aerosols'. In Houghton, J.T., Jenkins, G.J. and Ephraums, J.J. (eds) *Climate Change: The IPCC Scientific Assessment.* Cambridge University Press, Cambridge, pp. 1–40.

World Meteorological Organization, 1989. *Scientific Assessment of Stratospheric Ozone: 1989.* v. 1. WMO Global Ozone Research and Monitoring Project Report No. 20, Geneva, 486p.

III

IMPACTS

8

Health Impacts of Climate Change

WORLD PERSPECTIVE

World demography
(Haub, 1995)

One of the marked features of the twentieth century has been the sudden and unprecedented expansion of the world's population. At the beginning of the nineteenth century, the world's population stood at six hundred million (Figure 8.1). This reached 1.6 billion people by the turn of the twentieth century. The next billion was added by 1950, with 80% of the growth taking place in the world's poorer and underdeveloped countries. The last fifty years of the twentieth century have witnessed a further increase of three billion people, with no indication that this exponential growth in population is slackening. Most of this increase has occurred in Asia, Africa and Latin America. At present, a billion people are being added to the world's population every eleven years. None of this increase was accurately forecast, even in the last twenty years, in spite of an awareness that a population explosion was occurring. Moreover, the increase has occurred despite national initiatives to slow the birth rate. For instance, the United Nations predicted in 1951 that the world's population in 1980 would be between 3.0 and 3.6 billion. The actual figure was 4.4 billion, exceeding the estimate by about 22%.

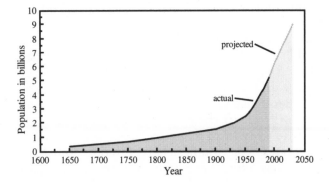

Figure 8.1 Actual and projected global population, 1650–2030 (data from the United Nations, 1992).

Whereas the population increases of the first half of the twentieth century were driven by increases in birth rates and decreases in infant mortality, the recent population growth has been caused by an unprecedented decrease in the death rate. In the early part of the twentieth century, death rates declined mainly in richer, industrialised countries, concomitantly with increased urbanisation, industrialisation and a reduction in family size. After 1950, global population exploded because of advances in public health and medicine in underdeveloped countries. The majority of population, however, was still agrarian-based, and children were deemed an economic asset. Some underdeveloped countries achieved birth rates of three or four percent per year, enough to double their population every twenty years. Since 1950, 89% of global population growth has occurred in developing countries, a figure that climbed to 97% by 1995.

Global projections
(United Nations, 1992; World Resources Institute and United Nations, 1992; Haub, 1995)

The ability to sustain present growth rates in many developing countries depends upon maintaining high fertility levels. In 1995, women in developing countries bore an average of 3.6 children in their lifetimes. India approximated this average despite decades of government-sponsored family planning. If China is excluded from the calculations, because of its stricter policy on reproduction, the global rate climbs to 4.2 children. The highest rate currently is in Mali with 7.2 children. While the global rate is high, it is substantially lower than the 6.1 figure of the early 1950s. If the fertility rate remains constant at its present level, world population will rise from 5.7 billion today, to 22 billion by 2050.

A medium population projection puts the world's population at 9 billion people by the year 2030, assuming a fertility rate of 2.1 children (Figure 8.1). However, there is presently no reason to believe that fertility rates in developing countries will fall to these levels. Certainly there are limits to growth in many countries with high fertility rates. For instance, both Ethiopia and Mali have levels which have been checked over the past thirty years by high death rates caused by major famines. If Ethiopia's present birth rate continues, its current population of 57 million would increase to 225 million by the year 2050. The latter figure is unsustainable.

In terms of climate impact, there will be significant changes to the age structure of the world's population by the year 2030. In 1975, 36.8% of the world's population was under the age of fifteen, while only 5.7% was over the age of sixty-five. By 1995, only 32% of the world's population was under the age of fifteen, and 6.5% over the age of sixty-five. By the year 2030, these figures will change to 22% and 11.5% respectively. Eventually, the under-fifteen age group will stabilise at around 18% of the total population, but the number of elderly people will continue to increase throughout the twenty-first century, until it eventually reaches a value over 20%. This high number of aged will have significant ramifications in terms

of human health, because the elderly have a higher incidence of climate-related skin cancers, eye diseases, respiratory afflictions, and heart and vascular problems.

In addition, the world's population will shift slowly towards cities. In 1960, only 34.2% of the world's population lived in urban centres. By 1990, this figure had risen to 45.2%, and is projected to reach over 65% by the year 2030. By then, 83% of people in developed countries will live in urban areas. The greatest change will occur in less developed countries, where the proportion of people living in cities will increase from 35% in 1990, to over 60% by the year 2030. However, these projections are subject to considerable uncertainty, because of the inherent variability over time in migration patterns. In 1990, there were 276 cities with populations of over one million inhabitants. Of these, forty-eight had populations of over four million people, and twenty had populations of more than ten million. The number of cities in each of these latter categories was increasing at the rate of ten and eight cities respectively per decade. Increased urbanisation, and its concomitant overcrowding, in developing countries is causing a resurgence in old diseases such as tuberculosis, influenza, meningitis, pneumonia and diarrhoea. Slum housing provides the conditions for the proliferation of disease vectors such as fleas, bedbugs, ticks, mites, flies, and mosquitoes. Urban encroachment upon natural environments is allowing insects to adapt to new conditions with alarming success. For instance, mosquitoes that once required clear water for breeding, now find the muddy puddles of cities just as livable. Poverty, amidst a background of ongoing diseases, exacerbates malnutrition and poor health, while overcrowding enhances the rate of transmission of disease. Apart from these economic and social problems, the growing cities are creating warmer climates because of their urban structure. Finally, although affluence and good health, rather than adversity (war, famine, pestilence, AIDS), have been the mechanisms for achieving zero population growth in the twentieth century, new diseases, increased natural hazards and climate impacts may become the means by which population growth rates will be lowered in the twenty-first century.

Marginalisation
(Bryant, 1991)

The health impact of climate change will affect those people, social or ethnic groups or nations, who either live in areas of greatest risk, or are marginalised within an increasing internationalised world. For example, the Sahelian droughts of the past thirty years affected marginalised people. The droughts occurred in an area with limited rainfall, and were induced by significant climate change in the adjacent Atlantic Ocean. The resulting famines were exacerbated by civil war, intertribal strife, and breakdowns in central government control and management. Marginalisation is the relegation or confinement of people to the edge of society, commerce, or transport and communication networks. These conditions pose different problems in terms of human health.

Such people consist of the 'forgotten few', the 'downs and outs', people in subcultures, those at the lower end of the socioeconomic scale, ethnic or racial minorities, oppressed cultural groups, or segments of society with little economic or political relevance. It may also include people who move frequently, or do not have a permanent address, the type of people who may be missed in a national census and excluded in government planning. Or they could be people whose religion, ethnicity, literacy level, occupation, or social habits isolate them, not just from the mainstream of society, but from frequent social interaction and communication. Marginalised people in western societies can include itinerant workers, gypsies, the homeless, young unemployed, the illiterate, the handicapped, those with physical disabilities, travellers, the hospitalised, the elderly, those with diminished mental functioning, those whose main language is not that of their surrounding community, native peoples, students, or people in occupations such as travelling sales, fishing and remote mineral exploration.

For example, marginalised people could consist of students travelling to and from colleges of residence, indigenous North Americans on isolated reservations, people in mental institutions, drug users in red-light districts, drop-outs living a nomadic or subsidence existence on communes in desert regions or in isolated valleys, seasonal workers harvesting crops, or newly arrived refugees from a foreign country with a different language. In scenarios where climate impacts on health, students could be travelling during the approach of a tropical cyclone; isolated reservations could be plagued by deer ticks carrying Lyme disease, unaware of the cause or the cure; commune inhabitants could be cut off or forgotten during a flash flood; seasonal workers could be subject to more frequent mosquito infestations

during an epidemic of Ross River fever or encephalitis; or newly arrived migrants could fail to understand what a state emergency worker is yelling at them as a bushfire bears down on them at thirty kilometres an hour. The total number of such people, even in developed countries, can consist of more than 25% of the population. For instance, it is estimated that two million people were not counted in the last American census because they could not be found at a permanent address. The health of marginalised people will probably be one of the major challenges facing governments under any changing climate.

HEALTH AND CLIMATE CHANGE

Figure 8.2 summarises the many health consequences of projected 'greenhouse' warming and ozone depletion. Climatic change will have both direct and indirect effects on health, as well as short- and long-term social effects. Direct health effects include increased heat stress, asthma, cataracts, skin cancer and other skin diseases. Indirect health effects include increased incidence of communicable diseases, increased mortality and injuries (morbidity) due to increased natural disasters, and change in diet and nutrition due to changed agricultural production. Most important, will be the adaptation required in response to opportunistic vector-borne diseases, which will expand in area and increase in magnitude under a warmer and wetter climate. The health conditions summarised in Figure 8.2 are addressed in the remainder of this chapter, and related to the climate scenarios presented in Chapter 7. For simplicity, the discussion focuses on health impacts related only to temperature and precipitation changes.

Definition of human health
(Ewan et al., 1993)

Human health encompasses the physical, emotional and mental wellbeing of an individual, free of disease or abnormality, and sustained by balanced and adequate nutrition, a nurturing physical environment, supportive social contacts and spiritual security. Good health permits an individual to live a life that is mentally fulfilled and functioning, and physically unrestricted until a time of dying that is natural, and not shortened by genetic impairment, physical abuse of the body or accidents. For a population, good health is measured by life expectancy, calorie intake, the incidence of

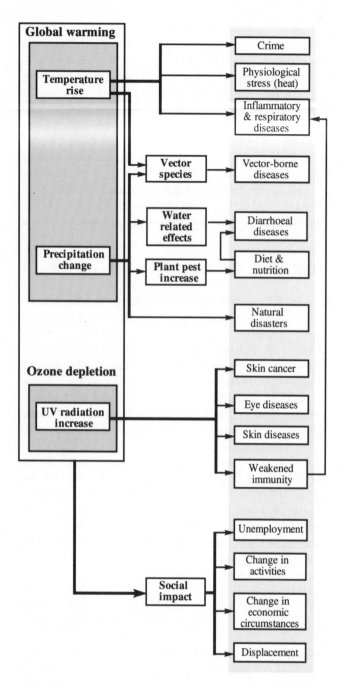

Figure 8.2 Schema of potential health effects resulting from climate change due to anthropogenically enhanced 'greenhouse' warming and ozone depletion (based on Marthick and Bryant, 1990).

have been obtained by controlling nutritional deficiencies, infectious diseases, social deprivation and exposure to environment pollutants and their effects.

Climate change can affect most of these factors in some way. Urban dwellers are very dependent upon services and infrastructure, as exemplified recently by the vagaries of civil strife (Bosnia, 1993–1995), terrorism (Tokyo nerve gas poisoning, 1995), ageing of infrastructure (Paris), and economic downturns (the former Soviet Union). Human health is influenced by the quality of water supplies and sanitation, and the prevention of disease. Increased rainfall, rising temperatures, coastal inundation and the frequency of extreme weather events can easily disrupt this infrastructure. The internationalisation of trade and the growth of cities also make health dependent upon the availability of food, the underlying transport system and food safety. Climate extremes can affect harvest yields, disrupt transport networks and jeopardise food safety. Human health is affected by the size and distribution of disease carrier populations, which can be modified by urban expansion and altered dramatically by climate change. In addition, changing climate may enhance domestic and industrial atmospheric pollution, causing increased respiratory problems. Poor working conditions or overcrowding amplify the deleterious effects of warmer temperatures and humidity, especially during summer nights. Finally, exposure of individuals to repetitive illnesses or stresses induced by climate change can culminate in psychosomatic and mental illness, leading to institutionalisation, increased violence, suicide and crime.

History of disease
(Lamb, 1985; McMichael, 1993; Garrett, 1995)

The history of human diseases is closely linked to stages in the development of civilisation. Three major developments are noteworthy: the development of cities about 3,000 years ago, the development of trade routes between Europe and Asia beginning mainly with the Roman Empire, and the formation of the global village of the twentieth century linked by rapid and voluminous international air traffic. Before the advent of cities, humans were affected by war, periodic famine and vector-borne infections. Some of the earliest diseases were related to bones and the skin. Bone diseases such as osteoarthritis appeared during cooler, damper times. Skin infections are spread by

heart disease, and other health indicators linked to major impairments or causes of death. While the preservation of personal health rests with the individual, population health is the domain of public health authorities. Achievements in public health contact and are related to the presence of clothing. Treponematoses and yaws are spread amongst children wearing little clothing, conditions favoured by a warm climate.

With the advent of cities came the arrival of contagious diseases that were virulent in crowded settlements. In many respects, the growth of cities constituted a distinct ecosystem that could amplify infectious diseases. Plagues appeared in human history during the development of the most prosperous city states, over 3,000 years ago. Athens in 430 BC was affected by an outbreak of epidemic typhus, a disease spread by the body louse *Pediculus humanus* under conditions of overcrowding, cold and squalor. Chinese records going back to 243 BC note massive epidemics that killed millions in China's cities. The wars of Europe, under the cold climate of the Little Ice Age, also witnessed innumerable typhus epidemics that often destroyed armies and changed the course of history.

The opening of trade routes, under expansive empires, witnessed the spread of new diseases. Smallpox arose in Asia in the year 50, and devastated the Roman Empire beginning in 165. Up to 35% of the population died. Six other notable epidemics occurred across Eurasia over the next half-millennium. When smallpox reached the Americas with the Spanish conquest, the death toll amongst Amerindians rose to 90%. The first great plague, the plague of Justinian of 543–547, also occurred within the Roman Empire. This plague, consisting of the pneumonic form of bubonic plague, swept through Europe and the Middle East via the trade routes established by the Romans. Similar pathways for disease occurred with the Black Plague in the fourteenth century; with influenza, which began its first of many global pandemics on the newly established European trade routes in 1580; and with yellow fever, which was introduced to the Americas via the slave trade in the seventeenth century.

The pace of human health afflictions has quickened since the Industrial Revolution, two hundred years ago. Mega-cities have developed worldwide, with many acquiring slums in their early stages of development. These conditions have seen the spread of epidemics and, at the same time, the implementation of modern health standards that have dramatically reduced death tolls. However, with the acceleration of urbanisation in developing countries in the last two decades, these health standards have not been maintained. Old diseases that were once under control are reappearing; new diseases find a virulent environment for dispersal; and

new hazards such as industrial chemical toxins and environmental degradation have arisen. Many of these recent changes are climatically affected.

Historical effects of climate change
(Gimpel, 1979; Lamb, 1982, 1985)

The historical record shows many fluctuations in temperature, rainfall and storms that have impacted upon human health. The Industrial Revolution almost broke out in mediaeval Europe in the thirteenth century. However after 1250, Europe witnessed abrupt deterioration of climate that substantially altered agriculture, settlement and trade across the North Atlantic. Most significant were the deaths due to great storms. In the twelfth and thirteenth centuries, famine had all but disappeared from Europe. This was reversed by the appalling weather conditions of 1315–1317. Torrential summer rains ruined harvests, and the resulting famine killed 10% of the population. The tales of survival were gruesome. In Ireland, people waited at the fringes of funerals, then dug up the bodies from churchyards, and ate them for food. Cannibalism became rife throughout Europe. Executed criminals were snatched from the gallows, to be eaten for food later by the watching populace. The economic conditions triggered a depression that lasted for one hundred and fifty years until the Renaissance. Widespread famine again occurred in the 1420s–1430s. These famines weakened the population for one of history's most infamous plagues, the bubonic plague, or Black Death.

Bubonic plague, caused by the bacterium *Yersinia pestis*, is spread by the flea *Xenopsylla cheopsis* whose reservoir is the black rat. The flea carrier is most active at temperatures between 20 and 32°C, and lives four times as long at relative humidities of 90% than at humidities below 30%. The bacterium invades the lymphatic system where it kills huge numbers of cells, leading to the formation of grotesque buboes and pus-filled boils. From there, the disease spreads to the liver, spleen and brain, causing wholesale haemorrhaging and dementia. The second pandemic began in China, under exceptional rains and floods associated with the early climatic cooling of the Middle Ages. Plague, and the ensuing famine, halved the Chinese population between 1200 and 1400. The Black Plague reached the city of Tana (now Feodosia) on the Crimean coast in 1347, where the besieging Tartars, decimated by the plague, catapulted their dead over the walls as a parting gesture to the enemy. Genoese merchants visiting the city carried the disease to Italy, where it spread between 1348 and 1350 with horrific consequences. Disease and famine decimated villages throughout Europe over the next two hundred years, with 15–60% of the population dying in different regions. Approximately 20%, 23% and 50% of villages disappeared in England, Germany, and southern Italy respectively.

Besides the Black Plague, another scourge arose. Wheat requires a higher summer temperature than barley or oats, and grows best when annual rainfall is below 900 millimetres. However, it can be harvested wet and dried indoors. Rye withstands colder winters and grows better on poorer soils. However, rye requires special harvesting and storage conditions to prevent the growth of ergot blight (*Claviceps purpurea*), which grows under damp conditions. Ergot blackens the kernels of rye and is highly poisonous. In humans, the resulting condition, known as ergotism or Saint Anthony's fire, can be induced by minute amounts of the blight, baked in bread. Because of increased rain and cooler temperatures, wheat could no longer be grown, so peasants switched to rye. However the cool, wet summers led to improper storage and occurrence of ergot blight. The mediaeval distribution system dispersed contaminated grain throughout whole villages. The consequences of ergotism were horrific. People suffered convulsions, hallucinated and underwent gangrenous rotting of the extremities. In the chronic stage, the extremities first developed an icy feeling, followed by a burning sensation. Limbs turned black, shrivelled and finally dropped off. The disease caused miscarriages in pregnant women. It also afflicted animals. Whole villages went mad as a result.

Climate change and crime
(Kaplan, 1960; DeFronzo, 1984; Kalkstein and Davis, 1989; McMichael et al., 1996)

There is considerable evidence relating climate to crime. One of the earliest studies, in New York City at the turn of the twentieth century, found that assault was more frequent when humidity and atmospheric pressure were low, temperatures high and winds mild. Subsequently, a strong relationship between crime and season was found in many countries. Most notable are the studies purporting to relate violent crime to temperature in American cities. In the United States, the FBI believes that rates of violent crime are higher in the west and south because the annual temperatures are higher

while precipitation is lower. Between 1967 and 1971, there was also a high positive correlation between high temperatures and rioting. Interestingly, rioting did not occur when temperatures exceeded 37.7°C. More detailed studies have found that crime rates against property (robbery, burglary, larceny, auto theft) and the person (homicide, assault and rape) are more likely to increase when it is hot. However, these relationships are now seen to be outweighed by socioeconomic factors.

Physiological stress
(Kreider, 1964; Lind, 1964; Linacre and Hobbs, 1977; Landsberg, 1981)

Physiological stress relates to physical changes within the human body due to temperature extremes. At low temperatures, evaporation from the skin and lungs can cause cooling that exceeds metabolic heat generation. The rate of cooling is enhanced by wind. For instance, at 15°C, an 80-kilometre-per-hour wind will produce an equivalent temperature on exposed skin of 4°C. At 0°C, the same reduction in temperature would be produced by only a 20-kilometre-per-hour wind. Exposed flesh freezes at equivalent wind chill temperatures below -14°C. Prolonged cold, evaporation and wind are life-threatening, leading to hypothermia, mental disorientation, and rapid shutting down of blood circulation to the extremities. If body temperature drops below 5°C, then death will result.

Cold weather can also cause mortality, mainly in the elderly, because of heart attacks due to strenuous activity such as snow shovelling. In cold climates, deaths due to heart disease, stroke and respiratory diseases are around 10% higher in winter than in summer. Cold spells, especially blizzards, are a common winter hazard in the United States. For instance, in March 1888, a spring blizzard, with wind gusts of 110 kilometres per hour, left four hundred dead in New York City. In recent times, the 28 January 1977 blizzard in Buffalo, where winds exceeded 134 kilometres per hour, resulted in over one hundred deaths. Of the 5,403 deaths from extreme temperatures in the United States between 1979 and 1985, 62% were from excessive cold.

The human body maintains a temperature of 37°C through metabolic heat generation internally in the body. Cooling is achieved through longwave radiation emission, and evaporation from the lungs and the skin. Sweating aids evaporation, but at high temperatures, can generate more than a litre of water per hour under extreme conditions. Rates of sweating above 0.5 litres per hour can lead to sweat gland fatigue, and a rapid elevation in core body temperature. Prolonged sweating, without adequate replacement of salt, will lead to heat exhaustion characterised by fatigue, nausea, vomiting, giddiness and muscle cramps. Rates of sweating above 1.0 litres per hour can cause life-threatening dehydration. Eventually, under prolonged heat, the circulatory system can fail, producing heat stroke. Heat stroke is most likely to occur above effective air temperatures of 33°C, and body temperatures of 40°C. Effective temperatures are reduced by wind; however, when the air temperature exceeds body temperature, increased wind velocity accelerates dehydration. Heat stress affects the old, the young and the infirm more than other segments of the population. In general, the death rate during a heatwave is six times greater for eighty-five-year-olds than for sixty-year-olds. Babies are five to ten times more likely to die than are older children.

In the United States in summer, warm humid conditions, especially at night, lead to the highest mortality. The relationship is strongest in regions where hot weather is uncommon. Hotter conditions, early in the summer, are also more life-threatening than later in the summer. Heatwaves were more deadly before the advent of air-conditioning in the late 1950s. Between 1900 and 1932, the death rates from heat in the United States and Australia were 1.78 and 1.49 per 100,000 people respectively. The greatest death toll in the United States occurred during the summer of 1936, when 4,700 excess deaths were recorded. Other notable heatwaves in the United States occurred in New York in September 1948 (1,400 excessive deaths in five days), the northeast at the end of July 1975 (2,800 excessive deaths in five days); Dallas, Texas in the summer of 1980 (1,200 excess deaths); and Chicago in July 1995 (566 extra deaths). In Australia, during the summers of 1895–1896 and 1938–1939, 431 and 438 people respectively died due to excessive heat. In Europe, heatwaves have had similar effects. In Vienna, Austria in July 1957, the death rate increased by 86–116%, while the hottest summer in a century in London, in 1976, increased the death rate to 50% above the seasonally expected rate.

Heatwaves will increase under global warming. The most affected segments of the population will be those less able to reduce or avoid the effects of heat. These include the physically and mentally handicapped; individuals with existing chronic conditions such as cardiovascular disease, diabetes and hypertension; people whose low incomes oblige them to live in substandard housing without adequate insulation, ventilation or air-conditioning (for example, the hot inner suburbs of large cities); the obese; the unfit; the alcoholic; those taking tranquillisers or certain other widely prescribed medications; and people whose occupations oblige them to perform strenuous work, often in heavy protective clothing (for example fire fighting and blast furnace workers). The impact of heatwaves will depend to a large degree on the rate of climatic change, and upon the ability of a population to acclimatise to higher temperatures. More worrisome is the fact that the increased use of air-conditioning to mitigate the effects of enhanced 'greenhouse' warming will exacerbate the incidences of Legionnaires' Disease, 'sick building syndrome', and infectious airborne diseases such as influenza and tuberculosis. The latter two diseases will become particularly opportunistic unless the present legislated recirculation rates for air in buildings are increased.

Vector-borne disease
(Knight, 1974; Learmouth, 1988; Marthick and Bryant, 1990; McMichael, 1993; Haines et al., 1993; Garrett, 1995; McMichael et al., 1996)

Diseases consist of an agent (viruses, bacteria or protozoa) which infects disadvantageously a host (humans). Some diseases are transmitted by an intermediary organism or vector, mainly insects, which are not affected by the agent. In some cases, a disease may be transmitted within a reservoir species without affecting humans. In this case, the disease is endemic (naturally occurring) in a region, with low rates of human infection. However, under favourable environmental conditions, vector populations can explode, and the resulting diseases may spread to humans as an epidemic. Diseases can be classified as two- (host, agent), three- (host, agent, vector) or four- (host, agent, vector, reservoir) factor diseases. Cholera is an example of a two-factor disease, being spread amongst humans by a bacterium. Malaria is an example of a three-factor disease; an intermediary vector transmits the malarial parasite amongst humans. Bubonic plague is an example of a four-factor disease, with the disease being harboured in a reservoir species, the black rat.

Arbovirus (*ar*thropod-*bor*ne virus) infections

cover a wide disease spectrum, and include asymptomatic infection and mild illness, as well as severe forms of inflammatory brain disease (encephalitis), and haemorrhagic fever with shock or circulatory breakdown. The most common insects involved in the transmission of arboviruses are mosquitoes, ticks and blood flukes (schistosomes). Vectors require adaption to specific ecosystems for survival and reproduction. Insects can breed over a wide range of climatic zones, and will invade uninfested areas opportunistically under increased temperature and humidity. There are approximately 102 arboviruses that can produce disease in humans, and about 50% of these have been isolated from mosquitoes. Table 8.1 summarises some of the main diseases that are climatically controlled.

Mosquito-borne diseases

The main mosquito-borne diseases are malaria, dengue and yellow fever; inflammatory brain diseases (encephalitis); and relapsing fevers such as epidemic polyarthritis. Malaria is a protozoal infection, while the others are virus diseases. The main mosquito vectors for these diseases are *Aedes* (dengue virus, epidemic polyarthritis), *Culex* (Australian encephalitis, epidemic polyarthritis) and *Anopheles* (malaria, filariasis). *Culex* and *Aedes* are able to use organically polluted water for egg-laying and the larval stages, whereas *Anopheles* prefer clean, still water. An increase in rainfall in summer will increase the breeding area for mosquitoes, and the incidence of disease. Mosquito populations in urban areas will increase due to a combination of increased rainfall; lack of predators; and more open drains, pools and damp backyards.

Malaria is a recurring infection produced in humans by four species of protozoan parasites (*Plasmodium*). The most virulent agent is *Plasmodium falciparum* which kills one in ten people it infects. The infection is transmitted through the bite of an infected female mosquito belonging to one of the twenty-seven species of the genus *Anopheles*. In Africa, malaria is associated with Burkitt's lymphoma, a cancer that attacks the lymphatic tissues in early life. Malaria is especially worrisome for three reasons. First, it once had a much greater geographical distribution than present, and could be reintroduced to many countries (Figure 8.3). Second, malaria is now resistant to most drugs. Third, the mosquito vector is very amenable to the warmer and more humid conditions that will characterise enhanced 'greenhouse'

warming. The optimum temperature for breeding by the *Anopheles* mosquito lies between 20–30°C; however, it will breed at temperatures as low as 16°C. Mosquitoes are very active at relative humidities over 60%. Mosquitoes die at temperatures above 35°C and at relative humidities less than 25%. The *Plasmodium* parasite usually fails to develop within the lifetime of the mosquito at temperatures below 19–20°C. However, even a small increase in temperature will accelerate the incubation period of the parasite from three weeks to one. A rainy season of three or more months is likely to enhance breeding by the different species of *Anopheles*. At present, these conditions put at risk 40% of the world's population, and kill over one million people annually.

Historically, places like the English fenland and the north German marshes have had temperatures favourable for the breeding of *Anopheles*. The disease was established in northern Europe in the tenth century, and reached its peak during the Mediaeval Warm Period in the 1150s. Malaria also was very prevalent in Sweden between 1847 and 1856 during warm years. One malaria parasite, *Plasmodium vivax*, has been recorded at Ottawa, Canada and in New York State. Globally, warming will permit many species of the *Anopheles* mosquito to reinvade these areas; major cities, such as Tokyo, Rome and New York, will be at risk. In addition, in tropical areas, *Anopheles* may become more active above 200 metres, putting at risk highland areas in central Africa presently free of the disease. However, malaria outbreaks can also be linked in some countries to ENSO events, which produce warm, wet conditions similar to those forecast for enhanced 'greenhouse' warming.

Dengue and yellow fever are related viral diseases transmitted mainly by the domesticated mosquito, *Aedes aegypti*, which is best adapted to the urban environment, and to the inside of buildings. Areas permanently colonised by this mosquito are limited by the 10°C isotherm, a value that can be enlarged globally under enhanced 'greenhouse' warming. Dengue or 'breakbone' fever results in acute febrile illness with severe generalised pains, headaches, sweating and a typical rash. While it can incapacitate for weeks, it rarely kills except with children, where the death toll can reach 10%. There are four different viruses that cause dengue. Dengue-2, which is the most lethal, has the unusual capacity to infect people that have been previously exposed to a weaker strain. Dengue is endemic to Southeast Asia, but is now spreading to the southwest Pacific (Figure 8.4A). While

Table 8.1 Climatically controlled diseases and their vectors (based on May, 1950).

Disease	Agent or pathogen	Intermediate vector (reservoirs)	Climate factors
Two-factor:			
Amoebic dysentery	*Entamoeba histolytica*		r,h,t
Anthrax	*Bacillus anthracis*		t
Ascariasis	*Ascaris lumbricoides*		r,h,t,l
Bacillary dysentery	*Shigellae*		r,h,t,l
Brucellosis	*Brucella* bacteria		r,h,t,l
Epidemic meningitis	*Neisseria meningitidis*		r,h
Epidemic cholera	*Vibrio cholerae*		r,h
Influenza	Orthomyxovirus		r,h,t
Legionnaires' disease	*Legionella pneumophila*		h,t,w
Leprosy	*Mycobacterium leprae*		t
Pinta	*Treponema carateum*		r,h,t
Poliomyelitis	polio viruses (3)		r,t
Tetanus	*Clostridium tetani*		t
Tuberculosis	*Tubercle bacilli*		r,h,t,l
Typhoid fever	*Salmonella typhi*		r,h,t,l
Yaws	*Treponema pertenue*		t
Three-factor:			
Australian encephalitis	flavivirus	mosquito (birds?)	r,h,t
Dengue	dengue viruses (4)	mosquito	r,h,t,l,w
Epidemic polyarthritis	Ross River virus	mosquito	r,h,t
Epidemic typhus	*Rickettsia prowazeki*	louse	t
Filariasis	Filariid roundworms	copepod, fly, tick mosquito, tabanid	r,h,t,l,w
Leptospiral diseases	*Leptospira*	rodent	r,h,t
Malaria	*Plasmodium*	mosquito	r,h,t,l,w
Relapsing fevers	*Pediculus, Ornithodoros*	louse (rodents?)	r,h,t,b
Four-factor:			
Bubonic plague	*Yersinia pestis*	flea (rodent)	r,h,t
Cesode diseases	*Diphyllobothrium latum*	copepod (fish)	r,h
Leishmaniasis	*Leishmania protozoa*	*Phlebotomus* (dog?)	r,h,t,l,w
Lyme disease	*Borrelia burgdorferi*	ticks (deer)	r,h,t
Q fever	*Colxiella burnetii*	ticks (animals)	r,h,w
Rocky Mountain spotted fever	Rickettsia	ticks (dogs, deer)	r,h,t
Scrub typhus	Rickettsia	mite (rodent)	r,h,t
Schistosomiasis	*Schistosoma*	blood fluke (snail)	r,h,t
Trematode diseases	blood fluke	snail (fish)	r,h,t
Trypanosomiasis	*Trypanosoma*	bug, lice, tsetse fly (various animals)	r,h,t,l,w
Yellow fever	yellow fever virus	mosquito (monkey)	r,h,t,l,w

r–rainfall, h–humidity, t–temperature, l–light, w–wind, b–barometric pressure

Australia has been periodically reinfected since the last century, the disease now appears to be endemic in northern Queensland, especially around Townsville. Dengue has also reinvaded the Americas where the *Aedes aegypti* mosquito had been virtually eliminated. In 1981, 344,000 cases of dengue occurred in Havana; 40% of those infected required hospitalisation. The hot summer of 1990, in Europe, also witnessed the proliferation of the *Aedes* mosquito across that continent.

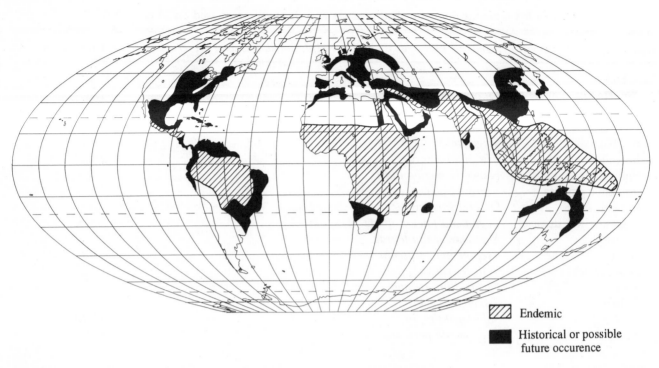

Endemic

Historical or possible future occurence

Figure 8.3 World incidence of malaria (based on May, 1951; Learmouth, 1988).

Yellow fever is a four-factor disease whose animal reservoir is the monkey. The disease causes acutely debilitating fever, chills, jaundice, internal haemorrhaging, coma and, in 5–15% of cases, death. The incubation period for yellow fever varies from three days to several weeks, depending upon temperature. The virus lives in mosquitoes when temperatures exceed 24°C under high humidity. Epidemics occur where mean annual air temperatures exceed 20°C. Yellow fever is prevalent across Africa and Latin America (Figure 8.4B). Humans pick up the disease when they enter forest areas harbouring the monkey reservoir. When these infected people return to settlements, the *Aedes* mosquito then retransmits the disease from person to person in the community. Yellow fever is thus a disease both of forests, and of adjacent urban settlements. This tropical disease has been detected even at mid-latitude ports such as Bristol, Philadelphia and Halifax, because the *Aedes* mosquito can survive in a ship's water tanks. Yellow fever can be spread in the southern United States and along the eastern seaboard, although it is doubtful if it could become endemic. Both yellow fever and dengue can be controlled easily by prohibiting standing water in urban communities.

Encephalitis or sleeping sickness is an inflammatory brain disease. It includes American trypanosomiasis (Chagas' disease), African trypanosomiasis, Japanese encephalitis and Australian encephalitis (also known as Murray Valley encephalitis). Only the last two diseases have a mosquito vector, *Culex annulirostris*. Japanese encephalitis is a flaviviral disease that affects 50,000 people in Asia each year. Australian encephalitis is a rarer disease occurring throughout that continent, particularly around the Murray–Darling River system. Since 1917, there have been seven major epidemics of this disease in Australia, with a death rate around 20%. Approximately 30% of infected people end up permanently incapacitated. *Culex annulirostris* thrives whenever mean daily temperatures exceed 17.5°C. Australian encephalitis outbreaks in southeastern Australia have tended to occur in summer and autumn, following extended periods of above average rainfall. Such outbreaks are well correlated to La Niña events.

Epidemic polyarthritis is characterised by pains in the joints, and is accompanied by a rash on the trunks and limbs. The duration of the disease is usually a few days to several weeks, although the rheumatic effects have been known to persist for years. The major vectors of epidemic polyarthritis in Australia (where it is known as Ross River fever) are *Culex annulirostris* in the inland, and

Part A

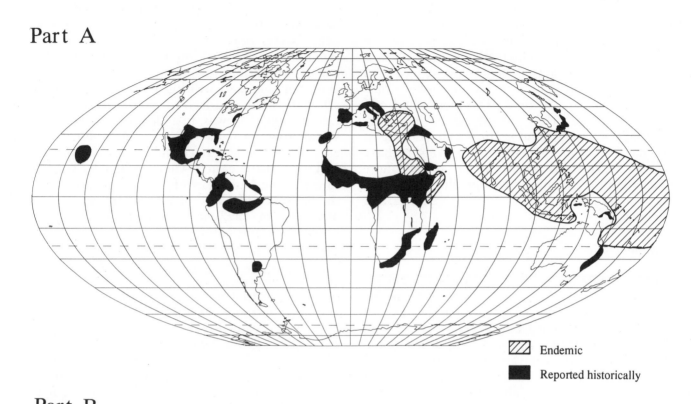

///// Endemic

■ Reported historically

Part B

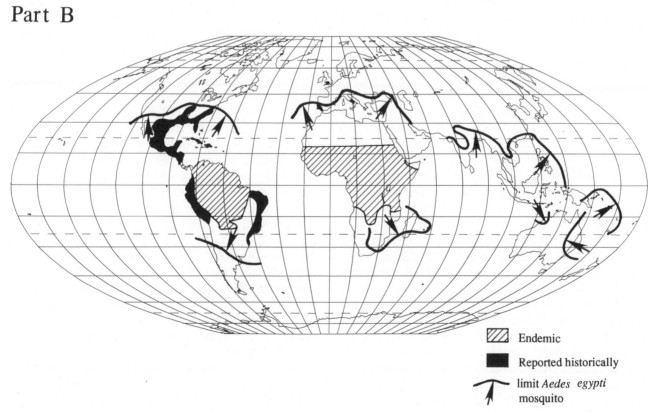

///// Endemic

■ Reported historically

↑ limit *Aedes egypti* mosquito

Figure 8.4 World incidence of (A) dengue and (B) yellow fever (based on May, 1952).

Aedes vigilax in coastal, tidally affected areas. The disease is now widespread throughout Australia. It is also endemic to Papua New Guinea and other islands in the western South Pacific. Because the disease is less severe in children, it may become less significant in the long-term because of acquired immunity. Those most at risk include people spending time in the open. The most rapid and intensive spread of the disease occurs in towns, especially if nearby land is subject to flooding in summer. Epidemic polyarthritis is likely to have higher background levels, to be spread more widely both seasonally and geographically, and to occur more frequently, during wetter and warmer summers.

Tick- and mite-borne (typhus) diseases

Tick- and mite-borne infections, known as rickettsial diseases or typhus fevers, are four-factor diseases. Rickettsiae are large bacteria that cannot live long outside living cells. Typhus causes high fever, chronic headache, depression, delirium and death. Specific diseases are usually spread by a single species of louse, flea, tick or mite. The most feared, and by far most ubiquitous typhus, is epidemic typhus caused by *Rickettsia prowazeki*, and spread by the body louse, *Pediculus humanus*. Like all rickettsial diseases, the vector is dependent upon a combination of environmental circumstances for survival. These include cold temperatures that force humans to wear heavy clothing and live close together. Epidemic typhus is a disease of war, concentration camps, unsanitary ships, squalor and urban living. Under certain conditions, many *Rickettsiae* species can survive in the excreta of fleas, lice and ticks. Under humid conditions, where wind or human activity stirs up dust, the pathogen can be inhaled to cause infection.

Tick-borne diseases, while globally widespread, are chiefly diseases of drier climates. Because they are vector specific, they tend to be given a regional name, such as Rocky Mountain spotted fever, Lyme disease and Q fever. The most affected areas are the United States, eastern Africa and central Asia. Rocky Mountain spotted fever is caused by *Rickettsia rickettsii* and occurs in the United States, mainly the east. It is spread to humans by at least three species of ticks, the wood tick *Dermacentor andersoni*, the dog tick *Dermacentor variabilis*, and the lone star tick *Amblyomma americanum*. The disease is characterised by chills, high fever, headache, backache and prolonged fatigue. Infections can also produce bleeding ulcers and lead to death. Lyme disease is caused by the bacteria *Borrelia burgdorferi* and is spread by deer tick, *Ixodes dammini*. The disease has similar symptoms to Rocky Mountain spotted fever, but causes skin lesions, meningitis, protracted muscle and joint pain, and arthritis. Left untreated, it can affect the brain and lead to fatal heart and respiratory failure. It was first described around the seaside town of Lyme, Connecticut, where its reservoirs include deer, mice and birds. With the revegetation of the eastern United States by scrubby forest in the latter half of the twentieth century, the incidence of Lyme disease has increased dramatically. In 1992, it became the main vector-borne disease for the United States with 10,000 cases reported each year. In Australia, Lyme disease appears to be a new disease whose reservoir is unidentified, although the tick *Ixodes holocyclus* has been implicated as the main vector, and other *Borrelia*-like bacteria as the pathogens.

Q fever is caused by ticks transmitting the bacteria, *Colxiella burnetii*, in their saliva to a wide variety of wild and domestic animals, apparently without harming them. The bacteria may infect milk, urine or dung. When livestock are slaughtered, the bacteria are transmitted via contact, or through the air. In humans, the disease causes severe influenza-like symptoms. Q-fever has appeared widely, being detected in Quebec, United States, Egypt, Panama, Turkey and most countries of Western Europe, including Great Britain. In northern Australia, notifications of Q fever peak in winter during the dry season, when dust is far more prevalent. Drought appears to be the favoured climatic condition for the spread of Q fever, although sufficient rain in the preceding wet season produces pasture growth that will carry increased numbers of livestock.

Because typhus infections are four-factor diseases, the effects of climatic change are more difficult to identify. The incidence of the disease depends more on the environmental conditions for the vector than for the pathogen. While temperatures must be warm enough to allow the life-cycle of ticks and mites to be completed, the reservoir population is also affected by the climate changes. In the United States, the ticks causing Rocky Mountain spotted fever would shift northwards under a warmer climate, and virtually be eliminated from the south, because the tick vector cannot survive hot temperatures. Wetter conditions would enhance the size of reservoir populations of deer, mice and birds, as well as enhancing the regrowth of forests. In Australia, the recent incidence of Lyme's disease may reflect generally wetter conditions over the last fifty years, despite the recurrence of droughts. Overall, enhanced 'greenhouse' warming should lead to a rise in the number of cases of typhus diseases, and their possible spread beyond present regional confines.

Other vectors

Other vectors for disease consist of blood flukes, the tsetse fly and the triatomid bug. The main disease carried by blood flukes is schistosomiasis or bilharzia. The incidence of this disease has doubled in the last fifty years, mainly because of the development of large irrigation schemes. Blood flukes infect the blood vessels of the intestines and bladder causing inflammation and tissue damage. They can lead to epilepsy, kidney failure and malignant cancers. The parasite's eggs are excreted in human faeces or urine, where they end up in near-stagnant surface water. The eggs then infect water snails where they hatch into larvae. The snails excrete the larvae, which attach to stems and leaves of underwater plants. Humans bathing in this water absorb these larvae through their skin, to continue the cycle. Schistosome infections are very debilitating and affect two hundred million people in Africa. In Brazil, the disease has become an urban disease having been carried from the northeast of the country, where it is endemic, into the cities with urban migration. The reservoir snail's ability to be infected is highly temperature dependent. Snails survive between 10°- and 37°C, with the optimum temperature being 24°C. Development of the larva takes forty days at a temperature of 30°C. Climate warming would witness the disease becoming persistent throughout more of the year, while a wetter climate would witness the disease spreading.

The tsetse fly (*Glossina*) spreads sleeping sickness in tropical Africa. It is caused by the parasitic zooflagellate protozoan *Trypanosoma gambiense* and *T. rhodesiense*. Both animals and humans are the reservoirs for this disease. Sleeping sickness is characterised by fever and inflammation of the lymph nodes, brain and spinal cord. The disease leads to extreme lethargy and eventually death, sometimes several years after initial infection. Sleeping sickness has been the scourge of central Africa for centuries, and affects about 25,000 people each year. The tsetse fly requires high humidity, warmth and shade, and spreads only during the rainy season. The tsetse fly does its bloodsucking only in bright daytime conditions, and ceases all activity below 15.5°C. The parasite infection requires eighteen days to develop at 24°C. The greatest European death toll in Africa occurred in the 1860s and 1870s, which were wet years associated with the summer monsoon; the average life expectancy of European immigrants in west Africa over this period was six months. In some locations, human habitation is all but prohibited. While the tsetse fly requires trees and shrubs to survive, increased rainfall and the 'fertiliser' effect of carbon dioxide would provide conditions for accelerated revegetation in those endemic areas where clearing was not maintained. A temperature increase of 1–2°C would witness a spread of the disease, as the activity of the tsetse fly increases with temperature.

Water-borne diseases
(Marthick and Bryant, 1990; McMichael, 1993)

Water-borne diseases are caused by organisms living in drinking water, or water coming into contact with humans. There are three major diseases of note: diarrhoeal diseases, cholera and typhoid. Diarrhoeal diseases may be caused by a variety of infective agents that are presently checked by public health standards. Diarrhoeal diseases consist of a variety of microorganisms including bacteria (*Staphylococcus aureus*, *Salmonellae*, *Shigellae*), viruses (*Echovirus*, *Adenovirus*, *Rotavirus*) and protozoa (*Giardia lamblia*, *Balantidium coli*). Many of these organisms may remain viable in water for months. The existence of surface and ground waters related to heavy rainfall offers opportunities for such pathogens not only to survive, but also to be transported from place to place (and person to person). Diarrhoeal disease may be promoted by the inability of sewers, septic tanks and drains to cope with higher rainfalls, soil moisture and runoff under enhanced 'greenhouse' conditions. Risks are greater in poorer or marginalised sections of society. Natural disasters always pose problems because drinking water may become contaminated, and alternative sources may be difficult to obtain. In developed countries, rising sea-levels could compromise coastal drainage systems, and cause flooding in swampland and low-lying areas. Combined with rising watertables, these conditions would adversely affect sewage disposal systems. One clinically significant water-borne disease, amoebic meningoencephalitis, could become more prevalent under warmer climate.

This disease is caused by the transport of water in pipes exposed to high temperatures.

Typhoid, or enteric fever, is an acute, highly contagious disease caused by the bacterium *Salmonella typhi*. It is transmitted chiefly by contaminated food or water, and characterised by prolonged high fever, headache, coughing and intestinal haemorrhaging. Death can result from the many complications associated with the fever. The typhoid bacillus can be transmitted in polluted water, through inadequate sewage or unhygienic waste disposal, by flies and by human carriers. The prevention of typhoid depends upon adequate water chlorination, sewage treatment, and the exclusion of carriers from the food industry. The first two conditions are a threat during flooding if untreated sewage contaminates drinking water.

Presently, the world is in the throes of a thirty-year global cholera pandemic that began in the Celebes Islands in 1961. Cholera produces a toxin that damages the gut lining, leading to diarrhoea and rapid, often fatal, dehydration. It is spread by the bacterium *Vibrio cholerae* whose survival is dependent upon its ability to hibernate, shrivelled to one hundred and fifty-times smaller than its normal size, beneath the mucous coating of algae. It is endemic around the shores of Bangladesh and India for this reason. The present epidemic is occurring for the first time under conditions of global urban squalor. The disease has also become endemic to the unsanitary slums and shanty towns of South America. Cholera spread to Peru in January 1991, when a Chinese freighter discharged bilge water, originating from Asian waters, into Callao harbour. The pathogen infected shellfish and spread into Lima's unchlorinated water supply. At the same time, the 1990–1995 ENSO event was starting, and warm water allowed the contaminated bilge algae to proliferate. Currents associated with this event then carried the algae-clinging cholera bacteria northwards along the South American coast. Within a year, over 300,000 people had become infected, and by 1995, the ongoing epidemic had infected millions and had incurred costs of $US200 billion to upgrade water, sanitation and sewage facilities throughout Latin America. Had the 1990–1995 ENSO event not happened when it did, the opportunity for cholera to occupy an ecological niche in South America might not have occurred. The algal reservoirs for cholera are very responsive to temperature, rainfall and nutrient loading in wastewaters. As nutrient loads to the ocean increase through pollution from agricultural fertilisers and sewage, algal blooms will become more extensive. The warmer temperatures of enhanced 'greenhouse' warming and the carbon dioxide 'fertiliser' effect will exacerbate this growth.

Respiratory diseases
(Marthick and Bryant, 1990)

Respiratory diseases are the most common of all diseases, and in the case of the common cold, infect over 50% of the world's population each year. More serious is influenza which reaches pandemic proportions as the general population's immunity wanes with time. Global influenza outbreaks take on a yearly or biannual recurrence; however, the more notable events of the twentieth century occurred with the 1918–1919 and 1957 outbreaks. The death toll in the 1918–1919 pandemic reached twenty million. The disease is caused by a virus that infects the respiratory system leading to fever, headache, pain, lethargy and complications such as pneumonia. It is a serious disease for children, the elderly and the infirm. The spread of the disease is enhanced by cold temperatures and high humidities, forcing closer human contact. These conditions are unlikely to occur in a warming world.

Climatic factors, including those that affect atmospheric pollution, have an effect upon the occurrence and worsening of many respiratory diseases, including chronic bronchitis and asthma. Asthma is most significant because its incidence has reached epidemic proportions in developed countries since 1970. To some degree it now afflicts 20% of children between five and nine years of age, and one in ten adults. Asthma is caused by muscular constriction of the bronchi and swelling of the bronchial mucosa. This blocks the free passage of air to the lungs, causing shortness of breath, wheezing and coughing, and in severe cases can lead to death. There are few data relating asthma deaths to environmental pollution, or to exposure to allergen-provoking substances in the air. Nevertheless, many asthmatics have allergic responses to pollen, yeasts, moulds, animal and insect proteins, and dust mite or cockroach excreta. These pathogens are favoured by increases in rainfall and temperature, climatic factors most likely to increase under enhanced 'greenhouse' warming. Light, intermittent rainfall appears to stir up pollen deposited on vegetation during rainless periods, while the incidence of asthma attacks increases above 21°C. Heat stress, combined with high humidity, also exacerbates asthma.

CLIMATE-INDUCED NATURAL HAZARDS AND HEALTH

(Garcia and Escudero, 1982; Marthick and Bryant, 1990; Bryant, 1991; Mitchell and Ericksen, 1992)

While the world's population has doubled since 1960, disaster losses have risen thirty times, from $US3 billion in 1960, to $US100 billion in 1995. Climatic disasters are part of this rising trend; for example, Hurricane Andrew caused $US24 billion damage in the southern suburbs of Miami in 1992. The potential effect of natural hazards is likely to increase globally due to climate warming. The Middle Ages storm record (Figure 5.10) indicates that catastrophic, mid-latitude storms are more prevalent during changeovers in climate, especially in the transition from cold-to-warm climates. For reasons outlined in Chapter 7, the incidence and magnitude of tropical cyclones should not increase; however, the response of the ocean to climate warming is ten to fifteen times slower than that of the atmosphere. Until the ocean re-equilibrates with a changed atmosphere, there is the possibility that certain types of storms, such as east coast lows, will increase. Assuming that there is no change in the variance of temperature or rainfall distributions, global warming should lead to an increase in the frequency of floods and high temperatures. Because floods are often linked to the Southern Oscillation, increased droughts cannot be ruled out. These climatic hazards will induce more bushfires and landslides. The health and social effects of these hazards include death, physical injury, psychological trauma, social dislocation, and the reduced socioeconomic viability of affected communities.

Floods

Historically, floods have produced the greatest death tolls of any natural hazard. By far the worse affected area is the Hwang Ho River in China. In 1332, seven million people drowned and over ten million people died from subsequent famine and disease. The 1887 flood breached twenty-two-metre-high embankments, and drowned one million people. The most recent flooding occurred in 1938, when another one million Chinese drowned and a possible eleven million people died from the ensuing famine. In the United States, flooding of the Mississippi River has continually disrupted commerce. The largest recent floods occurred in 1973–1974 and in 1993. The Middle Ages storms also generated extreme floods on

rivers in northern Europe, with death tolls on several occasions exceeding 100,000. The flooding in the winter of 1995 was the worse flooding in northern Europe to occur in the past two hundred years. In Australia, large floods have become more common in the second half of the twentieth century. During the La Niña event of 1988–1990, severe flooding covered one million square kilometres of eastern Australia. It is debatable whether or not these floods are a global signature of enhanced 'greenhouse' warming; however, they provide benchmarks against which preparation for increased precipitation and flooding can be assessed.

Flash flooding in urban areas poses unique problems. First, the probability of increased convective rainfall, in tropical and subtropical regions, will be felt in cities presently undergoing some of the largest urban growth in the world. Second, urban flash flooding is exacerbated by the structure of the urban drainage system, which may be blocked, diverted or supplanted by an inadequate storm drain system. During intense rainfalls, roadways may take the place of the natural drainage system. Third, urban growth modifies local climate and enhances intense convective rainfalls. While climatologists may be aware of the shift to more extreme rainfalls in a city, local authorities are probably oblivious to the change. For example, in Sydney, Australia, the State government realised that urban flooding was becoming more prevalent, and imposed a $25 surcharge on ratepayers to cover clean-up expenses and relief costs. However, the surcharge was applied to inland suburbs when the evidence clearly showed that coastal suburbs and the central business district were experiencing the greatest rainfalls. Neither the significance of present-day flooding, nor its consequences, have been fully appreciated.

Dam collapses historically have led to the largest flash flood death tolls in small catchments. The failure of dams during heavy rains can be attributed to neglect, inadequate design for high magnitude events, geological factors or mischance. Heavier than predicted rainfall may test the capacity of dam spillways, especially those that have been designed using rainfall records based upon drier periods. This latter situation pervades dam construction in Australia, where many dams in New South Wales were built using rainfall records before 1950. After 1950, rainfall increased by 30%, and many spillways have now reached the end of their design lifetime. It is estimated that, if urban expansion occurs in Sydney, up to one million people could be flooded if the main dam supplying drinking water to the metropolitan area were to fail. This situation will not be unique if rainfall increases globally because of warming.

Drought

Drought is by far the greatest and most widespread of all climate-related natural hazards. Droughts on the Indian subcontinent in 1769, 1790, 1866, 1876–1877 and 1943 killed millions, while in China in 1878, 10–13 million people died because of drought-related famine. Many of these droughts were linked to ENSO events. If famine occurs during drought, malnutrition and disease can lead to permanent intellectual and physical impairment. Not all famines are caused by drought, nor do all droughts lead to famine. Droughts are ubiquitous, and have an impact upon the economies of both developed and developing countries. While droughts can lead to long-term morbidity, they have little effect on population growth. Because of the quasi-cyclic nature of drought over large sections of the globe, it is doubtful if climatic warming will decrease their intensity or frequency. For example, while eastern Australia has become 30% wetter since 1950, droughts have not diminished in their frequency or intensity. In fact, the worst droughts recorded have occurred recently, in 1982–1983 and 1991–1995.

Bushfire

Recent fire frequencies and intensities in Australia, California and southern Europe conform with a globally warming climate. However, these fires should be set within a historical context. Much larger fires, with higher death tolls, have been experienced over the past two centuries, in the deciduous and mixed forests of Europe and North America. In North America, at least thirteen fires in recorded history have burned more than 400,000 hectares. The Wisconsin and Michigan forest fires of 1871 consumed 1,700,000 hectares and killed 2,200 people. Areas of similar magnitude were burnt in Wisconsin in 1894, and in Idaho and northwestern Montana in 1910. The Michigan fires of 1881 killed several hundred; the Hinckley, Minnesota fire in 1894 took 418 lives; and the Cloquet fire in Minnesota in 1918 killed 551 people. Fires were so common that most residents took fire warnings nonchalantly, to their own detriment.

The dominant factors contributing to the occurrence of bushfires are a source of ignition, the availability of highly flammable fuels, and the existence of meteorological conditions conducive to combustion and spread of fire. It is the last two factors that will be affected by global warming. Rainfall is a major parameter in the months preceding a dangerous fire period. Good spring and early summer rains will reduce the fire risk in a forest, but provide abundant undergrowth that will ensure a high fire risk at a later time. Severe antecedent drought will dry out this vegetation, and increase the fuel load. The alternation between record floods and extreme droughts recently plaguing some countries, such as Australia, favour both circumstances. Low relative humidity and high wind speed also play a key role in determining the likelihood of a bushfire and its severity. The intensity of any fire is not solely dependent upon climate conditions, but also upon the availability of dried litter or debris generated by disease or insect infestation, windstorms, previous fires or land clearing. High fuel loads are assured with warmer and wetter conditions, under the 'fertiliser' effect of higher concentrations of carbon dioxide.

Preparedness and the consequences of more disasters

The severity of hazards may not relate so much to climate change, as to the degree of preparedness. For example, the Sydney bushfires of January 1994, occurred during a drought which was well predicted. Temperatures had been above average for weeks, rain was non-existent, humidities were low, and undergrowth in the dry sclerophyll forest had undergone prodigious growth in the previous La Niña event. In addition, urban expansion had encroached upon bushland in terrain that was very susceptible to wildfire. Despite the apparent threat of the fires, only 5% of the treed landscape succumbed to fire, and there were only four deaths. Through a monumental organisational effort, that witnessed urban firetrucks and bushfire brigades being mobilised at a national level, disaster was averted because of foresight and planning. If the 1994 bushfire conditions in Sydney were a signature of a warmer globe, then the predicted consequences of such warming did not eventuate because of the high level of preparedness.

However this example may be an exception. Much of society's response to hazards is reactive rather than proactive. The most quoted response to Hurricane Andrew, which struck southern

Florida in August 1994, was from the Dade County emergency management director, who, on national TV four days after the disaster, and with no federal response in sight, said "Where the hell is the cavalry on this one?". So much development, worldwide, has been permitted in areas where flooding, storms and bushfires occur, that policy-makers must now seriously address the question of how the inevitable disasters should be handled. It is imperative that disaster management plans be formulated, not just to handle future climate change, but to deal with the present risk of hazards in expanding urban areas, in both developing and developed countries.

EFFECT OF OZONE DEPLETION

Effects on UV radiation amounts
(Setlow, 1974; National Health and Medical Research Council, 1989; McKenzie et al., 1992; de Gruijl, 1995)

Ultraviolet radiation can be broken down into three components, UVC, UVB and UVA, with wavelengths between 100–280, 280–315, and 315–400 nanometres respectively. Ozone effectively absorbs all UVC radiation and about half of the UVB radiation. The more important component of ultraviolet radiation, in terms of biological effects, is UVB radiation at the 300 nanometre wavelength. Both UVC and UVB are absorbed in living matter by molecules such as DNA, which can be damaged. UVB radiation is strongly absorbed in the skin and outer layers of the eye. The amount of UVB radiation reaching the ground depends upon the season of the year, the latitude and the amount of cloud cover. For small ozone depletions of less than 5%, a 1% decrease in stratospheric ozone will lead to a 2% increase in biologically accumulated dosage of UVB. Ozone depletion could thus lead to a weakening of the immune system, and an increase in the frequency and severity of skin cancer, eye cataracts, viral infection, and chronic fungal and protozoal infections.

Unfortunately there are few reliable, globally representative, long-term records of UVB radiation suitable for assessing these outcomes. Trends are complicated by the fact that UVB radiation can be absorbed by clouds, which are known to be increasing over most of the globe (Figure 7.6). Additionally, UVB radiation can be absorbed by the increase in ozone that is occurring within the lower troposphere, especially around urban areas.

In the United States, the amount of UVB radiation reaching the ground actually decreased between 1974 and 1985. UVB values, globally, have been inferred from satellite ozone measurements. Changes, by latitude, in erythemal and DNA-damaging ultraviolet radiation, between 1980 and 1990, are shown for two seasons in Figure 8.5. Erythemal ultraviolet radiation is that component of ultraviolet radiation that is biologically reactive. It produces sunburn, or reddening of the erythema. DNA-damaging radiation is a more accurate method of assessing total biological effects. Erythemal UV radiation has increased by 5–10% at mid-latitudes in both hemispheres during the 1980s, with spring values being slightly higher in each hemisphere. Noticeable increases of up to 25% have occurred over the Antarctic because of the 'Ozone Hole'. DNA-damaging radiation has increased more, especially in the southern hemisphere. While the largest increases for both types

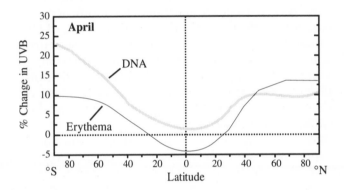

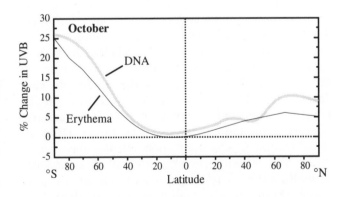

Figure 8.5 Change in erythemal and DNA-damaging ultraviolet radiation by latitude and season, 1980–1990. Values have been inferred using the TOMS, total column ozone measuring satellite (from McKenzie et al., 1992 and Madronich et al., 1991).

of UV radiation occur in winter and spring, these changes are relatively unimportant overall, because the total amount of solar radiation reaching the surface of the Earth at these times is small.

UV radiation and the skin
(Blum, 1964; Giles, 1989; Marks, 1989; Müller, 1989)

UV radiation is essential for human life. It triggers the formation of vitamin D3 essential for building and maintaining bones. Deficiencies in vitamin D lead to rickets, and the development of soft deformable bones. Excessive UV radiation causes both acute and long-term changes in the skin. Skin contains an outer layer called the epidermis, composed of squamous cells which develop from a basal layer of cells. Pigment or melanin is produced in the epidermis by melanocytes. If exposure to UVB radiation is slow, then melanin, produced within melanocytes, reflects ultraviolet radiation and neutralises its heat effects. It also neutralises free oxygen radicals generated in photochemical reactions within irradiated skin. Acute levels of UV radiation cause inflammation or sunburn, enhanced skin pigmentation and skin thickening. Long term exposure to ultraviolet radiation, usually over 20 or 30 years, can cause damage to the skin, leading to mutation and eventually skin cancer. UV radiation also alters elements of the immune system in the skin, leading to immunosuppression, and may affect the progression of several human diseases including cancer, the herpes simplex virus and some protozoal infections.

Sunburn is the best known low-level complaint of excessive ultraviolet radiation. Sunburn damages the epidermis, causing swelling, inflammation and leakage of plasma. Nucleic acids and proteins in the skin are also damaged. After a day in the sun, a typical cell in the epidermis can develop 100,000–1,000,000 damaged sites in its DNA. If the body's repair mechanisms cannot eliminate all damaged cells, then faulty DNA is reproduced. A series of such events may cause crucial tumour-suppressor genes, called oncogenes, to malfunction, leading to growth of a tumour and cancer. Skin cancers, mainly because of their visibility and death potential, have received the most publicity as a UVB-related health condition. The precursors of skin cancer consist initially of precancerous lesion solar keratoses, that are present in 40–50% of the total human population aged over forty. These can ultimately develop into the three main cancers, basal cell carcinomas (BCC), squamous cell carcinomas (SCC) and cutaneous malignant melanomas (CMM). The first two cancers are non-melanocytic skin cancers (NMSC), and tend to grow locally on those parts of the body that have been exposed to sunlight. Only 1–2% of these skin cancers become lethal. Cutaneous malignant melanoma is a tumour that grows from the melanocytes, the melanin pigment-producing cells in the skin. The mortality rate for cutaneous malignant melanoma approaches 25% for whites in the United States. There are four types of CMM: Hutchinson's melanotic freckle (HMFM), nodular melanoma (NM), superficial spreading melanoma (SSM), and acral lentiginous melanoma (ALM). HMFMs occur on skin surfaces exposed to the sun, and are related to cumulative sun exposure similar to non-melanocytic skin cancers. Superficial spreading melanomas occur on the body intermittently exposed to sunlight (the trunk in males and the thighs of females). Acral lentiginous melanomas occur on the soles and palms, and are restricted to black people.

The incidence of non-melanomas varies dramatically by skin type, UV exposure and age. South African blacks have one of the lowest rates in the world, with an annual incidence of one per 100,000 people, while Caucasians in Australia have the highest incidences of non-melanoma skin cancers and cutaneous malignant melanomas in the world. Figure 8.6 plots the mean daily erythemal dose of UV radiation against the incidence of skin cancers in Australia. The distribution of erythemal radiation across Australia is plotted in Figure 8.7. Mean daily erythemal dosage increases steeply towards the tropics, peaking at a value of

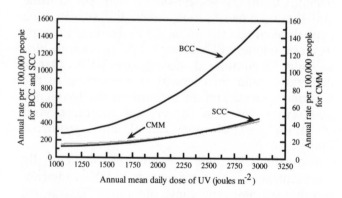

Figure 8.6 Average incidence per year of non-malignant (BCC and SCC) and malignant (CMM) melanomas plotted against annual mean daily erythemal dosage of ultraviolet radiation (in joules per square metre), in Australia.

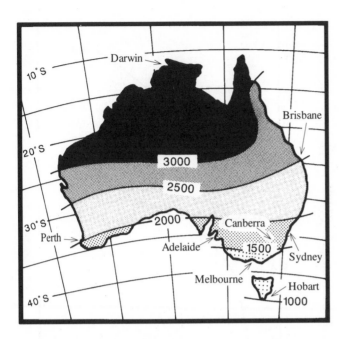

Figure 8.7 Annual mean distribution of daily erythemal radiation (in joules per square metre) over Australia (based on Barton and Paltridge, 1979).

3,000 joules per square metre over the north of the continent. The lowest values are recorded over the island of Tasmania at a latitude of 45°S. A similar pattern would exist across the United States under clear skies. There is a slight tendency for radiation to decrease towards the east coast as latitude decreases because of enhanced cloudiness. In the tropics, where the annual mean dose of UV radiation can be as high as 3,000 joules per square metre, the incidence of squamous cell carcinoma and basal cell carcinoma can reach 460 and 1,500 cases per 100,000 people per year, respectively. These rates decrease by a factor of two and four respectively at higher latitudes, where the annual UV dosage is only 1,000 joules per square metre. Cutaneous malignant melanomas have almost the same relationship to UV levels as squamous cell carcinomas; however, the incidence is an order of magnitude lower. Cutaneous malignant melanomas are also affected by behaviour and lifestyle. While non-melanomas are a cancer of the old, it is not uncommon for teenagers and young adults to develop CMMs.

Because skin cancers are dependent upon accumulated ultraviolet dosage, their incidence is age dependent. There is also a difference between the sexes. These relationship are plotted in Figure 8.8 for Australia. Generally, females suffer from a higher rate of non-melanomas early in life than men do. After the age of fifty-five, this incidence drops in females, but continues to rise exponentially in males. There is also a slight sex difference in the incidence of cutaneous malignant melanomas, and a tendency for the rate of increase to decelerate after the age of forty. The age differences can vary amongst countries. For instance, in England and Wales, females have twice the incidence of CMMs as males. The incidence of CMMs is also related to the number of sunburn events a person has experienced, especially as a child.

The United Nations has estimated that for every 1% decrease in ozone, the incidence of BCC and SCC will increase 2.0% and 3.5% respectively. In Australia, each 1% increase in biologically active ultraviolet radiation will result in 2,500 new cases of non-melanocytic skin cancer per year, at present rates of exposure. A 10% increase in solar ultraviolet radiation, here, would cause a 5 to 10% increase in mortality due to cutaneous malignant melanomas. In the United Kingdom, the number of cases of cutaneous malignant melanomas has doubled between 1980 and 1995. Future projections in this latter country suggest that the lifetime risk of suffering skin cancer will rise by 16%. Publicity of these rates has caused the people most at risk (fair-skinned, red-haired, easily sunburnt) to modify their exposure to the sun. In Australia, mortality from non-melanocytic skin cancers has decreased annually from 1930 to 1987 by 10% in males and 17% in females. However over the same period, mortality from cutaneous malignant melanoma has increased annually by 22% in males and 17% in females. There is little evidence that any of these recent changes have been caused by changes in stratospheric ozone levels.

UV radiation and the eye
(Hollows and Moran, 1981; Taylor et al., 1988; Favilla, 1989; Hollows, 1989)

Eye diseases are far more common and more serious an impairment than skin cancers. UV radiation has been shown to be harmful to the lens, retina, retinal pigment epithelium, ocular lens and cornea; and may be a factor in the development of cataracts, retinal degenerations and visual ageing. The damage is caused through photochemical reactions. In addition, presbyopia, often called age-related nearsightedness, and deformations of the anterior lens capsule, are affected by the degree of exposure to UV sunlight.

UVB radiation is directly, but not singly, linked

to the formation of cataracts. Over half of the world's estimated 25–35 million cases of blindness are caused by cataracts. Cataracts are an opacity of the lens affecting the cornea, mainly in the elderly. The cornea absorbs almost all radiation below 290 nanometres and a portion of UVB. There are two main types of senile cataract, of which 20–25% are nuclear (centred behind the pupil), and 75–80% are cortical (affecting the peripheral parts of the lens). Research has found an association only between UV radiation and cortical cataracts. In the United States, senile cataracts affect 12.3% of the population, with the incidence climbing from 3.5% for those under the age of sixty-five, to 41.4% for those over the age of seventy-five. In Australia, 1.4 million people had cortical cataracts in 1989, out of a population of seventeen million. Figure 8.9 plots the mean daily erythemal dosage of UV radiation against the incidence of cortical cataracts in Australia. The incidence is linearly related to the amount of exposure to this type of radiation, reaching a high of 12,000 cases per 100,000 people in the tropics, where the mean daily erythemal UV dosage is 3,000 joules per square metre. Cataracts are also a common cause of blindness amongst Australian Aborigines exposed continually to high levels of erythemal UVB. The incidence rises, for those over the age of sixty, from 13.6% of the Aboriginal population in the south to 30.5% in the north. In the United States, a doubling of UV exposure is associated with a 1.6-fold increase in the incidence of cortical cataracts.

Other common eye disorders include acute and chronic photokeratoconjunctivitis ('snow blindness'), and pterygium. Exposure to UV radiation is believed to be the major cause of these complaints. Pterygium is a common eye condition caused by reflected sunlight. It is a condition occurring in people exposed to reflected solar irradiance. Pterygium produces a wing-shaped vascular thickening of the conjunctiva, that spreads onto the cornea by the growth of subconjunctival fibroblasts. It results in discomfort, diminished vision, and, in a few rare cases, blindness. The incidence of pterygium in Australia is also plotted in Figure 8.9 against the mean daily erythemal dosage of UV radiation. Pterygium is rare where the mean, daily dosage of erythemal ultraviolet radiation falls below 1,400 joules per square metre. This occurs only in the southern part of Australia. However the incidence rises steeply towards the tropics where the incidence approaches 1,300 per 100,000 people. Again, the disease is more common in the

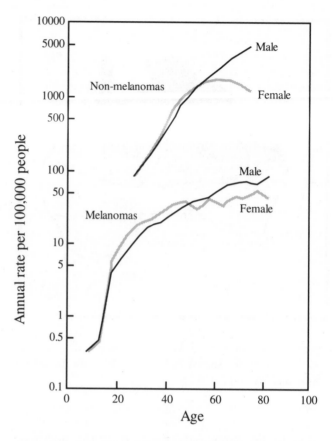

Figure 8.8 Incidence of non-malignant (BCC and SCC) and malignant (CMM) melanomas by age and sex, in Australia (from Giles et al., 1988 and Giles, 1989).

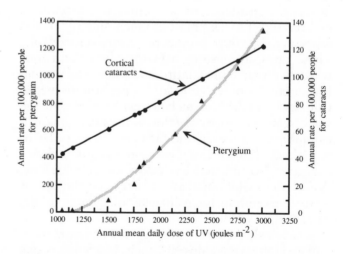

Figure 8.9 Incidence of pterygium and cortical cataracts plotted against annual mean daily erythemal dose of ultraviolet radiation (in joules per square metre), in Australia.

Aboriginal population, where a 1% increase in UV radiation would result in a 2.5% increase in its incidence. There is also a sex difference in the incidence of pterygium in Australia, with males of all ages having a higher incidence of affliction than females.

Photokeratoconjunctivitis results in inhibition of mitosis, epithelial loosening, and nuclear fragmentation. The acute form results in red and gritty eyes, photophobia and blepharospasm after 0.5–24 hours of exposure to sunlight, with some self-healing over the next 24 to 48 hours. In its chronic form, nodular opacities develop on the cornea. Nodular band-shaped keratopathy, climatic droplet degeneration of the cornea and Labrador keratopathy are all clinically similar eye complaints related to ultraviolet radiation. These latter conditions result in hazing of the cornea and are caused by prolonged exposure to high levels of reflected solar irradiance.

UV radiation and immunity
(Jose, 1989; de Gruijl, 1995)

UVB radiation not only increases the cancer potential in epidermal cells, but also initiates specific immunological events. These events lead to the appearance of T-suppressor cells that are linked to the development of UV-induced skin cancers. This immunosuppression occurs regardless of pigmentation. Irradiated skin also undergoes reduced sensitivity to contact with chemicals. People who have had kidney transplants, and who are put on immunosuppressive medications, dramatically increase their risk of skin cancer. The function of the immune system decreases up to two weeks after exposure to ultraviolet radiation. While interferon levels initially increase after exposure to sunlight, they decrease for four weeks afterwards. People who sun-bask in solariums show suppressed natural killer (NK) cell activity in blood, enhanced suppressor cell activity for immunoglobulin production, and delay of hypersensitivity in skin. However not all of these changes can be attributed solely to UVB radiation. UVA may also be responsible for NK cell suppression. UV-induced immunosuppression could lead to greater incidence or severity of infectious diseases, including viruses involved in herpes, papilloma and acquired immunodeficiency syndrome (AIDS). Skin cancers are prevalent on patients with AIDS, and people with latent herpes simplex virus can have the condition reactivated by ultraviolet exposure. UVB radiation may also trigger chronic fungal and protozoal infections. Ultraviolet radiation immune suppression has been linked, in the tropics, to increased leprosy, tuberculosis, and fungal and parasitic infections, although poor hygiene and malnutrition can also exacerbate the incidence of these diseases. Finally, exposure to UVB radiation may lower the effectiveness of some vaccinations such as tetanus.

Australia as a case study

Australia is one of the few countries to date affected by the dispersion of the Antarctic 'Ozone Hole' in summer. It also has some of the highest rates of UVB-induced diseases in the world. The effects of ozone depletion and ultraviolet radiation enhancement should peak in the year 2002 as the prohibitions on halocarbons take effect. Four medical conditions: non-melanocytic skin cancers, cutaneous malignant melanomas, pterygium and cortical cataracts, have been examined in detail in Australia in response to hypothesised changes in UV radiation to the year 2030. These four ailments are the most prevalent, publicised, life-threatening, debilitating and costly UVB-related health conditions in Australia today.

The forecast changes in ultraviolet radiation as the result of ozone depletion are shown for selected Australian cities, marked on Figure 8.7, in Table 8.2. No attempt has been made to forecast ultraviolet levels beyond 2010. Nor has any attempt been made to adjust the figures for the global effects of increased cloudiness (Figure 7.6). The 1% increase in cloudiness each decade in Australia should reduce ultraviolet radiation by 0.3%. Projected changes in ultraviolet radiation due to ozone depletion should be virtually non-existent close to the equator. Thus, in a tropical city such as Darwin there is little need to take extra precautions to mitigate the effects of exposure to sunlight because of ozone depletion. Erythemal ultraviolet radiation will increase progressively southwards across Australia. In Melbourne, a city of three million people, ultraviolet radiation will increase by up to 5.5% by the year 2000, while in Sydney, a city with four million people, it will increase by 4.7%. If DNA-damaging ultraviolet radiation is considered, then these estimates rise to 7.5% and 5.7% respectively for Melbourne and Sydney.

In this case study, the 1990 incidences of the four UVB-induced medical conditions have been applied to the existing age distribution in Australia, and summed across the continent, to

Table 8.2 Forecast percentage changes in ultraviolet radiation across Australia, 1990–2010
(all values are referenced to 1990).

Location	1980–1990	1990–1995	1990–2000	1990–2005	1990–2010
Darwin	−0.1	−0.1	−0.1	−0.1	−0.1
Brisbane	2.9	3.0	3.1	2.7	2.3
Perth	3.8	3.9	4.1	3.6	3.1
Sydney	4.3	4.5	4.7	4.1	3.5
Adelaide	4.7	4.9	5.1	4.4	3.8
Canberra	4.7	4.7	4.8	4.1	3.5
Melbourne	5.1	5.3	5.5	4.8	4.1
Hobart	6.6	6.8	6.9	6.2	5.1

give the present number of cases of each disease. Population projections to the year 2030 have been forecast, taking into account inward migration to Australia, and inter-state population movements. There is presently significant migration into Queensland because of the attraction of its sunnier and more benign climate. Queensland already has the highest rate of skin cancers in Australia, if not the world. The overall change in the demography of Australia by age and sex, to the year 2030, is shown in Figure 8.10. The most noticeable feature of these projections is the continued ageing of the Australian population. The result is significant, because most UVB-related diseases, especially non-melanocytic skin cancers, appear after the age of forty and increase in incidence with age.

The projected number of extra cases of non-melanomas (basal and squamous cell carcinomas), cutaneous malignant melanomas, pterygium and cortical cataracts are shown in Table 8.3. The percentage increases do not take into account the growth in the Australian population, but do include the effect of ageing. Changes in ultraviolet radiation due to ozone depletion by the year 2010 will increase non-melanomas and pterygium in Australia by over 10%, melanomas by 5.6%, pterygium by 8.1% and cortical cataracts by 3.6%. For comparison, the estimated increase in incidence of non-melanomas in the United States over this timespan is 16%. The medical costs of each health condition are also listed in Table 8.3. No attempt has been made to include inflation; but costs due to loss of income, transportation costs to hospital and funeral expenses have been included. The greatest cost will accrue because of the increased incidence of cortical cataracts which require hospitalisation. The present incidence guarantees that 61% of the Australian population

over the age of seventy-five will develop cortical cataracts. In rural Queensland and the Northern Territory, in the tropics of Australia, the incidence rises to 87% for this age group. Overall the four UVB-related diseases will cost the Australian economy over $100 million dollars each year. While the initial increase is due to enhanced ultraviolet radiation, much of the increase after the year 2000 will be due to the ageing of the Australian population. By the year 2030 the expenditure on these four UV-related medical conditions will rise to $185 million above 1990 levels. Over 70% of this cost will be attributable to cortical cataracts.

UNCERTAINTIES

Reappearance of old diseases

The reappearance of diseases that were controlled prior to 1970, is not necessarily linked to changing climate. For instance, malaria is now a major concern because it is known to have had a wider occurrence than at present, its vectors are very common, the mosquito vectors are resistant to DDT and its replacements, and the disease itself has now built up a resistance both to quinine and chloroquine. Since 1965, the incidence of malaria has been growing by about 50% per year in some Asian countries. For example, the incidence of malaria in India has grown from 100,000 cases in 1966 to over five million in 1975. In a country such as Australia, where the last case of endemic malaria disappeared in 1973, a reservoir population of 10,000 is all that would be required to reestablish the disease. At present 2,500 people who have been infected with malaria live in the receptive area for mosquitoes transmitting the disease.

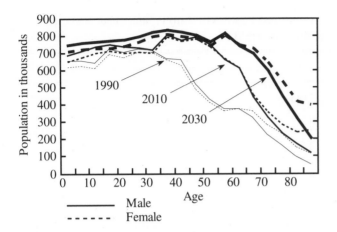

Figure 8.10 Demographic projections by age and sex, for Australia, for the years 1990, 2010 and 2030.

Similar conditions would apply for the reestablishment of the disease in North America and Europe.

Tuberculosis (TB) is a disease of cold climates, poverty and non-vigilance. It is more prevalent in rainy autumns, following dry summers. The disease is suppressed by ultraviolet light. The cold conditions of the fifteenth and seventeenth centuries aided the airborne transmission of tuberculosis, as people huddled indoors with little exposure to fresh air. In modern times, TB has become a disease of overcrowding and climatic deterioration. Wherever cities have expanded and developed slums, tuberculosis has occupied a new ecological niche, killing half the people it infects. TB declined in the twentieth century until, by 1970, it was no longer viewed as a disease of developed nations. But in the last twenty years of the twentieth century, tuberculosis has reemerged as a major disease of importance. In the United States, the incidence of the disease has increased 60% because of its ability to infect patients with AIDS, and because of drug resistance. In 1991, New York City was found to have a treatment compliance rate for TB of only 10%, lower than that of any developing country. In marginalised regions of that city, the infection rate is 51%, while in the general population it is 10.2%, an incidence worse than in many Third World countries. The disease is transmitted in the United States amongst prisoners in crowded jails, and amongst drug offenders. Its transmission is exacerbated by public health laxness in recognising the magnitude of the current epidemic.

The reemergence of malaria and tuberculosis as major diseases is not unique. It represents one more step in an evolving phase of interaction between disease and human ecology. In 1995, tuberculosis and malaria were responsible for two and one million deaths respectively, with annual rates of increase of 13% and 5%. Tuberculosis is now the number one killer. Of the more than seventeen million deaths in 1995 due to disease, 40% were caused by diseases that in 1980 would have been considered under control. More importantly, 76% were diseases where the virulence of the vector or pathogen is influenced by climate. Of these diseases only 40% would be transmitted more readily under global warming.

New diseases
(Garrett, 1995; Le Guenno, 1995; Budiansky, 1995)

The formation of the global village in the twentieth century has led to a third wave of infectious disease in human experience. Over the past two decades, as the world's population has grown, there has been increased human contact with species and organisms in areas that were once considered remote or isolated. The increased mobility of humans, the greater number of international travellers, the shortened time of travel over long distances, increased marginalisation of people, and increased multiple sexual partners have made the global community a reality, at least in terms of the potential for the spread of exotic microbes and mutated disease pathogens within human ecology. In addition, the lack of vigilance of the medical research community has often aided the spread of these diseases into human environments, mainly urban areas. No disease completely kills its host population; a proportion of the population is likely to build up an immunity. However, new diseases can wipe out over 30% of a population, as with the Black Plague when it first came into contact with human populations.

The source and vectors of new diseases appearing in the latter half of the twentieth century are listed in Table 8.4. The most feared new diseases are the haemorrhagic fever viruses: flaviviruses, arenaviruses, bunyaviruses (including hantaviruses) and filoviruses. These viruses have existed for millions of years, but have had little contact with humans. The flaviviruses are transmitted by mosquitoes or ticks. The arenaviruses and bunyaviruses circulate naturally in animal populations and are rarely spread by people. The animals serve as reservoirs for the virus, and sometimes as vectors for transmission to people.

Table 8.3 Projected increases in the number of cases of UV-induced health conditions, and their costs, for Australia, 1990–2010, relative to 1990.

Condition	Incidence/100,000	% increase in number of cases over 1990 levels			
	1990	1995	2000	2005	2010
Non-melanomas (BCC)	849	10.3	12.1	12.1	11.7
(SCC)	244	10.5	12.5	12.3	11.6
Melanomas	26	7.9	7.5	6.5	5.6
Pterygium	457	11.7	10.9	9.4	8.1
Cortical cataracts	7,960	4.4	4.4	3.9	3.6
		Estimated additional cost ($000,000) above 1990 values			
Non-melanomas (BCC)		27.4	32.5	32.1	30.2
(SCC)		16.1	19.0	19.0	18.4
Melanomas		7.4	8.4	8.0	7.6
Pterygium		9.2	10.6	10.1	9.4
Cortical cataracts		53.9	62.7	70.6	69.3
Total cost		114.0	133.2	139.8	134.9

Rodents are excellent reservoirs because they are not affected by the virus, yet they can shed viral particles through their faeces and urine. Hanta-viruses received widespread attention during the Korean War when over two thousand United Nations troops became infected between 1951 and 1953. The reservoir was identified as the Korean field mouse *Apodemus agrarius*. The virus is also found in Japan, Russia, the Balkans and Europe. In Europe, the virus is non-fatal and is known as Puumala virus. It is spread by wood-land animals, mainly the bank vole (*Clethriono-mys glareolus*). Since 1977, the disease appears to be on the increase mainly because of the inhalation of contaminated dust from the handling of wood, or from working in sheds and barns. Sabià and Muerto Canyon viruses are also hantaviruses, producing fever, health deterioration and eventually bleeding. Sabià was first isolated in 1990 in the state of São Paulo, Brazil, while Muerto Canyon appeared in 1993 in New Mexico. Filo-viruses have a filamentous appearance and an unknown method of transmission. They are all deadly. Two outbreaks of the Marburg filovirus, in Germany in 1967, and in Reston, Virginia in 1989, were traced to monkeys being used for research or the preparation of vaccines. One of the deadliest filoviruses is Ebola Zaire virus, which causes a breakdown of cell walls resulting in horrific and painful haemorrhaging. There have

been two outbreaks that were only stopped after international intervention. The latest outbreak occurred in January 1995 and killed 92% of people infected.

The primary cause of haemorrhagic fever viruses appears to be the intrusion of people into ecosystems where there is a prevalent reservoir animal population. For instance, the Guanarito arenavirus first appeared in Venezuela in 1989, in a rural community who were clearing forest. The animal reservoir turned out to be a species of cotton rat whose faeces and urine had been disturbed by the clearing. Both Machupo virus which appeared in Bolivia in 1952, and Junín virus which appeared in Argentina in 1958, are arenaviruses that are spread by vesper mice. Opening of large areas of pampas for maize cultivation led to a population explosion in vesper mice. Mechanisation in the harvesting of maize caused the airborne suspension of infective dust. Muerto Canyon or Sin Nombre virus is spread by the deer mouse *Peromyscus maniculatus* which lives on pine kernels. During the spring of 1993, heavy precipitation and high humidity, associated with the ENSO event of 1991–1992, led to a bumper crop of pine kernels, and a tenfold increase in mice in 1992–1993 in the southwest of the United States. This hantavirus infected 114 people, and subsequently killed fifty-eight in the southwest. Other hantaviruses are widespread across the United States with reservoirs

Table 8.4 New diseases of the twentieth century.

Disease	Type of virus	Vector or reservoir	First noted	Area detected
AIDS	HIV	monkeys?	1978	global
Ebola	filovirus	started in cotton mills	1976	Sudan & Zaire
Equine morbilivirus	morbilivirus	birds?	1995	Queensland
Guanarito	arenavirus	cotton rat	1989	Venezuela
Hantaan	hantavirus	field mice	1951	Korea
Japanese encephalitis	flavivirus	birds, mosquitoes	1990s	Australia
Junín virus	arenavirus	vesper mice	1958	Argentina
Lassa fever	arenavirus	rats	1970	west Africa
Machupo virus	arenavirus	vesper mice	1950s	Bolivia
Marburg	filovirus	laboratory monkeys	1967	Germany
Muerto Canyon	hantavirus	deer mouse	1993	southwest USA
Puumala virus	hantavirus	voles, field mice	1934	northern Europe
Rift Valley fever	bunyavirus	mosquito	1970	east Africa
Sabià	hantavirus	mice?	1990	São Paulo, Brazil

in deer mice, the voles species *Microtus*, and maybe the brown rat *Rattus norvegicus*. In Baltimore, 6.5% of people undergoing kidney dialysis treatment in the late 1980s tested positive for hantavirus antibodies. It is possible that many cases of acute respiratory distress syndrome (ARDS), kidney disease and hypertension in the United States could be related to hantaviruses.

Certainly some viruses are exacerbated by climate change. Some bunyaviruses are spread by mosquitoes linked to the building of dams and the expansion of irrigation, although wetter climates cannot be ruled out. The best known of these viruses is Rift Valley fever, known since 1931 mainly as an infection in sheep and cattle in South Africa. In 1977, seven years after the Assam Dam was constructed, not only were cattle infected, but so were 200,000 people with the death of six hundred. A similar epidemic followed the damming of the Senegal River in Mauritania in 1987. Rift Valley fever is spread by many species of mosquitoes, but commonly *Aedes*. In all cases, abundant rain or irrigation has allowed the populations to explode, leading to the spread of the disease. Legionnaires' disease, caused by the bacteria *Legionella pneumophila*, was only identified in 1975. Unless treated quickly, it kills approximately 20% of the people it infects. The bacterium grows in the humid conditions in centralised air-conditioning units, but can be dispersed easily on air currents. It is particularly virulent in old people, the very group most susceptible to heat stress, and who would require air-conditioning under enhanced 'greenhouse' warming.

CONCLUDING COMMENTS

Until the twentieth century, the history of disease has proceeded in two phases in relation to developing human civilisation. First, diseases infected humans under the new ecological conditions created by cities. Then, as civilisations grew, trade and the imperialistic expansion of western civilisation allowed global pandemics to develop. This second phase was supposedly brought to an end by the medical and public health advances of the twentieth century. As the twenty-first century approaches, this latter victory appears to be illusionary. With the formation of a global village, diseases are entering a third phase. A new human ecology has emerged that is being enhanced by subcultures, marginalised societies, crowded megacities, changing lifestyles, and increased contact with new disease reservoirs. The reemergence of old diseases that have acquired drug resistance, and the emergence of new diseases, such as haemorrhagic fever viruses, will dominate this third phase.

The impact of climate change upon human health is being played out against this background. Warming climate will have different impacts upon various aspects of health. In addition, the very health system that will be affected by climate change has permitted the development of an ageing population that is unique to human history. This ageing population will be most affected by climate changes and UVB-related diseases. Table 8.5 summarises the health impacts of climate change discussed in this chapter, presents the most likely climatic factors responsible for change, and attempts to rank the severity of each impact. The possibility that changes in the incidence of each health condition could be interpreted as a signature of enhanced 'greenhouse' warming is also indicated. From this table, seven diseases or hazards impacting on health: asthma, cholera, drought, floods, Lyme disease, malaria and tuberculosis, are undergoing, or are predicted to undergo, major increases. Of these, only four:

asthma, floods, Lyme disease and malaria, can be linked to present global warming. However other climatic factors cannot be ruled out. For instance, extremes in the Southern Oscillation have been linked to flooding and malaria (as well as to outbreaks of cholera, typhoid, Australian encephalitis). The causes of disease are presently too varied for any observed increase to be attributed to a single factor such as enhanced 'greenhouse' warming.

Western society emphasises death over injury and psychological effects. For example, it is common knowledge in Australia that the Ash Wednesday bushfires in southern Australia killed seventy-six people. Very few would know that 3,500 people were injured, many with burns that

Table 8.5 A summary of health-related climate effects.

Disease/condition	Climate variables controlling disease	Degree of impact (1 = most, 5 = least)
Asthma	increased rain, temperature	1*
Australian encephalitis	increased rain	5*
Bushfires	increased warming, rain, drought	5
Cataracts	increased UVB	4
Cholera	increased rain	1
Crime -violent	increased warming	5*
Crime -property	increased warming	5*
Dengue	increased warming, rain	2*
Diarrhoeal diseases	increased rain	3*
Drought	decreased rain	1
Epidemic polyarthritis	increased rain, warming	2*
Epidemic typhus	increased cooling	5
Floods	increased rain	1*
Haemorrhagic fever viruses	increased rain?	4
Heat exhaustion, stroke	increased warming	3*
Hypothermia	increased cooling, wind	5
Immunity	increased UVB	4
Japanese encephalitis	increased warming	2
Legionnaires' disease	increased warming	5*
Lyme disease	increased rain, warming	1*
Malaria	increased warming, rain	1*
Pterygium	increased UVB	4
Q fever	increased drought	5
Rocky Mountain spotted fever	increased warming, rain	3
Schistosomiasis or bilharzia	increased rain	2*
Skin cancers	increased UVB	4
Typhoid	increased rain	4
Trypanosomiasis	increased rain, warming	4*
Tuberculosis	increased cooling	1
Yellow fever	increased warming, rain	2*

*signature of enhanced 'greenhouse' warming

would cripple them for the rest of their lives. No one can put a figure on the number of people emotionally and psychologically traumatised, to the extent that their lives have been permanently disrupted or incapacitated, people who simply abandoned their property after the fires and left the region, who fear going outdoors, are terrified of fire, suffer insomnia, experience flashbacks, or are affected by a plethora of other mental effects. The subtle changes in health now being experienced at the end of the twentieth century, due both to a changing climate and to changes in the nature of disease, will increase such psychological stress.

A propensity for alarmism pervades our thinking on the possible health consequences of enhanced 'greenhouse' warming. However, despite the fact that our climate has been warming over the past century, whether globally or just in urban areas, it is difficult to detect increases in disease, reinfections or new epidemics specifically linked to such warming. Consider the background for heatwaves in the United States. New York City medical authorities in 1896, before any enhanced 'greenhouse' warming had taken place, were alarmed by the high death tolls during the heatwave of August that year. Air temperatures as high as 44.4°C were recorded in the centre of the city. There were 670 excess deaths from sunstroke alone in a city much smaller than present. Our 'record' heatwaves of the 1980s and 1990s are part of a continuum in urban living affected by such events.

Finally, climate related disease may not be the biggest threat to human health. Whereas contamination of food in the past could be linked to one restaurant, fast-food outlet or wedding reception, today the prospect exists for a pathogen to get into food at a central point of preparation, and then be spread rapidly throughout a nation. This was how ergotism spread through whole villages in the Middle Ages. Similar, rapid spread of disease can occur if domestic water supplies are contaminated, as was the case in 1993 in Milwaukee, when the parasite *cryptosporidium* entered the municipal water supply and infected 400,000 people. The state of human health in future is more likely to be determined by the recurrence of old diseases, our degree of affluence, the preparedness and funding of our public health authorities, and the nature of our sanitation, than by climate change.

REFERENCES AND FURTHER READING

Barton, I.J. and Paltridge, G.W. 1979. 'The Australian climatology of biologically effective ultraviolet radiation'. *Australian Journal of Dermatology* v. 20 pp. 68–74.

Blum, H.F. 1964. 'Effects of sunlight on the human body'. In Licht, S. and Kamenetz, H.L. (eds) *Medical Climatology*. Elizabeth Licht, New Haven, pp. 229–256.

Bryant, E.A. 1991. *Natural Hazards*. Cambridge University Press, Cambridge, 294p.

Budiansky, S. 1995. 'Plague fiction'. *New Scientist*, 2 December, pp. 28–31.

DeFronzo, J. 1984. 'Climate and crime: tests of an FBI assumption'. *Environment and Behavior* v. 16 pp. 185–210.

de Gruijl, F.R. 1995. 'Impacts of a projected depletion of the ozone layer'. *Consequences* (**http://www.gcrio.org/CONSEQUENCES/introCON.html**) v. 1 no. 2.

Ewan, C.E., Bryant, E.A., Calvert, G.D. and Garrick, J.A. (eds) 1993. *Health in the Greenhouse: The Medical and Environmental Health Effects of Global Climate Change.* Australian Government Publishing Service, Canberra, 237p.

Favilla, I. 1989. 'Ocular effects of ultraviolet radiation'. In *Health Effects of Ozone Layer Depletion.* National Health and Medical Research Council, Canberra, pp. 96–113.

Garcia, R.V. and Escudero, J. C. 1982. *Drought & Man. v. 2 The Constant Catastrophe: Malnutrition, Famines and Drought.* Pergamon Press, Oxford, 204p.

Garrett, L. 1995. *The Coming Plague: Newly Emerging Diseases in a World out of Balance.* Virago Press, London, 750p.

Giles, G.G., Marks, R. and Foley, P. 1988. 'Incidence of non-melanocytic skin cancer treated in Australia'. *British Medical Journal (Clinical Research)* v. 296 pp. 13–17.

Giles, G.G. 1989. 'Possible effects of increased exposure to ultraviolet radiation on the incidence of cutaneous malignant melanoma'. In *Health Effects of Ozone Layer Depletion.* National Health and Medical Research Council, Canberra, pp. 82–95.

Gimpel, J. 1979. *The Mediaeval Machine.* Futura, London, 255p.

Haines, A., Epstein, P.R. and McMichael, A.J. 1993. 'Global health watch: monitoring impacts of environmental change'. The *Lancet* v. 342 pp. 1464–1469.

Haub, B. 1995. 'Global and U.S. national population trends'. *Consequences* (**http://www.gcrio.org/ CONSEQUENCES/introCON.html**) v. 1 no. 2.

Hollows, F.C. 1989. 'Ultraviolet radiation and eye diseases'. *Transactions of the Menzies Foundation* v. 15 pp. 113–117.

Hollows, F.C. and Moran, D. 1981. 'Cataract-the ultraviolet risk factor'. *The Lancet* v. 2 pp. 1249–50.

Jose, D.G. 1989. 'Immunological effects of ultraviolet radiation'. In *Health Effects of Ozone Layer Depletion*. National Health and Medical Research Council, Canberra, pp. 114–126.

Kalkstein, L.S. and Davis, R.E. 1989. 'Weather and human mortality: an evaluation of demographic and interregional responses in the United States'. *Annals of the Association of American Geographers* v. 79 p. 44–64.

Kaplan, S.J. 1960. 'Climatic factors and crime'. *The Professional Geographer* v. 12 pp. 1–5.

Knight, C.G. 1974. 'The geography of vectored diseases'. In Hunter, J.M. (ed.) *The Geography of Health and Disease*. University of North Carolina at Chapel Hill, Department of Geography Studies in Geography no. 6 pp. 46–80.

Kreider, M.B. 1964. 'Pathologic effects of extreme cold'. In Licht, S. and Kamenetz, H.L. (eds) *Medical Climatology*. Elizabeth Licht, New Haven, pp. 428–468.

Lamb, H.H. 1982. *Climate, History and the Modern World*. Methuen, London, 387p.

Lamb, H.H. 1985. *Climatic History and the Future*. Princeton University Press, Princeton, 835p.

Landsberg, H.E. 1981. *The Urban Climate*. Academic, New York, 275p.

Learmouth, A. 1988. *Disease Ecology: An Introduction*. Basil Blackwell, Oxford, 456p.

Le Guenno, B. 1995. 'Emerging viruses'. *Scientific American* v 273 pp. 30–37.

Linacre, E. and Hobbs, J. 1977. *The Australian Climatic Environment*. Wiley, Sydney, 354p.

Lind, A.R. 1964. 'Physiologic responses to heat'. In Licht, S. and Kamenetz, H.L. (eds) *Medical Climatology*. Elizabeth Licht, New Haven, pp. 164–195.

Madronich, S., Bjorn, L., Ilyas, M. and Caldwell, M. 1991. 'Changes in biological active ultraviolet radiation reaching the Earth's surface'. In *Environmental Effects of Ozone Depletion:*

1991 Update. United Nations Environment Programme, Nairobi, pp. 1–13.

Marks, R. 1989. 'Possible effects of increased ultraviolet radiation on the incidence of non-melanocytic skin cancer'. In *Health Effects of Ozone Layer Depletion*. National Health and Medical Research Council, Canberra, pp. 70–81.

Marthick, J. and Bryant, E. 1990. 'Health implications of long term climatic change: literature review'. In Ewan, C., Bryant, E. and Calvert, D. (eds) *Health Implications of Long Term Climate Change. v II Survey Report and Commissioned Papers*. National Health Medical Research Council, Canberra, pp. 36–71.

May, J.M. 1950. 'Medical geography: its methods and objectives'. *The Geographical Review* v. 40 pp. 9–41.

May, J.M. 1951. 'Map of the world distribution of malaria vectors'. *The Geographical Review* v. 41 pp. 638–639.

May, J.M. 1952. 'Map of the world distribution of dengue and yellow fever'. *The Geographical Review* v. 42 pp. 283–286.

McKenzie, R., Frederick, J., Ilyas, M. and Filyushkin, V. 1992. *Ultraviolet Radiation Changes in Scientific Assessment of Ozone Depletion: 1991*. WMO Global Ozone Research and Monitoring Project, NASA/NOAA/UKDOE/ UNEP/ WMO Report no. 25, pp. 11.1–11.4.

McMichael, A.J. 1993. *Planetary Overload: Global Environmental Change and the Health of the Human Species*. Cambridge University Press, Cambridge, 352p.

McMichael, A.J., Haines, A., Sloof, R. and Kovats, S. (eds) 1996. *Climate Change and Human Health*. WHO, Geneva, 297p.

Mitchell, J.K. and Ericksen, N.J. 1992. 'Effects of climate change on weather-related disasters'. In Mintzer, I.M. (ed.) *Confronting Climate Change: Risks, Implications and Responses*. Cambridge University Press, Cambridge, pp. 141–151.

Muller, H.K. 1989. 'Ultraviolet radiation and the skin'. *Transactions of the Menzies Foundation* v. 15 pp. 91–94.

National Health and Medical Research Council 1989. *Health Effects of Ozone Layer Depletion*. Australian Government Publishing Service, Canberra, 139p.

Setlow, R. 1974. 'The wavelengths in sunlight effective in producing skin cancer: a theoretical analysis'. *Proceedings of the National*

Academy of Science v. 71 pp. 3363–3366.

Taylor, H., West, S., Rosenthal, F., Munoz, B., Newland, H.S., Abbey, H. and Emmett, E.A. 1988. 'Effect of ultraviolet radiation on cataract formation'. *New England Journal of Medicine* v. 319 pp. 1429–1433.

United Nations 1992. *World Population Monitoring 1991: With Special Emphasis on Age Structure*. Department of International Economic and Social Affairs Population Studies no. 126, 241p.

World Resources Institute and United Nations 1992. *World Resources: 1992–93*. Oxford University Press, New York, 385p.

9

Ecosystem Impacts of Climate Change

PRESENT DISTRIBUTION OF BIOME TYPES

An ecosystem is a system of interactions of a community of organisms with its environment. Involved in this interaction is mineral cycling, energy flow and population dynamics. The largest fundamental unit of an ecosystem is the biome, which consists of similar plants and animals that can be distinguished from each other by vegetation and climate. The biota (plants and animals) of each kind of biome have similar characteristics worldwide. Common biomes include grasslands, tundra, northern coniferous forests and tropical rainforests. Similar biomes comprise a biome type. About two-thirds of biome types relate chiefly to climate factors characterised mainly by temperature and rainfall, while the others relate to marine or wetland environments, for example the rocky littoral zone. The boundaries, or tolerance limits, of climatically determined biome types are controlled by two laws. Justus von Liebig's law of the minimum states that plants have minimum requirements for growth and reproduction, controlled by a range of factors. The factor that is least available has the greatest effect on plants, such that small changes in that parameter can exert a profound influence on a plant's survival. The law of the maximum, Shelford's law of tolerance, states that too much of a certain factor also limits a plant's existence. As a result of both these laws, and other environmental factors, biome types show up spatially as distinct zones of vegetation, mainly spaced parallel to the temperature gradient that exists either latitudinally between the equator and the poles, or with elevation on mountains. Climate change affects the spatial distribution of biome types through both of these laws.

The present, generalised distribution of the world's major biome types, without the influence of human activities, is shown in Figure 9.1. Regions of polar ice devoid of vegetation occupy a very small portion of the globe. Because these areas consist of massive icesheets, they will change little in area under a warmer globe. Extensive areas of tundra exist at high latitudes in the northern hemisphere, and along major mountain chains. While tundra is constrained by low temperatures, it also accumulates carbon because of low rates of decomposition. At the other temperature extreme, where moisture is not a constraint, extensive tropical forests and savanna have formed, not only in the equatorial regions, but also in the subtropics of southern Africa, Southeast Asia and southeastern South America. Where water availability in these regions becomes significant, an equivalent area of desert vegetation exists. In some of these deserts, for instance in Australia, plant vigour increases dramatically during wet periods associated with La Niña events. Extensive forests also dominate the northern and eastern parts of landmasses in the northern hemisphere. The taiga (boreal forest) of northern Asia and Europe forms the largest continuous belt of forest in the world. Where climate is more temperate, many of these forests have been cleared by Europeans. Both the taiga and

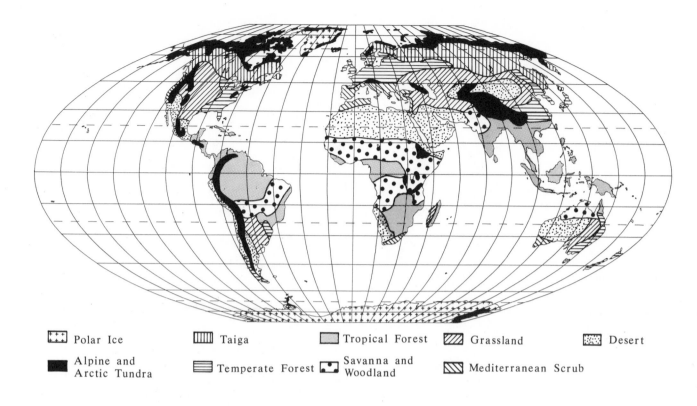

	Polar Ice		Taiga		Tropical Forest		Grassland		Desert
	Alpine and Arctic Tundra		Temperate Forest		Savanna and Woodland		Mediterranean Scrub		

Figure 9.1 Global distribution of major biome types of the world grouped into nine categories (various sources).

temperate forests are presently believed to be significant sinks for the missing carbon dioxide discussed in Chapter 7. In addition, significant areas cleared for agriculture have been abandoned, and have regrown forests over the latter part of the twentieth century. Finally, a small but well defined biome type consists of Mediterranean scrubland. This vegetation is especially fireprone, and occupies climates that are wet in summer and seasonally dry in winter. This biome type is located around the Mediterranean Sea, in California, South Africa and Western Australia.

MODELLING THE EFFECT OF CLIMATE ON ECOSYSTEMS
(Holdridge, 1947; Sellers et al., 1986; Calder, 1991; Xue et al., 1991; Gates, 1993; Monserud et al., 1993; Townshend et al., 1993)

Many schemes have been devised to depict the climatic limits to biome types. One of the earliest schemes was the Holdridge Life-Zone Classification, in which relationships amongst biome types, mean annual biotemperature, potential evapotranspiration, and mean annual precipitation are presented in the form of a triangular nomogram. A modification of this schema is presented in Figure 9.2, using the nine biome types mapped in Figure 9.1. Mean annual biotemperature refers to the mean of temperatures between 0–30°C. Outside these limits plants do not undergo a net gain in carbon. Potential evapotranspiration is the ratio between the amount of water that a plant can theoretically transpire and the amount of precipitation. Various biome types in Figure 9.2 can be related clearly to latitude or elevation. For instance, most tundra is constrained by a maximum mean annual biotemperature of 3°C. At the equator, this biome type occurs above elevations of 4,500 metres. If climate warms globally, then there would be both poleward and altitudinal shift of this temperature boundary, and the extent of tundra would decrease. In contrast, the existence of temperate woodlands is controlled both by temperature and rainfall. If rainfall increases, then woodlands could expand into semi-arid regions. However, if temperature increases, then woodlands would intrude into areas of tundra and their southern boundary would shift poleward. The areal extent of woodlands would not be detrimentally affected by enhanced 'greenhouse' warming.

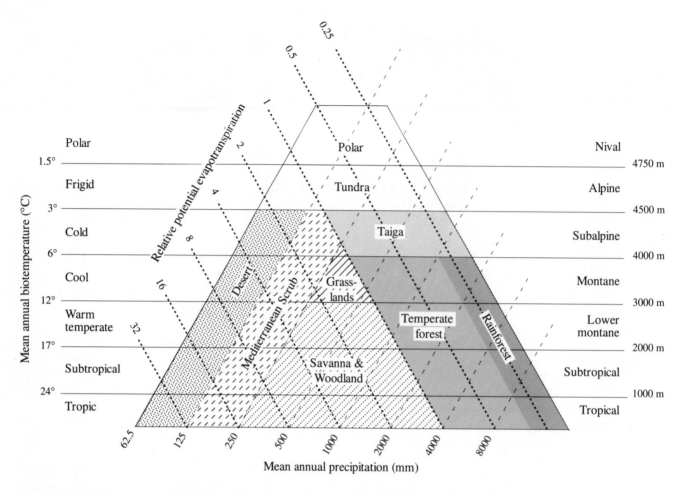

Figure 9.2 Classification of biome types based upon climatic classification of Holdridge (1947).

Alternative models of climatic effects on vegetation have been developed in Russia, where traditionally the boundaries of biome types have been linked to precipitation and potential evaporation. This early work led to the development of the Köppen and Thornthwaite classifications of climate. Budyko extended this work and showed that the influence of climate upon vegetation can be summarised by two equations as follows:

$$PE = B/L \qquad \text{9.1}$$

$$DI = PE/p \qquad \text{9.2}$$

where   PE = potential evapotranspiration
      DI = the dryness index
       B = radiation balance
       L = latent heat of evaporation
       p = precipitation

The radiation balance is the difference between incoming solar radiation at the ground and net longwave radiation. Potential evapotranspiration is the amount of water that can be evaporated from a surface with an unlimited water source. This value is determined from the radiation balance, and assumes that there is little heat advected by air masses. It also does not include the resistance of plant leaves or canopies to evaporation. The dryness index separates vegetation into five basic zones: tundra, forest, steppe, semidesert and desert, while differences in the radiation balance define biome types. The latter value is calculated only for the growing season when mean monthly temperatures are above 10°C. The parameters in equations 9.1 and 9.2 have been the subject of further, elegant simplification that can make use of basic data collected at weather stations around the globe, and can describe more biome types.

Such models are only first approximations of the

response of biomes to climate. Many vegetation types are either not included in the models, or else poorly represented. With the advent of satellite scanners and the use of computer simulated General Circulation Models (GCMs), more complicated models have been developed that emphasise the density of transpiring, carbon fixing vegetation, rather than species and plant associations. These models are based upon leaves as the fundamental indicator of plant 'healthiness'. Leaves trap photons of light that are used by chlorophyll to drive photosynthesis, converting carbon dioxide and water into plant biomass. Infrared wavelengths are not absorbed by leaf pigments, and over 40% of solar radiation at near-infrared wavelengths between 0.8 and 1.4 µm is reflected (Figure 9.3). The reflectance of short wave radiation, at various wavelengths, characterises the overall 'healthiness' of vegetation. The degree of reflectance between 0.45 and 0.52 µm (visible blue) differentiates vegetation from soil, and can be used to measure vegetation density. Because chlorophyll reflects more green light than other wavelengths in the visible spectrum, reflectance between 0.52 and 0.60 µm measures the intensity of photosynthesis or vegetation vigour, while reflectance between 0.63 and 0.69 µm (visible red) measures chlorophyll content and differentiates between vegetation species. Reflectance between 0.76 and 0.90 µm is affected by the mesophyll structure of leaves and indicates biomass content, while reflectance between 1.55 and 1.75 µm measures vegetation and soil moisture content. Finally, reflectance between 10.40 and 12.50 µm indicates the degree of heat stress affecting vegetation.

Ratios amongst various spectral reflectance bands characterise a number of vegetation attributes including leaf area, percentage vegetation cover, and green leaf biomass. In particular, the normalised difference vegetation index (NDVI) closely reflects the capacity of specific vegetation types to photosynthesise:

$$\text{NDVI} = \frac{\text{near-infrared} - \text{red}}{\text{near infrared} + \text{red}} \qquad \textbf{9.3}$$

where NDVI = the normalised difference vegetation index

near-infrared = reflectance between 0.76 and 0.90 µm

red = reflectance between 0.63 and 0.69 µm

This index is synonymous with vegetation activity, and is currently measured at a resolution of between one and four kilometres using the Advanced Very High Resolution Radiometer (AVHRR), launched in 1979 on a NOAA-7 weather satellite. Realistically, the NDVI can range from -0.3 for water, to more than 0.6 over areas of high vegetation activity. Desert vegetation typically has values of 0.05. The index can overlap for different vegetation types, and vary seasonally and annually. It is most efficient for dense vegetation cover, under clear skies and low humidities, at low latitudes. The index is an excellent indicator of the seasonal variation in vegetation growth and can depict vegetation differences caused by increased rainfall and temperature, fires, land clearing, and drought. The NDVI also accounts surprisingly well for the urban heat island effect mentioned in Chapter 3. Cities have less vegetation than their surroundings and thus undergo less evaporative cooling. For each 0.1 difference in the NDVI between a city and its rural hinterland, minimum temperatures can differ by 0.9°C. Finally, the NDVI can be used to monitor the impacts of ENSO events and climate change upon vegetation growth.

The interrelationships between plant 'healthiness' and climate can be simulated using a complex range of biosphere models developed around 1986. These models define not only the impact of climate upon vegetation, but also the feedback of vegetation upon climate. The Biosphere–Atmosphere Transfer Scheme (BATS) was one of the first biosphere models to be integrated into a GCM in this manner. One of the best models is the Simple Biosphere Model (SiB). The Simple Biosphere model simulates

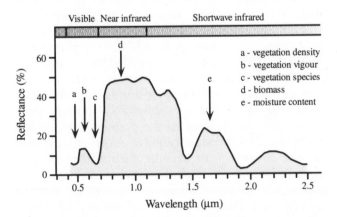

Figure 9.3 Relfectance of shortwave radiation by 'healthy' vegetation. The wavelengths characterising the main attributes of vegetation are marked.

the physical and biological aspects of the Earth's land surface more realistically than other models. For example, many biosphere models cannot mimic the surface warming caused by deforestation or desertification, because they do not adequately include vegetation components. The Simple Biosphere Model includes three soil layers and two vegetation layers. Crucially, the SiB models the vegetation itself and lets the leaves and canopy determine the means by which the Earth's land surface interacts with the atmosphere (Figure 9.4). Vegetation canopies increase evapotranspiration, intercept and store precipitation, increase ground roughness, friction or enhance turbulence, increase the transport of sensible and latent heat upwards from the ground, remove moisture from the ground, and shelter the ground surface from solar radiation. The SiB model mimics all of these characteristics. The model partitions surface energy transfers into sensible and latent heat components, and incorporates the fluxes of evapotranspiration and heat from the leaf to the vegetation canopy, through the canopy, and from the canopy to the surrounding atmosphere and the ground below (Table 9.1). Using simple weather data, the model can compute changes in leaf temperature, the wetting of leaves and soil, and the intake of solar energy. Because the rate of transpiration is controlled by stomatal resistance (stomata are microscopic pores on the surface of leaves that regulate gas exchange between the interior of the leaf and the adjacent atmosphere), the model accounts for differences in biome types from Arctic tundra to tropical rainforest. Of thirty-two different biome types representing global vegetation, the Simple Biosphere Model can identify thirty-one of them by combinations of just three parameters: percentages of bare soil, ground cover of grass and herbs, and canopy cover of shrubs and trees.

The third generation of GCMs incorporates the effect of the biosphere upon climate. Many of these models consider the transfer of momentum to the atmosphere from a vegetated landscape, and soil water and temperature components. The Simple Biosphere Model considers these aspects but has superior algorithms for simulating surface albedo, surface energy and soil moisture, and it can do this diurnally. Its scale of transfers between vegetation and the atmosphere is also more detailed than other models. Unfortunately, the SiB in its original form, while conceptually accurate, had forty-four parameters and 101 equations. Many of these parameters are poorly known for many different biomes. The model has since been simplified to twenty-one

parameters, fourteen describing the vegetation and seven describing the soil.

EFFECT OF ENHANCED 'GREENHOUSE' WARMING ON PLANTS
(Warrick et al., 1986; Gifford, 1988; Idso, 1989; Walker, 1991; Lockwood, 1995; Schimel, 1995)

Climate affects the entire life cycle of any plant species in five direct ways: temperature, atmospheric humidity, carbon dioxide concentration, soil moisture and wind speed. In addition, there are secondary factors, apart from climate, that influence plant growth. These include the nature of the fire regime, soil nutrient availability, the presence of herbivores as grazers, the availability of pollinators, the presence of plant pathogens, seed dispersal, and predation. Many of these secondary factors can also be affected by climate processes and change. While ecological models rely upon climate factors, usually just temperature and precipitation, to control change in vegetation, it should be realised that many secondary factors also have a crucial role in determining the boundaries of biomes and the presence of many plant species. For instance, the ability of many boreal forests to acquire a wide diversity in tree species after a major fire is not necessarily climatically controlled, but highly dependent upon the frequency of subsequent fires in regrowth.

The majority of plant species can be grouped into two categories, C_3 and C_4 plants, depending upon metabolic differences in the pathway of carbon dioxide uptake. C_3 plants under photosynthesis assimilate carbon dioxide into a three-carbon compound, phosphoglyceric acid. These plants have an active respiratory cycle triggered by light. Wheat, rice, beans, potatoes, citrus fruits and most trees are common C_3 plants. C_4 plants use the same pathway for carbon dioxide fixation as C_3 plants, but beforehand, they assimilate carbon dioxide into four-atom compounds such as oxaloacetic acid, in the mesophyll. Carbon is then passed into the interior of the plant through other reactions. This process requires more energy, but has the advantage of making carbon dioxide fixation independent of photorespiration. Hence, C_4 plants are more efficient at taking in carbon dioxide and fare better in hot or dry climates, where the carbon dioxide flux through stomata on leaves can be suppressed. Sugarcane, maize, sorghum and forage grasses in tropical regions, employing this pathway, can tolerate heat and

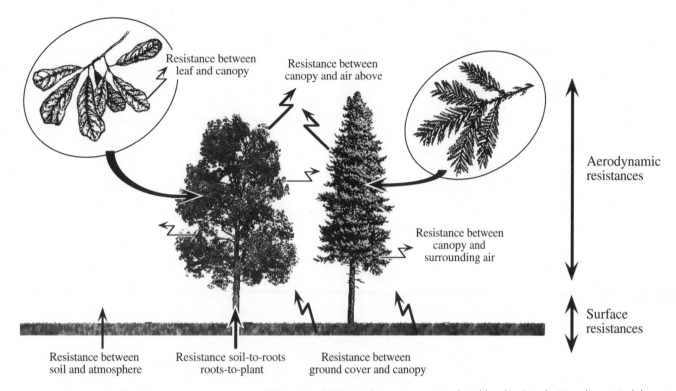

Figure 9.4 Various levels of resistance to water and heat transfer through a canopy, simulated by the Simple Biosphere Model.

water stress while producing the highest annual biomass yields of any species. C_3 plants account for over 95% of the world's plant species, however they are very sensitive to changes in the atmospheric concentration of carbon dioxide. C_4 plants are rarer, but their carbon dioxide fixation pathway makes them rather insensitive to changes in concentrations of carbon dioxide.

The effects of increased carbon dioxide upon plants are summarised in Table 9.2 Almost all experiments have shown increased plant biomass with increased levels of carbon dioxide. C_3 plants have shown greater increases than C_4 plants. However, this response is not necessarily long-lasting because some plant species become acclimatised to elevated carbon dioxide levels and stop their accelerated growth. This carbon dioxide 'acclimatisation' may be caused by limits in the availability of water and nutrients, rather than to any physiological adjustment to carbon dioxide levels. Under a doubling of carbon dioxide, plant biomass will increase on average 26% for C_3 plants in response to the increased carbon dioxide 'fertiliser' effect. For temperate cereals, the predicted mean increase in grain yield will be 36%. Biomass increases in C_4 plants will be smaller, if

not negative in some cases. Biomass increases appear to occur without any change in the efficiency of photosynthesis. Noticeable increases in the width of tree-rings, and by inference biomass, have already been measured for the 29% increase in atmospheric carbon dioxide that has occurred since industrialisation. In northern Europe, tree basal area has increased by 15–43% between 1950 and 1983.

Ecosystem carbon storage may be less responsive to the 'fertiliser' effect of carbon dioxide than plant growth. For instance, plants with low growth rates, characterising native plants, respond less to carbon dioxide increases than do crop species. Ecosystem modelling indicates that not only is the amount of biomass storage lower than for individual plants, but also that increases in carbon sequestering become limited eventually by nutrient availability. The process of fixing carbon in plants also enhances the absorption of nitrogen and nutrients. If these latter compounds are scarce, then plant growth will not respond any better to more carbon dioxide in the atmosphere. However, carbon dioxide 'fertilisation' could stimulate nitrogen fixation in soils, negating this restriction. Conservatively, a doubling of carbon dioxide should

Table 9.1 Parameters modelled in the Simple Biosphere Model (from Sellers et al., 1986).

Component	Parameter
Plant morphology	Area of canopy cover and ground cover
	Leaf orientations
	Height of canopy and thickness
	Density of leaves and stems in the canopy and on the ground
	Root depth and density
Soil characteristics	Thickness of three soil layers
	Soil albedo
	Soil pore space
	Soil moisture potential
	Rate of, and resistance to, moisture movement in soil
Plant physiology	Fraction of leaves and stems that are green
	Stomata resistance to fluxes dependent on light
	Temperature characteristics at which stomata function
	Stomatal response to water vapour pressure
	Stomatal response to leaf water
	Resistance to fluxes imposed by the plant vascular system
	Resistance to fluxes imposed by roots
	Reflectances of leaves in the canopy and on ground
	Transmission of leaves to light
	Resistance of ground surface to air movement
	Resistance of leaves to air movement
	Shelter factor of leaves in the canopy
	Heat and vapour transfer of canopy leaves
	Heat capacity of leaves and the ground
Environmental conditions	Canopy and ground temperature
	Amount of water stored on canopy and ground foliage
	Wetness of surface, root zone and soil
Atmospheric conditions	Wind speed at a reference height
	Air temperature at a reference height
	Vapour pressure at a reference height
	Direct and scattered solar radiation amounts
	Precipitation
Fluxes	Sensible heat flux between ground and the canopy
	Sensible heat flux between the soil and the ground cover
	Total sensible heat flux with the overlying atmosphere
	Latent heat transfer from canopy
	Latent heat transfer from ground surface
	Latent heat transfer from soil
	Total latent heat transfer to overlying atmosphere

result in a 1.8-fold increase in sustainable forest growth.

Morphologically, increased carbon dioxide concentrations reduce the density of stomata on plant leaves, because fewer pathways for carbon dioxide into the leaf are required. There is no difference in this regard between C_3 and C_4 plants. A 40% decrease in stomatal density has already been measured for plants over the past two hundred years. However, further increases in carbon dioxide, above present ambient levels, show no further reduction in stomatal density. A doubling of carbon dioxide will also result in a 40% reduction in stomatal conductance in all species. As a consequence, rates of plant evapotranspiration should decrease by a similar amount, improving

Table 9.2 Generalised effects of increased carbon dioxide on plant characteristics (based on Idso, 1989).

Characteristic	Effect
Morphology	Increased size, taller
	More branches and stems
	More and thicker leaves
	Smaller stomatal openings
	Decreased stomata density
	Increased seedling growth
	More and larger flowers
	More female flowers in some species
	Increased biomass
Physiology	More chlorophyll
	Reduced stomatal conductance
	Decreased dark respiration
	Decreased evapotranspiration
	Greater water-use efficiency
	Increased seed germination for most species
Soil	Greater cation exchange capacity
	Greater absorption of minerals by plants
	Greater absorption of trace elements
	Fungal suppression of toxic metal absorption
	Greater nitrogen fixation by bacteria
	More plant roots
Pathogens and hazards	Decreased incidence of plant diseases
	Reduced soil-borne pathogens
	Increased mortality of leaf-eating caterpillars
	More sustained growth under herbivore predation
	Better tolerance to climate extremes
	Greater salt tolerance
	Reduced damage caused by atmospheric pollutants
	More rapid decomposition of pesticide residues
Agronomy	Greater plant vigour for C_3 plants
	Increased dry plant matter
	Increased crop yields by 33% for a doubling of CO_2
	More and larger fruit
	Quickening of plant life cycle
	Greater potential for colonisation of severe environments

water-use efficiency. Plant transpiration will be reduced by 10–20% in C_3 plants, and 25% in C_4 plants. The presently measured 29% increase in carbon dioxide already should have reduced transpiration by 2.5–5.0% in C_3 plants and 6% in C_4 plants.

Reduced stomatal conductance implies that evapotranspiration from plant canopies should fall, reducing evaporative cooling as carbon dioxide levels increase. Air temperatures would rise as a result. However, this effect could be offset by the tendency for greater transpiration at higher temperatures. At mid-latitudes, each degree Celsius increase in temperature results in a 2% and 8% increase in transpiration for moist forest and pasture respectively. Below 25°C, carbon dioxide suppression approximately equals temperature enhancement for C_3 trees. However for C_3 grasses and all C_4 plants, carbon dioxide suppression of transpiration exceeds temperature enhancement. For temperatures above 25°C, carbon dioxide suppression dominates all vegetation types. Thus enhanced carbon dioxide, together with warming, will exacerbate warming over both forest and grass surfaces because of a reduction in evaporative cooling. Indirectly, a reduction in evapotranspiration should lead to an increase in runoff. This effect may already be detectable. Measurements in vegetated catchments in Wales between 1972 and 1988, show a 6.3–12.4% increase in precipitation, but a higher 19.2–29.4% increase in runoff. Calculation of evapotranspiration derived from these measurements indicates a 40–43% decline in transpiration. These results warrant further research, because they imply major alteration of the hydrological cycle, and potentially more stream erosion, as a result of increasing carbon dioxide.

Reduction in respiration enhances the uptake of carbon because plants lose less carbon dioxide back to the atmosphere (dark respiration). Reduced transpiration also increases the rate of plant growth, leading to a greater capacity for plants to act as carbon sinks. This aspect may explain why the amplitude of the annual cycle in atmospheric carbon dioxide has been increasing at a higher rate (0.5–0.7% per year) than the average concentration of carbon dioxide (0.4% per year) in the atmosphere. The annual fluctuation in carbon dioxide is a product of northern hemisphere vegetation respiration, dominated by trees. The net annual carbon gained by forests depends upon annual photosynthesis, minus annual dark respiration. If higher carbon dioxide concentrations are increasing this difference (photosynthesis remains unaffected while dark respiration is suppressed), then not only are forests taking in more carbon, but they are also increasing the amplitude of annual carbon dioxide fluctuations. This process probably explains why forests are the most likely source for 'The Great Northern Absorber' carbon sink described in Chapter 7. Overall plants should require less water, assimilate more carbon dioxide, and increase their biomass with increased atmospheric concentrations of carbon dioxide. These factors should nurture those plants where water stress presently limits growth.

Elevated levels of carbon dioxide in the atmosphere will not take place without attendant changes in temperature and precipitation. Both water- and temperature-stressed plants grow better under enhanced carbon dioxide, than unstressed ones. However, other climatic factors must be considered. In combination with increases in both precipitation and temperature, most plants will respond favourably to the 'fertilising' effect of carbon dioxide, and begin to occupy ecological niches that presently prohibit prolific vegetation growth. Where precipitation decreases, C_3 plants will have a distinct competitive advantage over C_4 plants under conditions of water stress.

While the growth of plants and their time of flowering and fruit-setting is dependent upon aggregated temperatures throughout a season, extreme events can also have an adverse effect on flowering and fruiting. Extreme heat can damage flowers, while heavy rain can adversely affect the filling of fruit. Many grains, under prolonged rain, will sprout while still attached to the parent stock. Elevated carbon dioxide levels appear to increase the flowering and fruiting prowess of plants. Mixed results have been found for the timing of flowering, and the number, size, weight and nutrient content of fruits and seeds. Increasing temperature also will augment plant yields. Of course a rise in temperature will tend to shift the latitudinal distribution of plants poleward. It is reasonable to assume that for each 1°C increase in temperature, plant species will move 100 kilometres poleward. Overall a 'greenhouse'- warmed world will be more benign for plant growth.

Climatic change will have profound effects upon agricultural production. However, these impacts are difficult to assign solely to anthropogenically induced climate changes, because the world's agriculture is also dependent upon the synchronous occurrence of droughts linked to ENSO events. A doubling of carbon dioxide concentrations will increase the growth and yield of C_3 plants (wheat, rye, barley) by between 10 and 50%, and of C_4 plants (sugarcane) by up to 10%. However, these higher yields could be offset by the effects of temperature and evaporation stress. For instance, the growing period between the flowering and grain development of wheat is particularly sensitive to temperatures above 35°C. In North America, it is estimated that a 2°C increase in average temperature would decrease grain yields by 3–17%. The effects of changing climate on other crops have been subjected to intense study, and have ramifications for national economies in many countries. For instance warming would probably shift the wheat belt in North America northwards, greatly benefiting Canada, but disadvantaging the United States. Many of these aspects are beyond the scope of this text.

EFFECT OF OZONE DEPLETION ON PLANTS
(Gates, 1993; Caldwell and Flint, 1994)

There are considerable differences in the response of plants and animals to the amount of biologically reactive ultraviolet radiation each receives. Many of the effects of enhanced UVB radiation have been determined in hothouse conditions. Plants growing in the field are much less sensitive to elevated UVB radiation than when grown in hothouses, or under laboratory conditions. Part of this response is due to the differences in the total amount of solar radiation being received in each setting. Exclusion experiments indicate that plants generally respond favourably to UVB radiation at current levels. There is thus only laboratory evidence suggesting that the physiological nature of some species may be distorted by enhanced UVB radiation. This may have an impact on the competitiveness of individual species within a plant community; however the effects are minor. Few studies have examined the interaction of elevated UVB levels and carbon dioxide at the same time. In those that have, unnaturally high UVB levels have often been used. Generally, higher UVB counterbalances the stimulation provided by enhanced carbon dioxide, while higher carbon dioxide ameliorates the impact of higher UVB levels.

Higher UVB levels have several explicit effects on individual plant species. At large dosages, UVB radiation inhibits photosynthesis, causes DNA damage, and generates changes in the structure of plants affecting light uptake. These can cause growth inhibition and decrease the degree of pollination, although the impact on overall plant vigour is difficult to establish, especially under natural conditions. If plants are growing under low sunlight conditions, these deleterious effects of enhanced UVB radiation are magnified. At present, the effect can be dampened, if plants are environmentally stressed by water shortages or nutrient deficiencies. When extrapolated to natural monocultures or plant communities, these results imply little reduction in primary productivity. Plants under enhanced UVB tend to produce phototoxins that may protect them against some insects, although other evidence suggests that increased UVB subjects plants to more fungal infection. Because plants exposed to more UVB radiation develop thicker leaves, there may be repercussions on the health of grazing animals.

The marine environment may be particularly sensitive to enhanced UVB, especially in the Antarctic. The springtime, when the 'Ozone Hole' is most developed, is the beginning of the reproduction cycle for many species. Phytoplankton, at the base of the food chain for the Southern Ocean, are concentrated in the surface waters of the Antarctic in spring. They are particularly vulnerable to UVB penetration of water at this time. Increases in UVB radiation inhibit photosynthesis in phytoplankton, although most species have a number of protective strategies that they can use to compensate for excess radiation at different wavelengths. Laboratory and field evidence suggests that UVB enhancement under the 'Ozone Hole' might reduce photosynthesis by 12–15% and primary productivity by 6–12% in the upper twenty-five metres of the ocean. In summary, present predictions of increasing UVB due to ozone depletion will have minimal global impact on plant or phytoplankton growth and viability. This is not surprising considering that all plant species have been subjected, repetitively in their evolutionary development, to natural UVB increases that have been much greater than those which are now predicted.

ECOSYSTEM CHANGE RELATED TO CLIMATE

Past record
(Walker, 1991; Friedlingstein et al., 1992; Woodward, 1992; Gates, 1993)

The most dramatic difference between present and past vegetation should have occurred during the Last Glacial Maximum. Various schemes, including a modified version of the Simple Biosphere Model, have been used to simulate the global distribution of vegetation 22,000 years BP. While the continental landmasses were up to 23 million square kilometres greater in extent, the amount of carbon locked in vegetation was lower than present. The change in area of eight major biome types is summarised in Table 9.3. Tundra covered much of the area presently occupied by taiga and deciduous forests in the northern hemisphere. The aridity that dominated climate at the time accounts for the 136% greater area of deserts than exists at present. Deciduous tropical forests were replaced by grasslands and shrublands; however, these forests expanded onto the present continental shelves of Southeast Asia and Oceania. In all, during the Last Glacial Maximum, forests shrank in extent by 20% while grasslands expanded by 20%. Tropical rainforests virtually disappeared from their present locations, and took on a refuge status. The actual amount of carbon in global vegetation was 23% lower during the Last Glacial Maximum than at present, despite 120 gigatonnes (Gt) of carbon being locked into vegetation covering continental shelves. Note that this lower carbon store does not explain the lower concentrations of carbon dioxide in the atmosphere during the Last Glacial Maximum (Figure 6.5). The removal of carbon dioxide from the carbon cycle was most likely due to the fixing of carbon by phytoplankton, the consequent fallout of dead organisms to the ocean bottom, and the lack of significant upwelling of deep bottom water to return this carbon to the ocean surface. In this regard, the oceans played an important role in carbon storage during glacials.

There is considerable evidence under warmer climates regarding the rate of species migration, colonisation and zonal shifts for vegetation. For instance, the warming of 0.6°C between 1880 and 1940 affected only the elevation of the treeline and secondary growth forest. The warming of the past 180 years, which includes the 29% increase in carbon dioxide, has changed the species composition of canopy trees. It would appear that changes in temperature must not only be large (probably greater than 1°C), but also prolonged over timescales of centuries, to affect the distribution of individual species and the boundaries of biome types. Supplanting of one biome type by another tends to occur at timescales of centuries, depending upon the longevity of declining species. However, colonisation of a particular biome type by different

Table 9.3 Changes in areal coverage of eight major biome types between the Last Glacial Maximum and the present (based on Friedlingstein et al., 1992).

Biome type	Area ($10^6 km^2$)		Carbon store in vegetation (Gt)	
	Present	Last Glacial Maximum	Present	Last Glacial Maximum
Ice/polar	3	23	0	0
Tundra	10	21	25	34
Taiga	24	12	236	91
Deciduous forest	16	14	237	204
Grassland and shrubland	36	34	30	27
Deciduous tropical forest	20	25	225	276
Tropical rainforest	11	6	147	64
Desert and semi-desert	16	24	1	1
Total	136	159	901	697

species may begin relatively quickly, as these intruders opportunistically compete for vacated space. This transition is very rapid for tree species encroaching upon tundra, where species invasion is noticeable in as little as ten years for a temperature increase of only 1°C.

The warmest period of temperature, often proposed as an analogue for future global warming, occurred during the Holocene Climatic Optimum. At this time, carbon dioxide concentrations were similar to pre-industrial levels, but temperatures were up to 2°C warmer than present, contemporaneously with greater precipitation. One of the better records of vegetation change during the Holocene has been constructed for eastern North America. Only six major biome types have dominated this landscape since the end of the Last Glacial Maximum: tundra; boreal forest; mixed conifer–northern hardwood forest; prairie; cool temperate, deciduous forest; and warm temperate, southeastern evergreen forest. At the beginning of the Holocene about 11,000 years ago, tundra was restricted to a 100-kilometre wide band in front of the retreating icecap in northern Ontario and Quebec. Boreal forest grew as far south as 42° N latitude, and the mixed conifer–northern hardwood forest grew about 200–250 kilometres further south than its present range. The boundaries of other forest biome types were only marginally different from the present. With the onset of warming and the peaking of the Holocene Climatic Optimum, vegetation boundaries expanded northward, with the greatest changes occurring in the location of boreal forest and tundra. These latter biome types shifted rapidly in response not only to deglaciation, but also to temperatures that in general were 1.5°C warmer than present. The mixed conifer–northern hardwood forest reached its maximum extent, growing to 49° N (Figure 9.5). Following the warmest peak of the Holocene, both cool and warm temperate forests south of 43° N latitude continued to expand northwards. However north of 43° N, forest boundaries shifted southwards by 100–150 kilometres. At the same time, cooler temperatures and lower potential evaporation allowed the cool temperate, deciduous forest to move westward into prairie vegetation, and to colonise the Mississippi River valley south to Baton Rouge. The forests of eastern North America north of 43° N latitude are still unstable, and have been in disequilibrium with climate for much of the Holocene. From this evidence, future warming should have the most impact on the boundaries of boreal forest. Not only does boreal forest have the most rapid movement of any biome type in these forests, but also its northern and southern boundaries (ecotones) appear to be synchronous with the average winter and summer position of the Arctic polar front. These predicted changes are described later.

The historical shift in zonation of eastern North American forests appears at first inspection to support the 'Clementsian' view that entire biome types, rather than individual plant species, respond to climate change. However, closer examination of pollen records for individual species indicates that each species has a different rate of migration. For instance, following the Last Glacial Maximum,

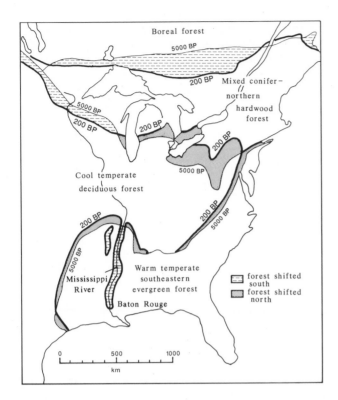

Figure 9.5 Change in extent of eastern North American forest biome types between the peak of the Holocene Climatic Optimum 5,000 years BP, and pre-European deforestation 200 years ago (based on Delcourt and Delcourt, 1981, 1983).

deciduous forest in Ohio State was colonised first by hickory, and then by beech 4,000 years later. This suggests that the rate of spread of species is not controlled as much by climate, as by the ability of each species to propagate opportunistically into favourable climate regimes, against the competition of species already growing at that site. The rates of spread by species throughout the Holocene ranged from a rapid 400 metres per year for Jack pine, to 250 metres per year for elm, to only 100 metres per year for chestnut. These rates of colonisation indicate that the type and method of seed dispersal were also important. Under enhanced 'greenhouse' warming, the boundaries of eastern North American forest types would not keep pace with the projected rate of warming in this region.

Finally, shorter climate changes should not be ignored as a factor shifting biomes. For instance, the Pilliga scrub in New South Wales, Australia is dominated by the conifer, *Callitris glaucophylla*. The Pilliga scrub now occupies more than 400,000 hectares of land, which over a hundred years ago, was open grazing country. The La Niña event that followed the severe ENSO event of 1876–1878

brought heavy rains which favoured the spread of this species. While human influences in terms of high grazing pressure and suppression of fires may have exacerbated the selective dominance of *Callitris*, climatic conditions certainly aided its spread.

Future change
(Smith et al., 1992; Gates, 1993; Monserud et al., 1993)

Except for the Simple Biosphere Model, the models described above have been used to forecast changes in biome types under enhanced 'greenhouse' warming. The Simple Biosphere Model has only been used regionally, because all the vegetation data required by the model are still lacking. Most zonation maps have been produced at a spatial resolution of 0.5° using the temperature and precipitation output from first generation GCMs. Note that at this resolution, the models underestimate the total landmass of the world by about 9%. This discrepancy may be significant given the magnitude of spatial shifts in some biome types. The results presented below for the Holdridge and Budyko models, each use the same four GCMs: Oregon State University, Geophysical Fluid Dynamics Laboratory, Goddard Institute for Space Studies (GISS) and United Kingdom Meteorological Office (UKMO). The temperature and precipitation predictions of the latter model have already been described (Figures 7.11 and 7.12).

The Holdridge model predicts that an equivalent doubling of carbon dioxide in the atmosphere will lead to a decrease in the extent of tundra and desert by 45% and 22% respectively (Table 9.4). These decreases will be caused by warming at high latitudes and greater precipitation. Overall present-day patterns of vegetation will remain stable for only 60% of the world's landmass. Such a large change implies that the existing levels of enhanced 'greenhouse' gases (as of 1996, 63% of the projected doubling) should already have produced changes in the distribution of biome types. Indeed, grasslands globally are presently being invaded by C_3 grasses and shrubs at the expense of C_4 species. Forest zones under an equivalent doubling of carbon dioxide will expand, although there is considerable variation depending upon the GCM model used. For instance, the results from some GCMs point to a reduction in the amount of moist forest. All models predict that boreal forest will colonise large sections of tundra. This change will be associated with a large increase in biomass carbon. Surprisingly, some tundra will revert to desert where warming is great enough to

Table 9.4 Changes in areal coverage of five major biome types predicted using the Holdridge nomogram, with a doubling of anthropogenic 'greenhouse' gases (based on Smith et al., 1992). Results have been averaged for four GCMs, and totals show weighted averages.

Biome type	Present area ($10^6 km^2$)	Forecast area ($10^6 km^2$)	% Common	% Changed
Tundra	9.4	5.1	11	−45
Desert	37.0	29.0	59	−22
Grassland	19.2	26.4	83	37
Dry forest	18.2	24.2	77	33
Moist forest	51.7	51.1	56	−1
Total	135.5	135.8	60	0

increase potential evaporation above any slight increase in precipitation. The area of grasslands will increase extensively, mainly at the expense of dry forest consisting of savanna. This shift is caused by drying mainly in the tropics. The Holdridge model predicts a substantial increase in areal coverage of tropical rainforest, ranging from 87–159%. The change is due mainly to increased temperature, even though such increases are projected to be small in the tropics.

Potential evaporation and dryness indices required by the modified Budyko model have been interpolated globally, for present conditions, using data from 9,000 stations measuring temperature and precipitation, and 4,000 stations measuring cloudiness and humidity. The Budyko model includes cloudiness and albedo, two parameters poorly modelled by GCMs. Hence, only changes in the temperature and precipitation aspects of the Budyko model have been simulated to date. Changes in the area of each biome type, and the distance and direction of shifts in type centroids can be calculated between present-day and projected vegetation maps. The results for a doubling of pre-industrial levels of carbon dioxide, and averaged for the four GCMs, are presented in Table 9.5. Overall present-day patterns of vegetation will remain stable for only 62% of the world's landmass. Again, such a large change implies that changes in the boundaries of biome types should be occurring already. The majority of biome types that expand in areal extent do so because of increases in precipitation. This expansion will occur in the form of a poleward shift in existing biome type boundaries at the expense of taiga, tundra and polar zones, which will decrease in extent by 25–49%. The southern limit of tundra will shift northwest by 350–600 kilometres, while the existing zone of taiga could shift eastward by

470 kilometres. The present area of tropical rainforest, subtropical vegetation, and temperate forest will increase by 39–88%. The zone of temperate forests will shift northeast by 420–1000 kilometres. This expansion is due to increased temperatures lengthening the growing season. In contrast, the potential for expansion of tropical rainforest is due to increased convective precipitation in the tropics. Paradoxically, while existing tropical rainforest is perceived as under threat from deforestation, global warming should favour its growth. This fact highlights the role that forest husbandry practices will have upon any change in the distribution of forests. Overall, these changes in vegetation will lower the planetary albedo leading to a positive feedback on warming. The increase in desertification occurs wherever increased warming is not offset by an increase in precipitation. This effect is most pronounced at temperate latitudes and excludes latitudes such as the Sahel region of Africa, that have experienced significant droughts during the last thirty years of the twentieth century.

Detailed analysis has also been performed on the distribution of forest biome types for eastern North America. If only temperature is considered, then the potential distribution of different forest types is dependent upon the annual number of degree days, defined by the time in days that the mean temperature exceeds 0°C multiplied by the mean temperature over that period. Under this formula, the boundary between boreal and mixed conifer–northern hardwood forest corresponds with the 1,100 growing degree day isopleth. Degree days were then calculated in a GCM for an equivalent doubling of carbon dioxide. No attempt was made to model the 'fertiliser' effect of enhanced carbon dioxide or precipitation changes. The shifts in forest boundaries between the

Table 9.5 Changes in areal coverage of sixteen biome types predicted using a modified Budyko model for a doubling of anthropogenic 'greenhouse' gases (Monserud et al., 1993). Results have been averaged for four GCMs, and totals show weighted averages.

Biome type	Present area (10^6 km^2)	Forecast area (10^6 km^2)	% Common	% Changed
Ice/polar	2.5	1.9	75	−25
Tundra	10.1	5.4	48	−46
Taiga	17.2	12.9	52	−25
Continental taiga	1.1	0.7	15	−36
Continental grassland	.2	0.1	5	−39
Temperate forest	10.4	16.0	72	53
Temperate grassland	7.6	7.0	51	−8
Subtropical forest	3.1	4.3	51	39
Subtropical desert	2.1	3.0	43	41
Subtropical grassland	0.7	1.3	29	79
Tropical rainforest	6.6	12.4	86	88
Tropical deciduous forest	13.6	12.9	42	−5
Tropical savanna	21.4	19.8	58	−7
Tropical scrubland	3.5	3.7	22	6
Tropical grassland	1.6	1.7	11	6
Desert	34.6	33.1	86	−4
Total	136.2	136.2	62	7

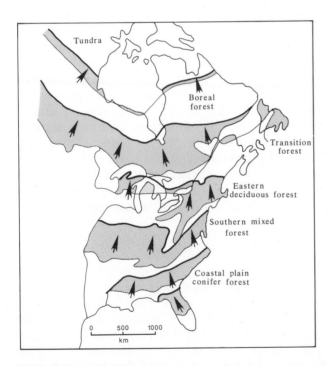

Figure 9.6 Modelled spatial changes in biome types of eastern North American forest, under warmer temperatures produced by an equivalent doubling of carbon dioxide (based upon Solomon et al., 1984). Shading shows the distances that the boundaries between major biomes have shifted northward.

present-day distribution and the forecast one are shown in Figure 9.6. All forest biome types shift northward, with boreal forest undergoing the greatest movement, 300 kilometres. The rapid change in potential distribution of different forests, when compared to the actual rate at which tree species and plant communities can opportunistically colonise new territory, shows major spatial disparities. It is doubtful if plant communities can perform this colonisation anywhere close to the rate at which predicted global warming will occur. Many species living under warmer temperatures, and a longer growing season, also could be environmentally stressed by excess heat and possible drought.

EFFECT OF ECOSYSTEMS ON CLIMATE
(Smith et al., 1992; Sahaglan et al., 1994; Henderson-Sellers and McGuffie, 1995; Lockwood, 1995)

The results of modelling, in cases where vegetation is assumed to adjust to an enhanced 'greenhouse' warmed climate, show that evaporation increases over continental land surfaces with a concomitant increase in absorbed solar radiation and surface temperature. The amount of energy transferred to the atmosphere by this increased evaporation is on the order of 5 watts per square metre, which is similar in magnitude to the radiative forcing of a doubling in 'greenhouse' gases. The differences are greatest in the northern hemisphere in summer. Results for more detailed modelling are shown in Table 9.6. The inclusion of vegetation, as a feedback on climate in GCMs, globally shows no added increase in temperature, a 0.5% increase in precipitation and evaporation, a 0.2% decrease in planetary albedo, and no change in cloudiness. While these values were determined at a gross scale, taken together with the concepts presented above for the Simple Biosphere Model they indicate that vegetation changes can have an impact measurable upon global climate. The magnitude of these feedbacks needs further research.

Although evaporation may increase globally because of enhanced warming and the spread of vegetation into tundra and desert areas, the decrease in transpiration, stomata size and density, with elevated levels of carbon dioxide, could locally reduce atmospheric moisture levels. While higher temperatures increase evapotranspiration, the effect is not unlimited. Above 25°C plants close their stomata sufficiently to prevent any increase in moisture loss. Over the past fifty years, under increasing carbon dioxide, evaporation by plants has probably decreased by 5% over tropical vegetation, by 2% over mid-latitude forests and by 2–4% over temperate grasslands. Hence, anthropogenic enhanced 'greenhouse' warming could conceivably make the air over vegetated surfaces drier. This might have a greater attendant risk of more frequent and intense fires in these landscapes. Analysis of the overall effect of vegetation upon climate under 'greenhouse' warming is conflicting at this stage.

As described in Chapter 7, there are four main reservoirs in the carbon cycle: fossil carbon, the atmosphere, the oceans and the terrestrial biosphere. In all biosphere modelling of the impact of enhanced 'greenhouse' warming, the amount of biomass locked into vegetation increases. A 10% increase in plant growth globally would soak up 0.5 gigatonnes (Gt) of carbon per year. If this growth increases to 40%, then the storage increases to 4.0 Gt per year. A best guess estimate is that a doubling of anthropogenic gases would sequester 1.2–1.5 Gt of extra carbon per year due to enhanced plant growth. These estimates equate with the amount of carbon hypothesised as being absorbed by forest growth each year in the northern hemisphere at present. Overall global warming could witness the absorption of up to 200 Gt of carbon by the terrestrial biosphere. This would impact upon the concentration of carbon dioxide in the atmosphere, or at least upon its rate of change. For the Holdridge model, the accelerated plant growth due to the carbon dioxide 'fertilisation' effect will lead to a 4–85 part per million reduction in atmospheric carbon dioxide levels. However, there is still a high degree of uncertainty about this value amongst different GCMs and biosphere models. Some approaches forecast that the reduction in atmospheric carbon dioxide could be as high as 128 parts per million. The increase in carbon storage will be due to a poleward shift in boreal and tropical rainforests, and to a decrease in the global extent of deserts. None of these estimates takes into account changing land use practices that may occur over the next 50–100 years as the result of continued human population growth, or as a response to a warming globe.

Table 9.6 Differences in GCM simulations of climate change, for a doubling of anthropogenic 'greenhouse' gases, with and without biosphere feedback (based on Henderson-Sellers and McGuffie, 1995).

Parameter (annual)	With biosphere feedback	Without biosphere feedback	Difference
Temperature	+2.5°C	+2.5°C	0.0°C
Precipitation	+6.1%	+5.6%	+0.5%
Evaporation	+6.1%	+5.6%	+0.5%
Planetary albedo	−4.5%	−4.3%	−0.2%
Cloud amount	−2.4%	−2.4%	0.0%

CONCLUDING COMMENTS

All biosphere models are static and assume that biome structures move as a fixed unit in time and space. While dynamic models such as the Simple Biosphere Model exist, to date they have only been applied to a limited number of individual biome types. The distances hypothesised for shifts in many important biome types require the rapid movement of species across the landscape, and the opportunistic invasion of new species into existing biomes. When compared to the timescales of evolutionary development and palaeo-environmental response, it is uncertain if such migration can take place in such a uniform or simple manner. More importantly, the concept of global biome types ignores the fact that a significant proportion of landmasses are cultivated, used for grazing, or logged by clear-felling or sustainable harvesting. All shifts in biomes must take place against this overwhelming human impediment.

The views presented in this Chapter, because they summarise a voluminous literature, are inherently simplistic. There is a lack of a consensus on many points. For instance, the effects of carbon dioxide summarised in Table 9.2 are mainly favourable. Some researchers would contend that they are overly so. Additionally, the results of GCM modelling of feedback to climate from the effects of the biosphere (presented in Table 9.5) show that evapotranspiration will increase globally, yet studies on individual plants and crops indicate that it will decrease. Finally, the increase in carbon dioxide and its impact on the biosphere are not without complications. The eruption of Mount Pinatubo in June 1991 caused both carbon dioxide and methane levels in the atmosphere to stabilise. The effect was due neither to the impact of volcanism on the biosphere, nor the impact of the biosphere upon the atmosphere. Clearly, as emphasised elsewhere in this book, there are other factors which can affect climate, and they probably do so to a greater degree than any change in the biosphere.

REFERENCES AND FURTHER READING

Calder, N. 1991. *Spaceship Earth*. Penguin, London, 208pp.

Caldwell, M.M. and Flint, S.D. 1994. 'Stratospheric ozone reduction, solar UV-B radiation and terrestrial ecosystems'. *Climatic Change* v. 28 pp. 375–394.

Delcourt, P.A. and Delcourt, H.R. 1981. 'Vegetation maps for eastern North America: 40,000 yr B.P. to the present'. In Romans, R.C. (ed.) *Geobotany II*. Plenum, New York, pp. 123–165.

Delcourt, P.A. and Delcourt, H.R. 1983. 'Late Quaternary vegetational dynamics and community stability reconsidered'. *Quaternary Research* v. 19 pp. 265–271.

Friedlingstein, P., Delire, C., Müller, J.F. and

Gérard, J.C. 1992. 'The climate induced variation of the continental biosphere: a model simulation of the Last Glacial Maximum'. *Geophysical Research Letters* v. 19 pp. 897–900.

Gates, D.M. 1993. *Climate Change and its Biological Consequences*. Sinauer, Sunderland, 280p.

Gifford, R.M. 1988. 'Direct effects of higher carbon dioxide concentrations on vegetation'. In Pearman, G.I. (ed.) *Greenhouse: Planning for Climate Change*. Brill, Leiden, pp. 506–519.

Henderson-Sellers, A. and McGuffie, K. 1995. 'Global climate models and 'dynamic' vegetation changes'. *Global Change Biology* v. 1 pp. 63–75.

Holdridge, L.R. 1947. 'Determination of world plant formations from simple climatic data'. *Science* v. 105 pp. 367–368.

Idso, S.B. 1989. *Carbon Dioxide and Global Change: Earth in Transition*. Institute for Biospheric Research Press, Tempe, 292p.

Lockwood, J.G. 1995. 'The suppression of evapotranspiration by rising levels of atmospheric CO_2'. *Weather* v. 50 pp. 304–308.

Monserud, R.A., Tchebakova, N.M. and Leemans, R. 1993. 'Global vegetation change predicted by the modified Budyko Model'. *Climatic Change* v. 25 pp. 59–83.

Sahaglan, D.L., Schwartz, F.W. and Jacobs, D.K. 1994. 'Direct anthropogenic contributions to sea level rise in the twentieth century'. *Nature* v. 367 pp. 54–57.

Schimel, D.S. 1995. 'Terrestrial ecosystems and the carbon cycle'. *Global Change Biology* v. 1 pp. 77–91.

Sellers, P.J., Mintz, Y., Sud, C. and Dalcher, A. 1986. 'A simple biosphere model (SiB) for use within general circulation models'. *Journal Atmospheric Science* v. 43 pp. 505–531.

Smith, T.M., Leemans, R. and Shugart, H.H. 1992. 'Sensitivity of terrestrial carbon storage to CO_2-induced climate change: comparison of four scenarios based on general circulation models'. *Climatic Change* v. 21 pp. 367–384.

Solomon, A.M., Thorp, M.L., West, D.C., Taylor, G.E., Webb, J.W. and Trimble, J.L. 1984. *Response of unmanaged forests to CO_2-induced climate change: available information, initial tests, and data requirements*. U.S. Department of Energy Report TR009, Washington.

Townshend, J.R.G., Tucker, C.J. and Goward,

S.N. 1993. 'Global vegetation mapping'. In Gurney, R.J., Foster, J.L. and Parkinson, C.L. (eds) *Atlas of Satellite Observations related to Global Change*. Cambridge University Press, Cambridge, pp. 301–311.

Walker, B.H. 1991. 'Ecological consequences of atmospheric and climate change'. *Climatic Change* v. 18 pp. 301–316.

Warrick, R.A., Gifford, R.M. and Parry, M.L. 1986. 'CO_2, climatic change and agriculture'. In Bolin, B., Doos, B.R., Jager, J. and Warrick, R.A. (eds) *The Greenhouse Effect,*

Climatic Change and Ecosystems. Chichester, Wiley, pp. 393–473.

Woodward, F.I. 1992. 'A review of the effects of climate on vegetation: ranges, competition, and composition'. In Peter, R.L. and Lovejoy, T.E. (eds) *Global Warming and Biological Diversity*. Yale University Press, New Haven, pp. 105–123.

Xue, Y., Sellers, P.J., Kinter, J.L. and Shukla, J. 1991. 'A simplified biosphere model for global climate studies'. *Journal of Climate* v. 4 pp. 345–364.

10

Epilogue

RELATIVE IMPORTANCE OF CHANGES IN ATMOSPHERIC RADIATIVE FORCING
(MacCracken, 1983; Harrison et al., 1993; Charlson and Heintzenberg, 1994; Jones, 1994; Houghton et al., 1995)

Cooling from anthropogenic aerosols, mainly due to industrial activity, but also including natural components, is changing the heat and energy balance of the atmosphere. The balance in radiation is being shifted between night and day, winter and summer, land and oceans, the tropics and mid-latitudes, the northern and southern hemisphere, and finally between the Earth's surface and the upper atmosphere. The impact of humans upon climate change cannot be attributed to a single factor weighted towards warming or cooling. It includes many factors operating with varying magnitudes, at different temporal and spatial scales. For instance, globally enhanced 'greenhouse' warming is being shielded regionally by the effects of industrial sulphates, and exacerbated by urban heat islands. There may also be undefined impacts that science has neither measured nor thought about. For example, humans may be enhancing dimethylsulphide production by phytoplankton in the oceans, through the overuse of agricultural fertilisers, and by discharging more iron in urban runoff and industrial effluent.

Human-induced climate change is occurring against a background of natural change. Anthropogenic radiative forcing and the inherent variability in natural fluxes of energy in the atmosphere are summarised in Table 10.1, and displayed in Figure 10.1. Values are based on references already quoted or listed in this text. While some anthropogenic forcings are supposedly known to a high degree of accuracy, our ability to measure the overall radiation components of the atmosphere is still subject to considerable error. Currently, global measurements are performed by satellites with an error in these measurements of at least 6 watts per square metre (W m^{-2}). This is greater than the total of all anthropogenic radiative forcings. The natural system has an interannual, global variability of 12.3–22.0 W m^{-2}, with the greatest variation, 22.0 W m^{-2}, occurring in the longwave flux. The total net radiation budget in the atmosphere fluctuates from year to year by 20.0 W m^{-2}. While part of this 'noise' can be attributed to differences in albedo, atmospheric moisture and cloudiness between La Niña and ENSO years, much is simply the background 'noise' against which climate change has to be assessed. Note that this 'noise' is greater than the radiative forcing of any natural or anthropogenic component invoked in the literature as controlling climate change over the last two hundred years.

Anthropogenic effects consist of two types of components, those enhancing the downward long-wave flux, and those shielding the surface of the Earth from solar radiation. The former are due to 'greenhouse' gases and, to date, total 2.76 W m^{-2} at the top of the atmosphere. The breakdown for individual gases is described in Chapter 7. Note that the 'greenhouse' effect of ozone is still considered uncertain, with stratospheric ozone depletion producing a decrease in radiative forcing of 0.15

Table 10.1 Variation in the components of natural radiation and radiative forcing of anthropogenic 'greenhouse' gases and atmospheric pollutants.

Factor	Radiative forcing ($W\ m^{-2}$)
'Greenhouse' gases:	
total	2.76
carbon dioxide	1.56
methane	0.50
nitrous oxide	0.10
CFCs & chlorine compounds	0.30
halocarbon substitutes	0.05
ozone	0.25[1]
Aerosols and dust:	
total	−2.10
due to sulphates	−0.50[2]
due to volcanoes	−0.50[3]
due to other sources	−1.10
Cloud:	
longwave variation by cloud	2.10
shortwave variation by cloud	7.60
increased cloudiness	−1.30
Radiation components:	
range in solar radiation	
11-year sunspot cycle	0.34
200-year sunspot cycle	3.40
interannual variation	
in albedo	12.30
in longwave	22.00
in net radiation budget	20.00
satellite measurement error	6.00
Recent positive forcings	2.76
Recent negative forcings	−3.40
Recent net change	−0.56

[1] stratosphere −0.15, troposphere +0.40
[2] human −0.28, other −0.26
[3] for comparison, Mount Pinatubo −4.0

$W\ m^{-2}$, while tropospheric ozone pollution produces a 'greenhouse' enhancement of 0.40 $W\ m^{-2}$. Shielding, or the reduction in atmospheric opacity, has received less attention. The overall shielding effect of aerosols in the atmosphere is −2.1 $W\ m^{-2}$. This value is based upon the fluctuation in direct solar radiation measured at the Earth's surface on cloudless days between 1890 and 1960. When the known forcing values for sulphate aerosols and volcanic dust are subtracted from this value, there may be a forcing of −1.1 $W\ m^{-2}$ at the Earth's surface that is currently unexplained. The most likely reasons for this latter forcing are smoke and dust production due to human activities. Little attention has been given to the atmospheric effects of either parameter.

An attempt is made in Chapter 7 to summarise the importance of anthropogenic dust (microscopic particles of soil, regolith, and industrially or agriculturally derived fine material) in forcing climate change. However, no recent literature has been found showing trends in regional or global dust production, or the degree of that radiative forcing. Unfortunately, dust suffers the adverse publicity of the 1970s, when it was invoked as the major cause of cooling between 1945 and 1978. In addition, dust has a residence time in the atmosphere of about one week, a value that has been considered of little consequence to long-term radiation forcing. A similar assumption was once made about sulphate aerosols. It is time that dust was 'rediscovered', and serious attempts were made to define not only a long-term, historical time series that is up-to-date, but also dust's spatial contribution to the radiation budget of the atmosphere.

This text has emphasised four human impacts on climate change: enhanced greenhouse warming, sulphate aerosols, dust and urban warming. Biomass burning rightly can be considered the fifth human impact, although its influence is often included under the effects of dust or sulphate aerosols. Statements by the Intergovernmental Panel on Climate Change, while they highlight the importance of biomass burning as a contributing factor in increasing 'greenhouse' gases, fail to define its contribution to radiative forcing. The historic role of fire in temperate forests is emphasised in Chapter 8. The degree of burning in the tropics is even greater. About 1.8–4.7 gigatonnes of biomass carbon, consisting of firewood, dry savanna grass, agricultural wastes, slash and burn agricultural activities, and the dead vegetation from deforestation, are burned each year. This is an enormous amount when compared to the average increase of 3.2 gigatonnes of carbon in the atmosphere each year. Recent assessments indicate that biomass burning may also contribute 8–10% to the production of anthropogenic methane in the atmosphere. Besides blocking incoming solar radiation, absorbing heat at elevation in the atmosphere, increasing atmospheric stability, and adding to sulphate aerosol production, smoke increases

Part A) Variation in radiation

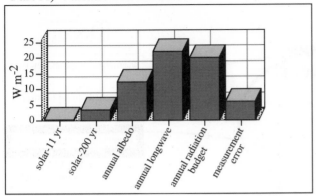

Part B) Enhanced 'greenhouse' gases to date

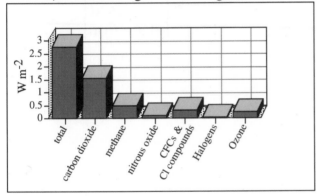

Part C) Aerosols and cloud

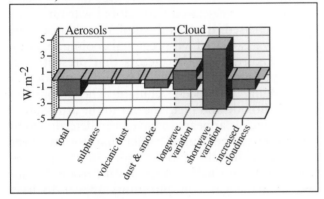

Figure 10.1 Radiative forcing: (A) natural variation in radiation components; (B) enhanced 'greenhouse' gases to date; and (C) aerosols and cloud.

cloud nucleation. Tropical smog resulting from human activities rivals the worst air pollution found in the skies of developing cities. And ozone levels can exceed 90 parts per billion by volume,

values that would trigger smog alerts in developed countries. Virtually nothing is known about the chemistry of this smoke, its spatial distribution, how it is changing over time, or its impact on radiative forcing. It is time that smoke received attention commensurate with its presence in the atmosphere.

The positive anthropogenic forcings of 'greenhouse' gases still would appear to outweigh the negative effects of aerosol loadings. However one additional parameter, namely cloud, must be considered. There is no doubt that the world is becoming cloudier (Figure 7.6). While natural effects cannot be ruled out, there is enough evidence to suggest that an anthropogenically enhanced warming globe, or one with sulphate pollution, is also cloudier. Little emphasis has been placed upon the radiative forcing of increased cloudiness. It probably has grown to -1.30 W m^{-2} over the past fifty years. When increased cloudiness is included in net radiative forcing calculations, then the climatic impact of human activities, to date, has been -0.56 W m^{-2}. The magnitude and role of clouds in cooling climate need further consideration, as they may be a crucial component of current climate change.

PERSPECTIVE ON FEEDBACK PROCESSES FAVOURING ICE AGES

The scenario on the causes of the Ice Ages, summarised at the end of Chapter 6, provides an alternative method for assessing the potential impact of human activities upon climate. Figure 10.2 presents a simplified version of this scenario that includes the positive feedback mechanisms translating orbit variations in solar radiation into an Ice Age. The interactions work in the opposite manner for deglaciation. Not all of the interrelationships are shown. For instance, no consideration is given to the slower acting impacts of changing sea-level, the North Atlantic salinity sink or Antarctic sea-ice formation. Certainly as icecaps develop, sea-levels drop, leading to more continental landmasses, greater continental aridity and a greater landmass subject to the effects of wind. More sea-ice also enhances the planetary albedo. All the linkages in Figure 10.2 lead towards an increased probability of glaciation. It should be emphasised that the human impact on climate will never produce this condition. However the linkages allude to the impact that human activities could have in cooling climate.

Processes Favouring Ice Ages

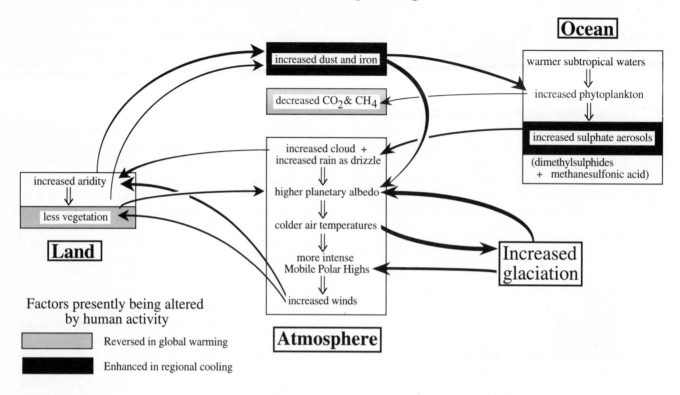

Figure 10.2 Positive feedback processes favouring the development of Ice Ages.

There are three different components involved in the schema: the atmosphere, the oceans and the land surface, each representing the three main parts of the Earth–atmosphere system involved in climate change. The most significant factor driving a global Ice Age, apart from latitudinal differences in solar radiation that are not shown, is a higher planetary albedo. This prevents radiation from reaching the Earth's surface, leading to cooler air temperatures and more glaciation. Increased glacial coverage enhances the planetary albedo. The planetary albedo is also augmented by the formation of cloud generated over the oceans by increased phytoplankton growth and dimethylsulphide nucleation. This latter process is further enhanced by the iron 'fertilisation' effect of dust. Increased glaciation leads to more intense Mobile Polar Highs and increased windiness. These impact upon elements of the land surface and change global climate by increasing aridity and decreasing vegetation coverage. Finally, the latter feedback to the atmosphere provides a continuous source of dust that also augments the overall planetary albedo, and marries into the ocean segment of the model.

There are four factors in the schema that are being altered by human activity. Two of them, vegetation and elevated levels of 'greenhouse' gases, are presently associated with global warming; and two, increased dust and sulphate aerosols, are presently associated with regional cooling. According to Figure 10.2, the factors that presently are leading to global warming play only a minor role in changing Pleistocene climate. The amount of carbon dioxide and methane gas in the atmosphere during glacials appears to have been determined by the cooler temperatures, rather than to have been responsible for the cooling roles in the positive feedback favouring Ice Ages. Both carbon dioxide and methane gas concentrations appear to have responded to cooler temperatures during glacials, rather than causing the cooling. Even during the Little Ice Age, when the globe cooled by up to 1°C, the concentration of these gases did not fluctuate much over time (Figure 7.1). Nor would increases in these gases have reversed any cooling that was already in train. Present-day increases in the extent of vegetation are probably being driven by the 'fertilisation' effect of enhanced carbon dioxide in the atmosphere, and the regrowth of mid-latitude forests in the northern hemisphere. However, vegetation also has a minor role in the feedback processes altering climate during the Pleistocene. Thus both of the factors in Figure 10.2 that are presently linked to global warming, have minimal effects in reversing the positive feedback processes leading to any cooling.

A different picture emerges with the two factors presently being enhanced by human activity, and which are currently cooling the globe regionally. Increased atmospheric dust is a prime mechanism in the model forcing the phytoplankton–dimethyl-sulphide–cloud linkage leading to glaciation. As well, sulphate aerosols, through industrial emissions, are exacerbating cloud formation over the ocean that produces cooling in the schema. Additionally, both dust and sulphate aerosols are increasing the planetary albedo. Human discharge of contaminated water into the oceans is also mimicking the iron 'fertilisation' effect that dust has on phytoplankton growth during glacials.

If enhanced 'greenhouse' warming is a reality, than its arrival may well be timely, because it has the potential to compensate for the strong cooling that may be a natural part of the ocean–atmosphere system, under the heavy atmospheric dust and sulphate aerosol loadings presently being stimulated by human activity. The resultant global effect of these human impacts may not be the sum of their parts. Instead, the changes in the radiation budget may be forcing the Earth's climate in some manner that has no past analogue. A general cooling or unprecedented warming may not be the final state of our climate. These scenarios seem too predictable. The new climate is probably too complicated to be defined or forecast yet. Storminess in the North Atlantic, droughts in Sahel, prolonged ENSO events, increased variability of severe events, hotter summers followed by colder winters at mid-latitudes, and urban heat island effects may be the present signatures of this new anthropogenically induced climate.

REFERENCES AND FURTHER READING

Charlson, R.J. and Heintzenberg, J. (eds) 1994. *Aerosol Forcing of Climate*. Wiley, Chichester, 416p.

Harrison, E.F., Minnis, P., Barkstrom, B.R. and Gibson, G.G. 1993. 'Radiation budget at the top of the atmosphere'. In Gurney, R.J., Foster, J.L. and Parkinson, C.L. (eds) *Atlas of Satellite Observations Related to Global Change*. Cambridge University Press, Cambridge, pp. 19–38.

Houghton, J.T., Meira Filho, L.G., Bruce, J., Hoesung Lee, Callander, B.A., Haites, E., Harris, N. and Maskell, K. (eds) 1995. *Climate Change 1994: Radiative Forcing of Climate Change and an Evaluation of the IPCC IS92 Emission Scenarios*. Cambridge University Press, Cambridge, 339p.

Jones, A., Roberts, D.L. and Slingo, A. 1994. 'A climate model study of indirect radiative forcing by anthropogenic sulphate aerosols'. *Nature* v. 370 pp. 450–453.

MacCracken, M.C. 1983. 'Have we detected CO_2-induced climate change? problems and prospects'. *Proceedings Carbon Dioxide Research Conference*, United States Department of Energy, 19–23 September, Berkeley Springs, West Virginia, pp. v.3–v.45.

Index